Structural Identification and Damage Evaluation by Integrating Physics-Based Models with Data

Structural Identification and Damage Evaluation by Integrating Physics-Based Models with Data

Guest Editors

Zhiming Zhang
Mingming Song
Qipei Mei

MDPI

Basel • Beijing • Wuhan • Barcelona • Belgrade • Novi Sad • Cluj • Manchester

Guest Editors
Zhiming Zhang
Research and Modeling Group
Verisk Analytics
Boston, MA
USA

Mingming Song
Department of Bridge Engineering
Tongji University
Shanghai
China

Qipei Mei
Department of Civil and Environmental Engineering
University of Alberta
Edmonton, AB
Canada

Editorial Office
MDPI AG
Grosspeteranlage 5
4052 Basel, Switzerland

This is a reprint of the Special Issue, published open access by the journal *Buildings* (ISSN 2075-5309), freely accessible at: https://www.mdpi.com/journal/buildings/special_issues/Models_Data.

For citation purposes, cite each article independently as indicated on the article page online and as indicated below:

Lastname, A.A.; Lastname, B.B. Article Title. *Journal Name* **Year**, *Volume Number*, Page Range.

ISBN 978-3-7258-4859-1 (Hbk)
ISBN 978-3-7258-4860-7 (PDF)
https://doi.org/10.3390/books978-3-7258-4860-7

Contents

buildings

MDPI

Article

Swarm Intelligent Optimization Conjunction with Kriging Model for Bridge Structure Finite Element Model Updating

Jie Wu [1,2], Fan Cheng [1], Chao Zou [3], Rongtang Zhang [1], Cong Li [1], Shiping Huang [4] and Yu Zhou [2,5,*]

1 School of Civil Engineering and Architecture, Wuhan Polytechnic University, Wuhan 430023, China; wujiemc@gmail.com (J.W.); 13409991545@163.com (F.C.); peterzhang@whpu.edu.cn (R.Z.); licong@whpu.edu.cn (C.L.)
2 National-Local Joint Engineering Laboratory of Building Health Monitoring and Disaster Prevention Technology, Hefei 230601, China
3 School of Civil and Transportation Engineering, Guangdong University of Technology, Guangzhou 510006, China; zouchao226@gmail.com
4 School of Civil Engineering and Transportation, South China University of Technology, Guangzhou 510640, China; ctasihuang@scut.edu.cn
5 School of Civil Engineering, Anhui Jianzhu University, Hefei 230601, China
* Correspondence: yuzhou923@outlook.com; Tel.: +86-13195619595

Abstract: For the simple bridge structure, the finite element model established by drawing and elastic mechanics method is accurate. However, when faced with large and complex long-span bridge structures, there are inevitable differences between the finite element model and the physical model, where the model has to be updated. It is problematic that the updating structural matrix cannot be fed back into the existing general finite element calculation software in the traditional structural matrix updating method. In this paper, a parameter-type updating method based on the "Kriging model + swarm intelligence" optimization is proposed. The Kriging model, based on Genetic Algorithm (GA), Bird Mating Optimizer (BMO), and Particle Swarm Optimization algorithm (PSO), is introduced into the finite element model, updating this to correct the design parameters of the finite element model. Firstly, a truss structure was used to verify the effectiveness of the proposed optimization method, and then a cable-stayed bridge was taken as an example. Three methods were used to update the finite element model of the bridge, and the results of the three optimization algorithms were compared and analyzed. The results show that, compared with the other two methods, the GA-based model updating method has the least time due to the small computation. The results of the BMO-based model were time consuming compared to the other two algorithms, and the parameter identification results were better than the GA algorithm. The PSO algorithm-based model updating method to solve the finite element model was repeated, which required a large amount of computation and was more time consuming; however, it had the highest parameter correction accuracy.

Keywords: model updating; Kriging model; swarm intelligence optimization; cable-stayed bridge

1. Introduction

During the bridge structural analysis, the finite element model of the structure is usually established according to the design drawings. In order to obtain the real dynamic characteristics of the structure, the parameters of the model must be set accurately [1–3]. By analyzing the structure with an accurate finite element model, the static and dynamic responses of the structure can be predicted, and the damage condition of the structure can be directly simulated [4,5]. Therefore, it is of considerable significance to obtain the real finite element model of the bridge structure [6–8]. However, in general, due to the complexity of the bridge structure itself, it is challenging for researchers to build models that match theory exactly with the experiment.

The errors between the calculated results of the theoretical model and the test results are attributed to three aspects: (1) model structural error, (2) model parameter error, and (3) model order error [9]. The model error is usually related to the selected mathematical model, which depends on the main characteristics of the actual structure. The error of the model order is caused by the finite element discretization, which cannot be avoided in finite element analysis. Therefore, for most of the bridge structure finite element model updating, the key is to use appropriate methods to reduce the second type of error, that is, the model parameter error [10].

At present, methods of bridge structural model updating mainly include: frequency response function method [11], neural network method [12], genetic algorithm [13], simulated annealing method [14–17], and statistical method [18,19].

The method of frequency response function [20] can provide adequate response information by itself; however, when using this method, response information needs to be obtained in a high-demand environment, which limits the application of this method to some extent. Because of its strong learning ability and nonlinear mapping ability, the neural network is suitable for finite element model updating. However, the method has some defects in robustness. The genetic algorithm [21] is a worldwide optimization search algorithm. It is natural and universal, robust, suitable for parallel processing, efficient, and practical. The selection of fitness function determines the speed and effect of the algorithm. The simulated annealing method [22] is a standard method of probability calculation. Because of its high quality, a strong initial robustness, simplicity, generality, and ease of implementation, it is often used to search for the optimal solution in an ample space. However, this method also has the disadvantages of a high initial temperature, a slow cooling rate, a low termination temperature, and a time-consuming optimization process. The model updating method based on the statistical method only uses the finite element analysis during the initial sample preparation stage. Therefore, it is suitable for all kinds of projects [23]. However, when using this method, feature extraction, parameter screening, and response surface fitting are also needed, and there are problems such as inappropriate feature extraction and multiple types of parameter selection.

Most of the above methods have problems such as too many parameters to be corrected, too much calculation, and low efficiency of model updating. Not all of the parameters to be updated have a significant influence on the structural model. Therefore, the method based on sensitivity analysis should be used to analyze the updated parameters. The problem with sensitivity-based model updating is that the impact of parameters with high sensitivity on the structure may be small [24], while the influence of parameters with low sensitivity on the structure may be critical. Therefore, it is more meaningful to adopt the structural model updating method, which is not based on sensitivity.

Genetic algorithm (GA) [21] is a calculation method based on "natural selection and survival of the fittest" in the theory of evolution. It is a parallel, random, and adaptive search algorithm. Because GA can effectively avoid the problem of local optimization in the search process, it developed as one of the principal artificial intelligence algorithms.

Bird mating optimizer (BMO) is a new kind of swarm intelligence optimization algorithm [25] which has fewer controlled parameters and can avoid local optimization in the optimization process. Therefore, the BMO algorithm has become a lively research direction in structural finite element model modification.

Particle Swarm Optimization (PSO) [26], an evolutionary computing technology developed in 1995, is derived from the simulation of a simplified social model and belongs to a swarm intelligence algorithm.

In this paper, we firstly studied the updating method of the Kriging model [27]. Three kinds of optimization algorithms, GA, BMO, and the PSO, were applied to the updating method of the Kriging model, respectively. Secondly, a truss structure verified the effectiveness of the three methods and compared the time consumption and calculation accuracy. Next, we applied the Kriging model updating method based on three optimization algorithms to a cable-stayed bridge structure for analysis.

2. Methods

2.1. Kriging Model

The Kriging model is based on the original data of information samples in a specific region [28,29], and carries out linear unbiased, minimum variance estimation for unknown data with the same characteristics in the selected region. It consists of two parts: the linear regression part and the non-parametric part. The non-parametric part can be regarded as a Gaussian stationary random process, and a polynomial and a random distribution can represent the Kriging model:

$$y(\mathbf{x}) = f^T(\mathbf{x})\beta + z(\mathbf{x}) \tag{1}$$

$$f^T(\mathbf{x})\beta = \beta_1 f_1(\mathbf{x}) + \beta_2 f_2(\mathbf{x}) + \ldots + \beta_m f_m(\mathbf{x}) = [f_1(\mathbf{x}) \ldots f_m(\mathbf{x})]\beta \tag{2}$$

where $y(\mathbf{x})$ represents the unknown function; β represents the regression coefficient; $f(\mathbf{x})$ represents the polynomial function of the variable $\mathbf{x}$; m is the number of $f(\mathbf{x})$.

In the design space, $f(\mathbf{x})$ can be used to represent the global approximation and $z(\mathbf{x})$ can be used to simulate the local approximation. $z(\mathbf{x})$ is a Gaussian stationary random process, the mean value of which is zero and the covariance non-zero, and it obeys normal distribution $N(0,\sigma^2)$. The covariance matrix of $z(\mathbf{x})$ can be expressed as:

$$E\left[z(\mathbf{x}_i)z(\mathbf{x}_j)\right] = \sigma^2 R(\theta, \mathbf{x}_i, \mathbf{x}_j)\ i, j = 1, 2, \ldots, n \tag{3}$$

where $R(\theta, \mathbf{x}_i, \mathbf{x}_j)$ represents the spatial correlation function between selected two points from samples, which directly affects the accuracy of simulation; θ represents the parameters of the correlation function; n represents the total number of points in the sample.

To build a completed Kriging model, assuming the test sample points in the region of $n \times p$ are $\mathbf{x} = (\mathbf{x}_1, \mathbf{x}_2, \ldots, \mathbf{x}_n)$, where $\mathbf{x}_i^k \in R^p$, p is the number of design variables. The corresponding output can be expressed as:

$$\mathbf{Y} = \{y_1(\mathbf{x}), y_2(\mathbf{x}), \ldots, y_n(\mathbf{x})\} \tag{4}$$

β and σ^2 can be estimated as follows:

$$\hat{\beta} = \left(\mathbf{F}^T R^{-1} \mathbf{F}\right)^{-1} \mathbf{F}^T R^{-1} \mathbf{Y} \tag{5}$$

$$\hat{\sigma}^2 = \frac{1}{\mathrm{n}}(\mathbf{Y} - \mathbf{F}\hat{\beta})^{\mathrm{T}} R^{-1} (\mathbf{Y} - \mathbf{F}\hat{\beta}) \tag{6}$$

where $\hat{\beta}$ and $\hat{\sigma}^2$ are estimated value of β and σ^2, $\mathbf{F}$ represents the estimated value vector containing the $f(\mathbf{x})$ for each test point, R is the correlation matrix of the test points, which can be expressed as:

$$R = \begin{bmatrix} R(\mathbf{x}_1, \mathbf{x}_2) & \ldots & R(\mathbf{x}_1, \mathbf{x}_n) \\ \vdots & \ddots & \vdots \\ R(\mathbf{x}_n, \mathbf{x}_1) & \ldots & R(\mathbf{x}_n, \mathbf{x}_n) \end{bmatrix} \tag{7}$$

To get the values of $\hat{\beta}$ and $\hat{\sigma}^2$, the parameter θ should be figured out due to the $\hat{\beta}$ and $\hat{\sigma}^2$ are related to parameter θ. The maximum likelihood estimation is used to solve the minimum value of the following equation:

$$\min_{\theta>0}\left\{1/2\left[n \ln\left(\hat{\sigma}^2\right) + \ln|R|\right]\right\} \tag{8}$$

Equation (8) is equivalent to an optimization problem, and θ can be obtained by taking the minimum value of $\min_{\theta>0}\left\{|R|^{\frac{1}{n}}\hat{\sigma}^2\right\}$.

When solving the optimization problem, GA, BMO, and PSO are adopted in the Kriging model to solve the optimization of θ, respectively. After the parameter θ in the correlation function is obtained, the optimal linear unbiased prediction result of the response can be expressed as:

$$\hat{y}(\mathbf{x}) = f^T(\mathbf{x})\hat{\beta} + r^T(\mathbf{x})\hat{\alpha} \tag{9}$$

where $r^T(\mathbf{x})\hat{\alpha}$ represents the difference value of residual error of the regression function $f^T(\mathbf{x})\hat{\beta}$, and the vector $\hat{\alpha}$ can be expressed as:

$$\hat{\alpha} = R^{-1}(\mathbf{Y} - \mathbf{F}\hat{\beta}) \tag{10}$$

$r^T(\mathbf{x})$ in Equation (9) can be expressed as the vector of the correlation function between the test point and the unknown point:

$$r^T(\mathbf{x}) = \{R(\mathbf{x}, \mathbf{x}_1), \mathrm{L}, R(\mathbf{x}, \mathbf{x}_n)\} \tag{11}$$

The Kriging model established can be used to predict the points to be measured. In Equation (3), the simulation of the random process includes a correlation function that can affect the stationary characteristics of the model. Then, express the correlation function of the test point as [30]:

$$R(\theta, \mathbf{x}_i, \mathbf{x}_j) = \prod_{\mathrm{k}=1}^{\mathrm{p}} R_{\mathrm{j}}\left(\theta, \mathbf{x}_i^k - \mathbf{x}_j^k\right) \tag{12}$$

where $\mathbf{x}_i^k$ and $\mathbf{x}_j^k$ represent the part k of the test point. Gaussian function can be used as kernel function of correlation function:

$$R(\mathbf{x}_i, \mathbf{x}_j) = \exp\left[-\sum_{k=1}^{p} \theta_k \left|\mathbf{x}_i^k - \mathbf{x}_j^k\right|^2\right] \tag{13}$$

The steps to update the finite element model by Kriging model are showed in Figure 1 and explained as follows:

(1) Determine the variables to be corrected in the finite element model, and then generate a certain number of samples according to the distribution form of the determined variables.
(2) Substitute the generated samples into the finite element model of the bridge for modal analysis, and extract the modal correlation information corresponding to the variables.
(3) The variables and the corresponding structural mode frequencies are taken as the input and output samples, respectively, and GA, BMO, or PSO is used to optimize the parameters θ in the correlation function of the Kriging model to complete the establishment of the Kriging model.
(4) Take the frequency of modal test of the bridge as the input, use the Kriging model to predict the structure modal, and extract the variables by finite element model updating.

2.2. Kriging Model Updating Method Based on GA

When dealing with a specific problem, the GA treats the possible solution to the problem as a solution space. The solution space is regarded as a population, and the solution or solution vector is taken as individuals in the population. When the possible solution to a problem is transformed from the solution space to the search space using the GA, the transformation process is regarded as coding. GA randomly generates the initial population and calculates the fitness value of each individual in the population according to the fitness function. Based on the first generation, according to the evolutionary theory of "natural selection, the survival of the fittest", each offspring will inherit a better solution than the previous generation. In the application of the theory of evolution, individuals with high fitness are selected and calculated, utilizing crossover and mutation operators in genetics to generalize new solutions into a new population. Finally, decoding the optimal individuals in the constrained population and the decoded optimal solution can be taken as the approximate optimal solution of the optimization problem [13].

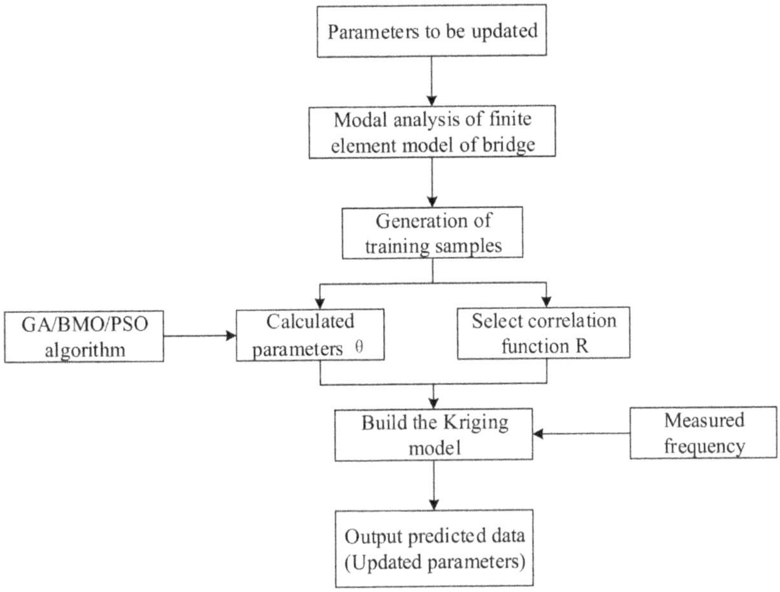

Figure 1. Kriging model-based finite element analysis model updating.

The specific steps of using GA to update the finite element model can be summarized as follows:

(1) Code the samples to generate the initial population;
(2) Set the fitness function of the population, which is the inverse of the difference between the modal frequency of the model and the measured field frequency. Then the fitness of the individuals in the population is calculated;
(3) Conduct individual selection, crossover, mutation to generate new progeny population;
(4) Decode the obtained new progeny population and input it into the finite element model to calculate and extract the modal information of the structure;
(5) Obtain the frequency error by comparing the finite element modal frequency with the measured frequency. When the error meets the requirement of accuracy, complete the correction process; if the accuracy cannot meet the requirements, then go back to step (1).

2.3. Kriging Model Updating Method Based on BMO

The BMO is an algorithm that imitates the mechanism by which birds produce superior genetic offspring. Consider a set of solution in the optimization problem as a bird. In nature, birds are divided into males and females, with females representing the better solution. The mating patterns of female birds are divided into two categories based on the breeding categories of the birds: polyandry and self-mating. The mating patterns of male birds fall into three categories: monogamy, polygamy, and hybridization. In the BMO algorithm, the number of birds in each category is determined by the user, while in the actual situation, the proportion of birds in monogyny and polygyny is relatively large. In contrast, the proportion of birds in polyandry and hybridization is relatively small [25].

2.3.1. Monogamy

If a male x_M and a female x_i mate, the equation for producing offspring is:

$$\begin{gathered} x_b = x_M + \mathrm{w} \times \mathrm{r_g} \times (x_i - x_M) \\ \text{if } r_1 > mcf, x_b(c) = l(c) + r_2 \times (u(c) - l(c)) \end{gathered} \quad (14)$$

where x_b represents the offspring; w is the time-varying weighting factor, which decreases linearly from 1.85 to 0.15; $\mathrm{r_g}$ is the vector of random distribution of elements, and the interval is from 0 to 1. $mcf = 0.85$ is the variation control factor. r_1 and r_2 represents the random number between 0 and 1. c represents the random number between the number of optimized solutions and 1; u and l represent the upper and lower critical values of the initial solution, respectively.

2.3.2. Polygyny

Polygyny refers to the male birds that mate with more than one female, and the formula for the offspring can be expressed as:

$$\begin{gathered} x_b = x_{Pg} + \mathrm{w} \times \sum_{j=1}^{n_i} \mathrm{r_g} \times (x_{ij} - x_{Pg}) \\ \text{if } r_1 > mcf, x_b(c) = l(c) + r_2 \times (u(c) - l(c)) \end{gathered} \quad (15)$$

where n_i represents the number of females attracted by the male birds; x_{Pg} represents the male birds; x_{ij} indicates that the jth female has good genes.

2.3.3. Promiscuity

If the solution of the worst fitness value is mutated then hybrid birds will appear. In the algorithm, a chaotic sequence generates hybrid birds, and a chaotic sequence generates a new feasible solution. The offspring of the hybrid birds are calculated by using Equation (14) with reference to the monogamous birds.

2.3.4. Parthenogenesis

Parthenogenesis birds represent the best result. Calculate the brood as following:

$$\begin{gathered} \text{if } r_1 > mcf_P,\ x_b(i) = x(i) + \mu \times (r_2 - r_3) \times x(i) \\ \text{if } r_1 \le mcf_P,\ x_b(i) = x(i),\ i = 1, 2 \ldots n \end{gathered} \quad (16)$$

where mcf_P represents the variation control factor of self-mating the bird population, which increases linearly from 0.10 to 0.75, r_2 and r_3 represents the random number between 0 and 1. μ is the step size and value is 9.1×10^{-3}.

2.3.5. Polyandry

In the case of polyandry, the female bird will choose several males with good genes for mating in order to obtain offspring with better genes. See Equation (15) for the calculation of the offspring birds.

2.3.6. Steps

In the selection of female or male mates, apply the roulette selection strategy. The higher the fitness value, the greater the probability of being selected. The specific steps of using BMO to update the finite element model can be summarized as follows:

(1) Assuming the size of the bird population is m, the probability of selecting the kth bird is calculated as:

$$p_k = \frac{1/f(x_k)}{\sum_{i=1}^{m} 1/f(x_i)} \quad (17)$$

where $f(x_i)$ represents the fitness function of the problem to be optimized; m is the size of the population.

(2) The selection probability $p(x_k)$ and accumulation probability $Q(x_i)$ of individuals are calculated according to their serial numbers.

$$Q(x_i) = \sum_{k=1}^{i} p(x_k) \tag{18}$$

(3) A number $r = rand(0,1)$ is randomly generated between 0 and 1 to determine which interval the number falls in. If $Q(x_{i-1}) < r < Q(x_i)$, the interval is selected.

(4) Repeat step 3, and stop when the number of selected offspring reaches the required number.

2.4. Kriging Model Updating Method Based on PSO

PSO is a method developed by simulating the predation behavior of birds. Regard the set objective function as the behavior of the birds searching for food. Regard the range of single or multiple variables in the objective function as the flight space of the birds. Regard the space where the solution of the objective function is located as the space where the birds are searching for food in flight. Regard the birds as particle groups and regard each bird as a particle. In the process of searching for and capturing food, carry out information exchange and transmission among individual birds. The flight trajectory and flight speed are adjusted continuously, and then prey is gradually approached. When birds are in the process of finding and capturing food, individual birds continuously share and transmit information. This sharing of information enables each individual to understand their location. The information is then updated continuously to determine whether the target they are locked on to and the flight status is the best combination, and they pass this information on to other individuals. The birds then search near the target and regard the process of finding the target as the process of finding the optimal solution in the optimization problem [26].

In the PSO algorithm, assume there is a group of particles with the total number of m, which belongs to D-dimensional space, and use the position of $x_i = (x_{i1}, x_{i2}, \cdots, x_{iD})$, $i = 1, 2, \cdots m$ for each particle in this group.

Then set the objective function, namely the fitness function. Through the objective function, the most satisfying position of each particle in the group in the space can be found and then expressed with $P_i = (p_{i1}, p_{i2}, \ldots, p_{iD})$. The velocity of the particle is expressed in terms of $V_i = (v_{i1}, v_{i2}, \ldots, v_{iD})$. $P_g = (p_{g1}, p_{g2}, \ldots, p_{gD})$ represents the optimal position of each particle in the group.

The updated position and velocity of each particle in the particle swarm can be calculated according to the following formula:

$$\begin{gathered} v_{(i+1)d} = wv_{id} + c_1 r_1 (p_{id} - x_{id}) + c_2 r_2 \left(p_{gd} - x_{id}\right) \\ x_{(i+1)d} = x_{id} + v_{(i+1)d} \end{gathered} \tag{19}$$

where $x_{(i+1)d}$ represents the position of the next generation of the particle, which is calculated by the current position of the particle x_{id} and the velocity of the next generation $v_{(i+1)d}$. w is the weight for speed update, or inertia weight; c_1 and c_2 is the acceleration factor, and they are generally taken as $c_1 = c_2 = [0,4]$; r_1 and r_2 represent the random number between 0 and 1. The individual optimal solution for particles can be expressed in terms of P_{best}, and the optimal solution for particle swarm can be expressed in terms of G_{best}.

The optimization process of the PSO algorithm can be summarized as follows:

(1) It is assumed that the total number of particles in the particle swarm is m, which belongs to D-dimensional space. The inertia weight is w , the acceleration factor is c_1 and c_2, and the maximum flight speed is v_{max}.

(2) Calculate the fitness value $f(x)$ of each particle according to the objective function, compare the current fitness value of each particle with the adaptive value of the optimal individual position, compare the current fitness value of each particle P_i with

the fitness value of the global optimal position P_{best}, and calculate the velocity $v_{(i+1)d}$ and position $x_{(i+1)d}$ of the particle at the next moment.

(3) Iterative calculation.
(4) Calculate the fitness value of each particle and calculate the optimal position of the individual. Calculate the optimal position of the population and the optimal fitness value of the population, and update the speed and position of the particle.
(5) Determine whether the iteration reaches the maximum number or the convergence threshold. If the iteration reaches the maximum number or the convergence threshold, the optimization will stop. Instead, go back to step 3.

2.5. Latin Hypercube Sampling

Based on the theory of small sample learning and prediction, the samples of the Kriging model are limited. The method of selecting sample points is particularly important to carrying out the experimental design. In this section, adopt the Latin hypercubic sampling method for sample experimental design.

Latin hypercubic sampling is a stratified sampling method proposed by Makay et al. [31]. This method can avoid the problem that the sampling points overlap in the local area, thereby ensuring that the sampling points are evenly distributed in the sampling space. Suppose there is a hypercube, the variable dimension of which is n, $\mathbf{x}^i \in [\ \mathbf{x}_l^i \quad \mathbf{x}_u^i\ ]$, i = 1,2 . . . n, $\mathbf{x}^i$ represents the i th dimension variable. $\mathbf{x}_u^i$ and $\mathbf{x}_l^i$ are the upper and lower bounds of the $\mathbf{x}^i$, respectively. The operation of generating m samples using the Latin hypercube sampling method is as follows:

(1) Determine the number of samples m to be taken.
(2) Divide the interval of each dimension $\mathbf{x}^i$ into m non-overlapping cells with the same probability, and divide the original hypercube into small hypercubes with the number of m^n.
(3) Generation matrix M, whose dimension is $m \times n$. Each column in the matrix M is formed by random arrangement of sequence $\{1, 2, \cdots, m\}$.
(4) Each row in the matrix M represents a small hypercube to be extracted, and a point is randomly selected from each small hypercube to obtain the required m samples.

3. Finite Element Model Updating for Two Bridges

3.1. Verification of Plane Truss

The Kriging model method based on GA, BMO, and PSO algorithms is used to update the finite element model of a 14-span plane truss structure in reference [32], as shown in Figure 2. There are two supports in the model with a fixed support at the left end and hinge support at the right end. Each member in the structure is a round steel tube with an inner diameter of 0.054 m and an outer diameter of 0.085 m. The initial elastic modulus is 210 GPa, mass density is 7800 kg/m^3, and Poisson's ratio is 0.3.

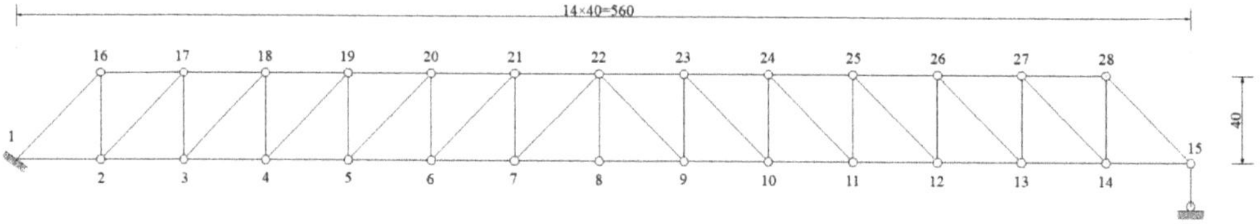

Figure 2. The diagram of plan truss (Unit: cm).

During the finite element model updating, the elastic modulus and density of the material are selected as the parameters to be corrected. Assume that the structural parameters to be corrected obey the normal distribution of X~N (μ, σ). μ is the initial value (i.e., the design value) of the parameters to be corrected. σ = μ × α is the standard deviation of the parameters to be corrected, and α represents the coefficient of variation of the parameters to be corrected. Set the coefficient of variation of the design value of the elastic modulus and density as 5%.

Use the Latin hypercube sampling method to sample the elastic modulus and density, and substitute 100 samples into the finite element model of the bridge structure, and calculate the first six modal frequencies of 100 truss models. Take the first three frequencies of the structure obtained in the model as training samples, and the Kriging model based on the GA algorithm, BMO algorithm, and PSO algorithm to train the samples, respectively, and obtain the corresponding training models of the three methods. According to the measured data, and using the best three groups of a training model to predict, the elastic modulus and density can be obtained. Use the obtained values to input into the finite element model based on three methods and compare with the reference [32]. The specific results are shown in Table 1.

Table 1. Comparison of optimization results of truss model (Unit: Hz).

Order	Measured	Ref. [32]		GA		BMO		PSO	
		Modal Frequency	Error	Modal Frequency	Error	Modal Frequency	Error	Modal Frequency	Error
1	8.79	8.83	−0.55%	8.79	0.00%	8.76	0.34%	8.75	−0.46%
2	29.60	30.18	−1.48%	29.78	−0.60%	29.98	1.28%	29.78	−0.60%
3	43.39	41.65	2.23%	42.66	1.69%	43.14	−0.58%	42.92	1.08%
4	59.10	59.62	−0.24%	59.55	−0.75%	59.77	−1.13%	59.19	−0.15%
5	90.62	91.34	−0.40%	91.09	−0.52%	90.75	0.14%	90.85	0.25%
6	119.81	120.84	−0.11%	120.86	−0.88%	119.95	0.12%	119.88	0.06%

As shown in Table 1, the results obtained by the Kriging model based on the GA algorithm, BMO algorithm, and PSO algorithm are close. The maximum modal frequency error of the reference [32] is 2.23%, and the minimum value is −0.11%. The maximum mean error of results based on the GA algorithm, BMO algorithm, and PSO algorithm is 1.69%, 1.28%, and 1.08%, respectively, and the minimum value is 0.00%, 0.12%, and 0.06%, respectively.

3.2. *Finite Element Model Updating for Cable-Stayed Bridge*

3.2.1. Description

The proposed optimization algorithm was applied to the finite element model updating for a cable-stayed bridge. The main bridge is 604 m in length. The half span of the bridge is 13.45 m, which consists of 2 × 0.5 m collision barrier, 2.5 m hard shoulder, 2 × 3.75 m lanes, 0.5 m curb, 0.25 m constant, and 1.7 m half column width. The main bridge is a prestressed concrete cable-stayed bridge with two towers and a single cable plane, as shown in Figure 3.

The cable tower of this bridge adopts the single-column tower, and the height above the bridge deck is 78 m. The cross-section of the tower column is a circular hollow-core. The longitudinal width is 6.9 m, transverse width is 3.4 m, and thickness is 0.8–1.1 m in the transverse direction and 1.5 m in the longitudinal direction. The central piers are round-ended piers with transverse width of 13.8–15 m and longitudinal width of 5.8 m and 7 m, and the thickness is 1–2 m. At the bottom of the pier, a substantial section is used. The foundation is made up of 16 bored piles with a diameter of 2.8 m, and a monolithic reinforced concrete bearing platform with a height of 6 m. The stay cables are dense cable systems and fan space layout with a total of 184. Considering the reference wind speed is up to 61.3 m/s of the design of the main girder, the high strength galvanized steel wire is used to make the parallel steel wire rope. The primary beam cable spacing is 6 m, and the tower cable spacing is 1.6 m. The main beam section adopts a triangular box with a single box and a double chamber. The full width of the box girder of this bridge is 26.9 m, the beam spacing is 6 m, the height of the main girder is 3.2 m, the thickness of the box girder roof is 27 cm, and the thickness of the bottom plate is 25 cm. The cross-section of the main girder is shown in Figure 4.

Figure 3. Photo and elevation view of the cable-stayed bridge. (**a**) The side elevation photo of the bridge during construction. (**b**) Elevation view.

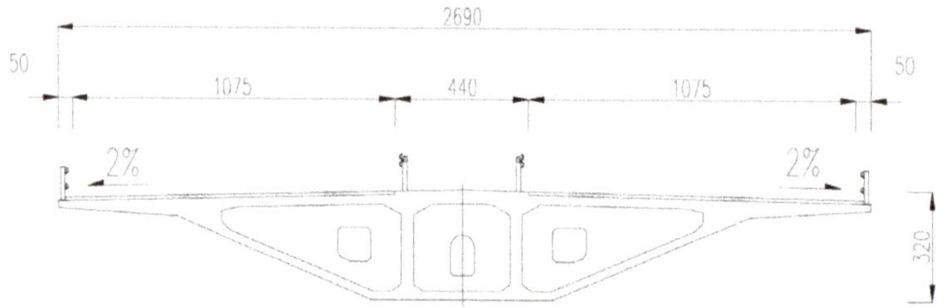

Figure 4. Cross section of the cable-stayed bridge (unit: cm).

3.2.2. Modal Test

Modal tests were carried out on the bridge before the traffic officially opened.

In order to obtain the vertical modal characteristics of the structure, vertical measurement points were arranged at each cable at both bounds of the main span of the main girder. The whole bridge had 24 measurement points which divide into four groups. Since the bridge was an asymmetrical structure, the sensors were arranged by using the symmetry of the structure when selecting the measurement points, as shown in Figure 5.

Figure 5 shows the layout of the 1/4 bridge, which contains two groups of test points. The first group contains V1, V2, V3, V4, V5, and V6 (test reference points), and the second group contains V7, V8, V9, V10, V11, and V6.

By collecting the modal information of the bridge structure, the time-history curves of the structure can be obtained, respectively (see Figure 6). The results of vibration mode and frequency of the cable-stayed bridge were also obtained, as shown in Table 2.

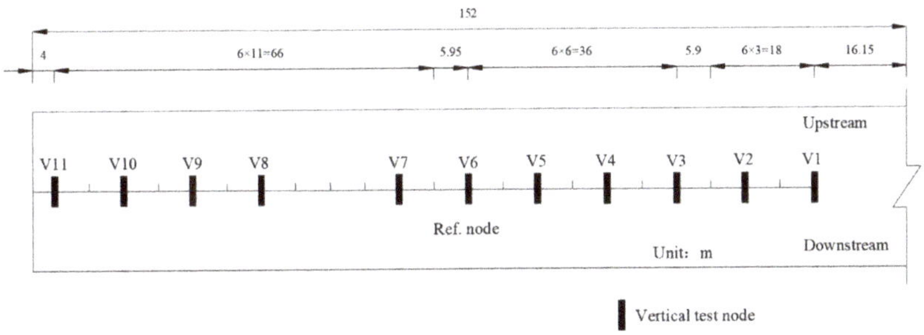

Figure 5. Measurement setup of the girder (1/4 span).

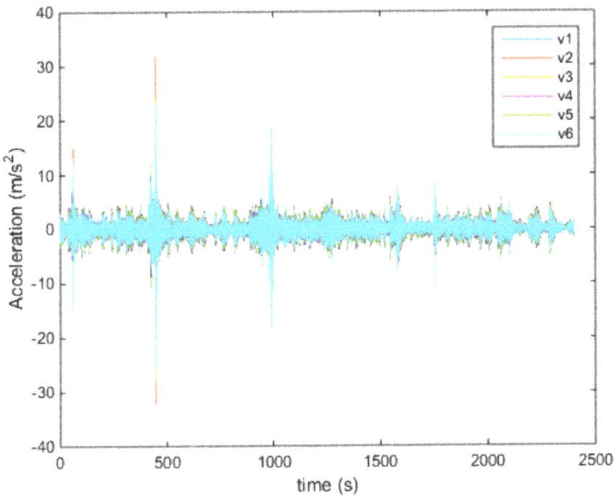

Figure 6. The typical time-history response.

Table 2. Comparison of measured and calculated data of truss model (Unit: Hz).

Order	Measured	Finite Element Model	Error (%)	Modal
1	0.4639	0.4582	1.23	Vertical bending
2	0.6507	0.5981	8.08	Antisymmetric vertical bending
3	1.1960	1.0517	12.07	Vertical bending
4	-	1.5123	-	Vertical bending
5	1.9814	1.9806	0.04	Vertical bending
6	2.6415	2.6222	0.73	Vertical bending

3.2.3. Finite Element Model of Cable-Stayed Bridge

In order to accurately analyze the structure, Midas/Civil software was used to build the model of the cable-stayed bridge, as shown in Figure 7.

In the process of model updating, the selected parameters were elastic modulus E_{c1} and bulk density γ_{c1} of concrete of main girder, and elastic modulus E_{c2} and bulk density γ_{c2} of concrete of the main tower.

By analyzing the acceleration response of the main girder, obtain the first six orders, as shown in Table 2. The measured modal frequency of the bridge is higher than that of the finite element model, and the error is between 0% and 13%. Hence, the parameters need to be updated to obtain accurate results. This section used the Kriging model to update the finite element model based on the proposed three algorithms.

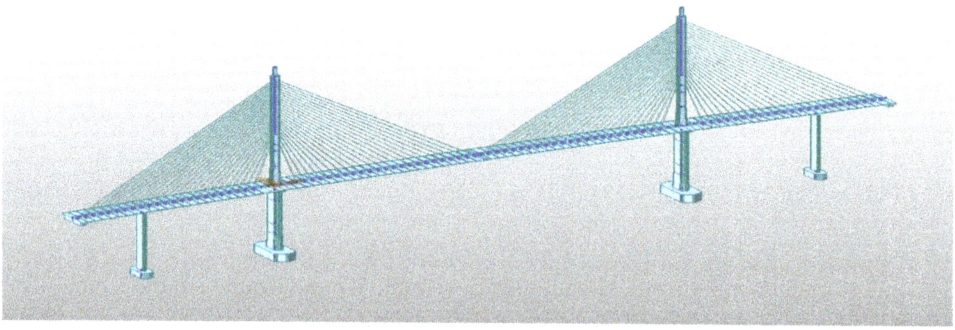

Figure 7. Finite element model of the cable-stayed bridge.

3.2.4. Results

From Table 2, it is easy to find that the vertical modes of the first order, second order, third order, fifth order, and sixth order of the main girder of the cable-stayed bridge were effectively identified according to the measured data. Therefore, select the frequencies corresponding to the vertical modes of the first, second, third, fifth, and sixth order of the main girder of the cable-stayed bridge as the training samples for calculation and analysis. According to GA, BMO, and PSO algorithms, train the Kriging model after obtaining the optimization results of theta. The measured results were substituted into the trained model to obtain the correction results of the parameters to be updated. Next, evaluate the accuracy of parameter correction by using the 2-norm of analysis frequency error as the evaluation index. The calculated frequency results after the model updating as shown in Table 3. Table 4 shows the modification results of the design parameters to be modified obtained in the three finite element model modification methods.

Table 3. Comparison of optimization results of the cable-stayed bridge (Unit: Hz).

Order	Measured Natural Frequency	GA Natural Frequency	BMO Natural Frequency	PSO Natural Frequency
1	0.4639	0.4595	0.4611	0.4622
2	0.6507	0.6351	0.6375	0.6398
3	1.1960	1.1462	1.1453	1.1538
5	1.9814	1.9785	1.9825	1.9811
6	2.6415	2.6209	2.6235	2.6285

Table 4. The updating parameters of finite element model.

Parameters	Not Updating	GA	BMO	PSO	Coefficient of Variation
E_{c1}/GPa	35.50	35.08	35.21	35.68	10%
$\gamma_{c1}/(N/m^3)$	26,000	26,121	25,874	26,225	10%
E_{c2}/GPa	36.00	35.21	35.87	36.33	10%
$\gamma_{c2}/(N/m^3)$	26,000	25,854	26,172	26,228	10%

Table 5 shows the comparison of the time consumed by three methods. By using the same laptop and the same structure, GA methods took half a minute, followed by the BMO method, and finally the PSO method, which took 90 s.

The Kriging model-based model updating method reduced the requirement for sample size in the optimization process and improved the calculation efficiency without sacrificing the calculation accuracy. According to the results of the Kriging model updating method based on the GA algorithm, the frequency error between the training samples in the finite element model and the updating model was the largest, which ranges from 0.15% to 4.39%.

Table 5. Comparison of time-consuming of three methods (Unit: min).

GA	BMO	PSO
0.5	1.1	1.5

When the Kriging model based on the BMO algorithm was adopted, the error range of the results compared with the measured data was smaller than that of the Kriging model based on the GA algorithm, ranging from 0.06% to 4.24%. The result was similar to the updating result based on the GA algorithm. However, in the process of calculation, in order to obtain high-precision results, it was necessary to set a sufficient number of population and number of descendants, which results in the updating method based on the BMO algorithm being more time-consuming than the updating method based on GA algorithm.

Compared with the optimization results based on the GA algorithm and BMO algorithm, the updating results of the Kriging model based on the PSO algorithm have the smallest error range from 0.02% to 3.53% compared with the measured results. Similarly, if high-precision optimization results were needed, sufficient population numbers and the number of descendants should be set in the algorithm, which also has the problem of being time-consuming. Compared with the other two methods, the Kriging model updating method based on the PSO algorithm was the most time consuming of the three methods.

Calculate the Modal Assurance Criterion (MAC) values of the analysis mode and the test mode to illustrate the degree of correlation between the updating model and the test mode. Since the cable-stayed bridge used in this example mainly collected the vertical frequency and mode information of the main girder during dynamic testing, the MAC values calculated here also focused on the relevant information of the main girder, as shown in Table 6. For the PSO algorithm, the MAC is the best, the second one is the BMO algorithm, and GA is the last one.

Table 6. The MAC comparison results of the cable-stayed bridge.

Order	Not Updating	GA	BMO	PSO
1	0.9325	0.9083	0.9245	0.9588
2	0.9182	0.9192	0.9199	0.9215
3	0.9923	0.9921	0.9925	0.9933
5	0.9322	0.9304	0.9338	0.9358
6	0.9159	0.9158	0.9164	0.9172

Table 7 shows the comparison of 2-norm of errors by using different model updating methods. When the finite element model was not updated, the error of 2-norm between the measured frequencies and the corresponding frequencies of the finite element model was the largest. Among the three Kriging model updating methods, the order of the error of 2-norm between the frequency of the updated model and the measured frequency was the GA algorithm, BMO algorithm, and PSO algorithm.

Table 7. Comparison of 2-norm of errors by using different model updating methods.

Not Updating	GA	BMO	PSO
0.1788	0.0506	0.0475	0.0394

4. Conclusions

This paper presented a finite element model updating method based on the Kriging model and intelligent group optimization. The method can optimize the parameters of the bridge structure and of updating the finite element model. Based on the proposed optimization algorithm, verify the effectiveness of the model updating methods by using a plane truss supported beam structure. Then, taking a cable-stayed bridge as an example, the proposed method was used to update the design parameters of some components in the bridge structure.

To update finite element model of bridge, using GA, BMO, and PSO methods can obtain consistent results and have good accuracy. This proves that the three methods used in this paper can be applied not only to simple structures but also to complex structures such as cable-stayed bridges.

Among the three updating methods, the PSO algorithm has the highest accuracy, and the GA algorithm has the lowest accuracy. In terms of computing efficiency, the GA algorithm takes the least time. The BMO algorithm requires relatively more sample data in the calculation process. Therefore, the time consumption was increased. The PSO algorithm is the most time consuming because it needs to be solved iteratively.

Author Contributions: Conceptualization, J.W.; methodology, J.W. and F.C.; software, C.Z. and C.L.; validation, C.Z. and J.W.; data curation, C.Z. and J.W.; writing—original draft preparation, J.W.; writing—review and editing, J.W., S.H., R.Z., Y.Z.; supervision, R.Z. and J.W.; funding acquisition, C.Z. and Y.Z. All authors have read and agreed to the published version of the manuscript.

Funding: The first author wishes to thank the support by National Natural Science Foundation of China Grants 51908139, National-local joint Engineering Laboratory of building health monitoring and disaster prevention technology Grants GG19KF003, Department of Transport of Hubei Province Grants 2020-2-5-1, Scientific research project of Wuhan Polytechnic University Grants Y047.

Institutional Review Board Statement: Not applicable.

Informed Consent Statement: Not applicable.

Data Availability Statement: The data presented in this study are available from the second author upon reasonable request.

Conflicts of Interest: The authors declare no conflict of interest.

References

1. Xu, X.; Huang, Q.; Ren, Y.; Zhao, D.Y.; Yang, J.; Zhang, D.Y. Modeling and separation of thermal effects from cable-stayed bridge response. *J. Bridge Eng.* **2019**, *24*, 04019028. [CrossRef]
2. Xu, X.; Huang, Q.; Ren, Y.; Zhao, D.Y.; Zhang, D.Y.; Sun, H.B. Condition evaluation of suspension bridges for maintenance, repair and rehabilitation: A comprehensive framework. *Struct. Infrastruct. Eng.* **2019**, *15*, 555–567. [CrossRef]
3. Islam, M.M.; Hu, G.; Liu, Q. Online Model Updating and Dynamic Learning Rate-Based Robust Object Tracking. *Sensors* **2018**, *18*, 2046. [CrossRef] [PubMed]
4. Xiong, W.; Cai, C.S.; Kong, B.; Zhang, X.; Tang, P. Bridge Scour Identification and Field Application Based on Ambient Vibration Measurements of Superstructures. *J. Mar. Sci. Eng.* **2019**, *7*, 121. [CrossRef]
5. Kim, S.; Kim, N.; Park, Y.-S.; Jin, S.-S. A Sequential Framework for Improving Identifiability of FE Model Updating Using Static and Dynamic Data. *Sensors* **2019**, *19*, 5099. [CrossRef]
6. Jan, B.; Pawel, H.; Marco, T. SHM system and a FEM model-based force analysis assessment in stay cables. *Sensors* **2021**, *21*, 1927.
7. Alberto, M.B.M.; Luís, M.C.S.; João, H.J.O.N. Optimum design of concrete cable-stayed bridges. *Eng. Optimiz.* **2016**, *48*, 772–791.
8. Farid, G.; Niloofar, M.; Hamed, E.; Ertugrul, T. Bridge Digital Twinning Using an Output-Only Bayesian Model Updating Method and Recorded Seismic Measurements. *Sensors* **2022**, *22*, 1278.
9. Xu, X.; Huang, Q.; Ren, Y.; Zhao, D.Y.; Yang, J. Sensor fault diagnosis for bridge monitoring system using similarity of symmetric responses. *Smart Struct. Syst.* **2019**, *23*, 279–293.
10. Asadollahi, P.; Huang, Y.; Li, J. Bayesian Finite Element Model Updating and Assessment of Cable-Stayed Bridges Using Wireless Sensor Data. *Sensors* **2018**, *18*, 3057. [CrossRef]
11. Cha, P.D.; Tuck-Lee, J.P. Updating structural system parameters using frequency response data. *J. Eng. Mech.* **2000**, *126*, 1240–1246. [CrossRef]
12. Zapico, J.L.; Gonzalez, M.P.; Friswell, M.I.; Taylor, C.A.; Crewe, A.J. Finite element model updating of a small scale bridge. *J. Sound Vib.* **2003**, *268*, 993–1012. [CrossRef]
13. Tran-Ngoc, H.; Khatir, S.; De Roeck, G.; Bui-Tien, T.; Nguyen-Ngoc, L.; Abdel Wahab, M. Model Updating for Nam O Bridge Using Particle Swarm Optimization Algorithm and Genetic Algorithm. *Sensors* **2018**, *18*, 4131. [CrossRef]
14. Kirkpatrick, S.; Gelatt, C.D.; Vecchi, M.P. Optimization by simulated annealing. *Science* **1983**, *220*, 671–680. [CrossRef]

15. Levin, R.I.; Lieven, N.A.J. Dynamic finite element model updating using simulated annealing and genetic algorithms. *Mech. Syst. Signal Process.* **1998**, *12*, 91–120. [CrossRef]
16. Astroza, R.; Nguyen, L.T.; Nestorović, T. Finite element model updating using simulated annealing hybridized with unscented Kalman filter. *Comput. Struct.* **2016**, *177*, 176–191. [CrossRef]
17. Guo, J.; Yuan, W.; Dang, X.; Alam, M.S. Cable force optimization of a curved cable-stayed bridge with combined simulated annealing method and cubic B-Spline interpolation curves. *Eng. Struct.* **2019**, *201*, 109813. [CrossRef]
18. He, S.; Ng, C.T. A probabilistic approach for quantitative identification of multiple delaminations in laminated composite beams using guided waves. *Eng. Struct.* **2016**, *127*, 602–614. [CrossRef]
19. Behmanesh, I.; Moaveni, B. Probabilistic identification of simulated damage on the Dowling Hall footbridge through Bayesian finite element model updating. *Struct. Control Health Monit.* **2015**, *22*, 463–483. [CrossRef]
20. Guo, N.; Yang, Z.; Wang, L.; Bian, X. A updating method using strain frequency response function with emphasis on local structure. *Mech. Syst. Signal Process.* **2019**, *115*, 637–656. [CrossRef]
21. Shabbir, F.; Omenzetter, P. Model updating using genetic algorithms with sequential niche technique. *Eng. Struct.* **2016**, *120*, 166–182. [CrossRef]
22. Li, B.; Yang, X.; Xuan, H. A hybrid simulated annealing heuristic for multistage heterogeneous fleet scheduling with fleet sizing decisions. *J. Adv. Transp.* **2019**, 5364201. [CrossRef]
23. Katahira, K. The statistical structures of reinforcement learning with asymmetric value updates. *J. Math. Psychol.* **2018**, *87*, 31–45. [CrossRef]
24. Bartilson, D.T.; Jang, J.; Smyth, A.W. Finite element model updating using objective-consistent sensitivity-based parameter clustering and Bayesian regularization. *Mech. Syst. Signal Process.* **2019**, *114*, 328–345. [CrossRef]
25. Zhu, J.J.; Huang, M.; Lu, Z.R. Bird mating optimizer for structural damage detection using a hybrid objective function. *Swarm Evol. Comput.* **2017**, *35*, 41–52. [CrossRef]
26. Qin, S.; Zhang, Y.; Zhou, Y.-L.; Kang, J. Dynamic Model Updating for Bridge Structures Using the Kriging Model and PSO Algorithm Ensemble with Higher Vibration Modes. *Sensors* **2018**, *18*, 1879. [CrossRef]
27. Wu, J.; Yan, Q.; Huang, S.; Zou, C.; Zhong, J.; Wang, W.W. Finite Element Model Updating in Bridge Structures Using Kriging Model and Latin Hypercube Sampling Method. *Adv. Civ. Eng.* **2018**, *2018*, 8980756. [CrossRef]
28. Wang, Z.; Shafieezadeh, A. Real-time high-fidelity reliability updating with equality information using adaptive Kriging. *Reliab. Eng. Syst. Saf.* **2020**, *195*, 106735. [CrossRef]
29. Krige, D.G. A Statistical Approach to Some Mine Valuation and Allied Problems on the Witwatersrand. Master's Thesis, Faculty of Engineering, University of the Witwatersrand, Johannesburg, South Africa, 1951.
30. Lophaven, S.N.; Nielsen, H.B.; Søndergaard, J. *DACE—A Matlab Kriging Toolbox*; IMM-TR-2002-12; Technical University of Denmark: Lyngby, Denmark, 2002; pp. 1–28.
31. McKay, M.D.; Beckman, R.J.; Conover, W.J. A comparison of three methods for selecting values of input variables in the analysis of output from a computer code. *Technometrics* **2000**, *42*, 55–61. [CrossRef]
32. Yan, G.R.; Duan, Z.D.; Ou, J.P. Application of genetic algorithm on structural finite element model updating. *J. Harbin Inst. Technol.* **2007**, *39*, 181–186. (In Chinese)

buildings

MDPI

Article

A Joint State-Parameter Identification Algorithm of a Structure with Non-Diagonal Mass Matrix Based on UKF with Unknown Mass

Shiyu Wang and Ying Lei *

School of Architecture and Civil Engineering, Xiamen University, Xiamen 361005, China; kaixing007wang@126.com
* Correspondence: ylei@xmu.edu.cn

Abstract: Inaccurate mass estimates have been recognized as an important source of uncertainty in structural identification, especially for large-scale structures with old ages. Over the past decades, some identification algorithms for structural states and unknown parameters, including unknown mass, have been proposed by researchers. However, most of these identification algorithms are based on the simplified mechanical model of chain-like structures. For a chain-like structure, the mass matrix and its inverse matrix are diagonal matrices, which simplify the difficulty of identifying the structure with unknown mass. However, a structure with a non-diagonal mass matrix is not of such a simple characteristic. In this paper, an online joint state-parameter identification algorithm based on an Unscented Kalman filter (UKF) is proposed for a structure with a non-diagonal mass matrix under unknown mass using only partial acceleration measurements. The effectiveness of the proposed algorithm is verified by numerical examples of a beam excited by wide-band white noise excitation and a two-story one-span plane frame structure excited by filtered white noise excitation generated according to the Kanai–Tajimi power spectrum. The identification results show that the proposed algorithm can effectively identify the structural state, unknown stiffness, damping and mass parameters of the structures.

Keywords: unknown mass; joint state-parameter identification; non-diagonal mass matrix; partial acceleration measurement

1. Introduction

Nowadays, more and more large-scale civil building structures, especially high-rise buildings and long-span bridges, are being built all over the world. Once these structures are constructed and used, functional degradation of the structures will become a concerning issue as the working-age of these structures increases. In order to ensure the safety and reliability of these structures, it is particularly important to obtain the information on the state and parameters of these structures accurately and timely. Structural health monitoring (SHM) has received increasing attention in recent decades with the increasing demand for effectively managing the health condition of these important infrastructures. Structural identification (SI) methods play key roles in structural damage detection, model updating and performance evaluation, which are the most important parts of structural health monitoring. Therefore, the proposal of efficient and reliable structural system identification algorithms is very important for the evaluation of the working performance of the structure and the assessment of post-event conditions after natural disasters.

Over the past decades, a great deal of research has been conducted for structural identification in either the frequency domain [1–5] or the time domain [6–10]. It should be noted that almost all of the studies reviewed above assume that structural mass is known when structural systems are identified. In practice, it is often difficult or even impossible to obtain a priori information about the mass of an engineering structure in service. The inaccurate structural mass estimation causes large errors in the identification of the structure, which endanger the reliability and even the safety of the structure.

In order to deal with this problem, many researchers developed some system identification algorithms under unknown mass. These system identification algorithms can be roughly divided into two categories, namely, the frequency-domain method and the time-domain method. In the frequency-domain system identification algorithm, Yuan et al. [11] proposed an iterative algorithm for identifying structural mass and stiffness matrix elements of chain-like structures based on the first two orders of structural mode measurement by combining modal expansion with the least-squares algorithm. Chakraverty et al. [12] refined the above method by using Holzer criteria to improve its computational efficiency and accuracy. Then, Mukhopadhyay et al. [13,14] also proposed a stiffness and mass matrix identification method based on modal expansion. However, in order to perform modal mass normalization processing, the structural mass, which the sensor is installed, must be known prior. In addition, a flexibility-based damage identification algorithm that does not require knowledge of structural mass is provided by Zhang et al. [15]. Farshadi et al. [16] developed a (Transfer ratio function) FRF-based finite element (FE) model updating algorithm. In this algorithm, the sensitivity equation between frequency response function and parameter change is constructed, and the change in stiffness and mass are identified by solving the equation. The above algorithms are all deterministic algorithms that cannot quantitatively describe the uncertainty of identified results. The Bayesian statistical probabilistic approach provides a method that can not only provide us with an optimal estimate of the state and parameters of a structural system but also quantitatively describe the uncertainty of this estimate. Mustafa et al. [17] proposed an efficient and robust Bayesian model updating to update mass and stiffness by introducing a new objective function to remove the coupling effect of stiffness and mass matrix to solve the unidentifiable problem of the traditional Bayesian method when stiffness and mass matrix identified simultaneously. Furthermore, Kim et al. [18] proposed a novel Bayesian model updating algorithm. In the algorithm, the additional mass or additional stiffness is added to the structure to decouple the coupling effects of mass and stiffness matrices of the identification algorithm so that the model mass and stiffness parameters can be updated by comparing the measured data of the reference model and the modified model.

On the other hand, various time-domain techniques were developed. Mei et al. [19] proposed an algorithm synthesis of the Auto-Regressive Moving Average model and structural dynamics equations to identify the changes in structural element mass and stiffness. However, in this algorithm, only one damage index is defined, so it cannot identify the changes in mass and stiffness simultaneously. The algorithm above is improved by Do et al. [20] by introducing two damage indicators for identifying the changes in the stiffness and mass in identifying the structure simultaneously. A restoring force identification method is provided by Marsi et al. [21] to identify chain-like dynamic structural systems under unknown mass. Based on his work, a time-domain identification algorithm of modal parameters to handle the case of chain-like dynamic systems with unknown ambient excitation under unknown mass was proposed by Nayeri et al. [22]. However, this algorithm can only identify the stiffness and mass coupling coefficients of the structure. Zhan et al. [23] generalized the approach by introducing the clustering algorithm to decouple the stiffness and mass coupling coefficients. Meanwhile, Nayeri et al. [24] provided an algorithm combining natural excitation technology and eigenvalue realization technology to identify the modal parameters of structures with unknown mass. Xu et al. [25] investigated a time-domain algorithm for simultaneous identification of mass and nonlinear restoring force based on the least square algorithm and verified the algorithm with a chain-like nonlinear structure of six degrees of freedom with a Magnetorheological (MR) damper mounted in the middle. Huang et al. [26] employed the Kalman filter (KF) technique together with energy equilibrium equations to develop a method that can identify the damping, stiffness and mass of the structure simultaneously online. However, these time-domain algorithms mentioned above all need to be employed under the condition of full measurement of structural acceleration, which may have limited the application of the algorithms. In order to address the aforementioned issues, an adaptive Extended Kalman Filter (EKF) was proposed by Reina et al. [27] to perform joint time-varying mass and state estimation for vehicles, which is simplified to a single degree of freedom model under partial acceleration measurements. Boada et al. [28] proposed a real-time locomotive Vehicle and Road Irregularities estimation algorithm based on dual KF simultaneously. In the dual KF, the first KF is used to estimate the Vehicle state, and the latter is used for mass estimation. Lei et al. [29] extended the method of Zhan et al. [25] by introducing EKF to replace the least squares to identify the mass stiffness coupling parameters to identify structural stiffness and mass changes in the case of partial acceleration response measurement. Zhang et al. [30,31] proposed a loop substructure identification method for chain-like structures, which can identify the mass and stiffness parameters of selected substructures under partial acceleration measurements. Then, Xu et al. [32] also investigated a method for the identification of nonlinearity restoring force of chain-like structural and mass simultaneously using partial acceleration measurements. However, the application of EKF with weighted global iteration (EKF-WGI) makes it impossible to implement the algorithm online.

It is noted that most of the methods reviewed in the aforementioned literature are only suitable for chain-like structures, in which both the mass matrix and the inverse matrix are diagonal matrices. This characteristic of a mass matrix for the chain-like structure greatly reduces the difficulty of structure identification under unknown mass. However, many civil structures cannot simply be simulated by a chain-like structure model, such as super high-rise frame shear wall structures, long-span bridges and industrial plants. To the best of the authors' knowledge, there are very limited studies on joint state-parameter identification algorithms of non-chain-like structures under unknown mass. The joint state-parameters identification, even for linear structure, is essentially a nonlinear problem due to the coupling effects between unknown structural parameters and unknown state variables. Compared with other schemes based on the nonlinear Kalman framework (e.g., EKF, particle filter (PF)), which can identify nonlinear systems, UKF becomes a better choice because it does not need to calculate the Jacobian matrix and has high computational efficiency. To this end, this paper provided an online joint state-parameter identification algorithm of a non-chain-like structure based on UKF under unknown mass using only partial acceleration responses. The content of the paper is organized as follows: Section 2 briefly introduces the calculation process of UKF; Section 3 includes two numerical simulation cases in the context of the beam-type model and plane-frame model used to assess the performance of the joint state-parameter identification algorithm of a non-chain-like structure under unknown mass. Finally, the conclusion and further research are given in the conclusions section.

2. Brief Review of the Unscented Kalman Filter

A generalized n-DOF structural system dynamics equation can be expressed as

$$\mathbf{M}\ddot{\mathbf{z}} + \mathbf{F}(\mathbf{z}, \dot{\mathbf{z}}, \theta) = \eta\mathbf{f} \tag{1}$$

In which $\mathbf{M}$ is the mass matrix; $\mathbf{z}$, $\dot{\mathbf{z}}$ and $\ddot{\mathbf{z}}$ are n-dimension vectors of displacement, velocity and acceleration, respectively; $\mathbf{F}(\mathbf{z}, \dot{\mathbf{z}}, \theta)$ is a generalized restoring force equation vector; θ is a q-dimension structural parametric vector containing the parameters that need to parameterize the restoring force function $\mathbf{F}(\mathbf{z}, \dot{\mathbf{z}}, \theta)$; θ_m is an l-dimension vector that to be identified in this study, which includes the structural mass. $\mathbf{f}$ is a p-dimension external excitation vector and η is the influence matrix corresponding to $\mathbf{f}$. Since the structural system in this paper is a time-invariant system, the time derivatives of unknown parameters in the structure $\dot{\theta}_{m(i)} = 0(i = 1, 2, 3, \ldots l)$. An augmented state vector is defined as $\mathbf{X} = \left\{\mathbf{X}_{\mathbf{z}}^T, \theta_m{}^T\right\}^T = \left\{\mathbf{z}^T, \dot{\mathbf{z}}^T, \theta_m{}^T\right\}^T$, which includes structural displacement, velocity and unknown structural parameters, including the structural mass, stiffness and damping coefficients. The state-space form of Equation (1) can be expressed as follows:

$$\dot{\mathbf{X}} = \begin{Bmatrix} \dot{\mathbf{X}}_{\mathbf{z}} \\ \dot{\theta}_m \end{Bmatrix} = \begin{Bmatrix} \dot{\mathbf{z}} \\ \ddot{\mathbf{z}} \\ \dot{\theta}_m \end{Bmatrix} = \begin{Bmatrix} \dot{\mathbf{z}} \\ \mathbf{M}^{-1}\left[\eta\mathbf{f} - \mathbf{F}(\mathbf{z}, \dot{\mathbf{z}}, \theta)\right] \\ 0 \end{Bmatrix} = \mathbf{g}(\mathbf{X}, \mathbf{f}) \tag{2}$$

where $\mathbf{g}(.)$ denotes system equations for the structural system.

The continuous system Equation (2) on the kth time step can be discrete to be the following form:

$$\mathbf{X}_{k+1} = \mathbf{X}_k + \int_{k\Delta t}^{(k+1)\Delta t} g(\mathbf{X}_{t|k}, \mathbf{f}_k)\boldsymbol{dt} \tag{3}$$

The observation equations for the structural system can be formulated as

$$\mathbf{y}_{k+1} = \mathbf{h}(\mathbf{X}_{k+1}, \mathbf{f}_{k+1}) + \mathbf{v}_{k+1} \tag{4}$$

where $\mathbf{y}_{k+1}$ is an m-dimension measurement vector at time $t = (k+1)\Delta t$ with Δt being the sampling time step, and $\mathbf{v}_{k+1}$ is the measurement noise vector modeled as Gaussian white noise with zero mean and a covariance matrix $\mathbf{E}(\mathbf{v}_{k+1}\mathbf{v}_{k+1}^T) = \mathbf{R}_{k+1}$.

The unscented Kalman filter is implemented in the following three steps:

(1) Sigma point generation step

Firstly, a set of 2N + 1 sigma points whose mean and covariance are $\mathbf{X}_{k|k}$ and $\mathbf{P}_{k|k}^{\mathbf{XX}}$, respectively, are reproduced as

$$\chi_{i,k|k} = \begin{cases} \mathbf{X}_{k|k} & ,i = 0 \\ \mathbf{X}_{k|k} + \left(\sqrt{(N+\lambda)\mathbf{P}_{k|k}^{\mathbf{XX}}}\right)_i & ,i = 1, \ldots N \\ \mathbf{X}_{k|k} - \left(\sqrt{(N+\lambda)\mathbf{P}_{k|k}^{\mathbf{XX}}}\right)_i & ,i = N+1 \ldots 2N \end{cases} \tag{5}$$

where N is defined as the dimension of the state vector $\mathbf{X}$, $\mathbf{X}_{k|k} = \mathbf{E}\{\mathbf{X}_k\}$, $\mathbf{P}_{k|k}^{\mathbf{XX}} = \mathbf{E}\left\{\left(\mathbf{X}_k - \mathbf{X}_{k|k}\right)\left(\mathbf{X}_k - \mathbf{X}_{k|k}\right)\right\}$, $\left(\sqrt{(N+\lambda)\mathbf{P}_{k|k}^{\mathbf{XX}}}\right)_i$ denotes the ith column of the matrix square root, $\lambda = \alpha^2(N+\kappa) - N$ is a scaling parameter, α is a scaling parameter used to incorporate higher-order information. It is often set to an extremely small positive value (e.g., 2×10^{-3}); $\boldsymbol{\kappa}$ is a secondary scaling parameter.

(2) The time updating step

The sigma points are propagated by structural system dynamic Equations as follows:

$$\chi_{i,k+1|k} = \chi_{i,k|k} + \int_{k\Delta t}^{(k+1)\Delta t} \mathbf{g}\left(\chi_{t|k}, \mathbf{f}_k\right)dt \tag{6}$$

and the a priori estimate of the state vector $\mathbf{X}_{k+1|k}$ and corresponding error covariance matrix $\mathbf{P}_{k+1|k}^{\mathbf{XX}}$ are calculated as

$$\mathbf{X}_{k+1|k} = \sum_{i=0}^{2N} w_i^{(m)} \chi_{i,k+1|k} \tag{7}$$

$$\mathbf{P}_{k+1|k}^{\mathbf{XX}} = \sum_{i=0}^{2N} w_i^{(c)} \left(\chi_{i,k+1|k} - \mathbf{X}_{k+1|k}\right)\left(\chi_{i,k+1|k} - \mathbf{X}_{k+1|k}\right)^T + \mathbf{Q}_{k+1} \tag{8}$$

where $w_i^{(m)}$ and $w_i^{(c)}$ are the weights for the predicted mean and covariance, respectively, and given by

$$w_0^{(m)} = \frac{\lambda}{N+\lambda}; w_0^{(c)} = \frac{\lambda}{N+\lambda} + (1 - \alpha^2 + \beta); \tag{9a}$$

$$w_i^{(m)} = w_i^{(c)} = \frac{\lambda}{2(N+\lambda)}(i = 1, 2 \ldots 2N) \tag{9b}$$

where β is a parameter used to contain prior information of the distribution of state variables, and for Gaussian distribution, $\beta = 2$ is optimal.

The predicted measurement $\mathbf{y}_{k+1|k}$ and its error covariance matrix $\mathbf{P}^{\mathrm{yy}}_{k+1|k}$ is computed as

$$\mathbf{y}_{i,k+1|k} = h(\chi_{i,k+1|k}, \mathbf{f}_{k+1});\ \mathbf{y}_{k+1|k} = \sum_{i=0}^{2N} w_i^{(m)} \mathbf{y}_{i,k+1|k} \tag{10}$$

$$\mathbf{P}^{\mathrm{yy}}_{k+1|k} = \sum_{i=0}^{2N} w_i^{(c)} (\mathbf{y}_{i,k+1|k} - \mathbf{y}_{k+1|k})\left(\mathbf{y}_{i,k+1|k} - \mathbf{y}_{k+1|k}\right)^T + \mathbf{R}_{k+1} \tag{11}$$

and the cross-covariance $\mathbf{P}^{\mathrm{Xy}}_{k+1|k}$ matrix is calculated as

$$\mathbf{P}^{\mathrm{Xy}}_{k+1|k} = \sum_{i=0}^{2N} w_i^{(c)} (\mathbf{X}_{i,k+1|k} - \mathbf{X}_{k+1|k})\left(\mathbf{y}_{i,k+1|k} - \mathbf{y}_{k+1|k}\right)^T \tag{12}$$

(3) The measurement updating step

Finally, the augmented state vector $\mathbf{X}_{k+1|k+1}$ and error covariance matrix $\mathbf{P}^{\mathrm{XX}}_{k+1|k+1}$ are updated with the measured output using the Kalman filtering Equations

$$\mathbf{X}_{k+1|k+1} = \mathbf{X}_{k+1|k} + \mathbf{K}_m(\mathbf{y}_{k+1} - \mathbf{y}_{k+1|k}) \tag{13}$$

$$\mathbf{P}^{\mathrm{XX}}_{k+1|k+1} = \mathbf{P}^{\mathrm{XX}}_{k+1|k} - \mathbf{K}_m \mathbf{P}^{\mathrm{yy}}_{k+1|k} \mathbf{K}_m^T \tag{14}$$

in which is the Kalman gain matrix $\mathbf{K}_{k+1}$ given by

$$\mathbf{K}_{k+1} = \mathbf{P}^{\mathrm{Xy}}_{k+1|k}\left(\mathbf{P}^{\mathrm{yy}}_{k+1|k}\right)^{-1} \tag{15}$$

By implementing the identification algorithm based on the unscented Kalman filter, the augmented state vector of the structural system, which contains unknown structural parameters including structural mass, stiffness and damping coefficients, can be identified.

3. Numerical Validation

In this section, two numerical simulation cases are given aimed at verifying the effectiveness of the UKF algorithm for joint state-parameter identification of a structure with a non-diagonal mass matrix under unknown mass in the context of two types of non-chain-like structural models: beam-type model and plan-frame model.

3.1. Identification of a Beam-Type Structure Subjected to White Noise Excitation

Considering only chain-like structures were used to identify stiffness, damping and mass simultaneously in most previous studies, a beam-type structure is investigated in this case. The structure under consideration is a simply supported Euler beam, shown in Figure 1. The beam is modeled using a two-dimensional finite element (FE) model and is equally discretized into six beam elements. For these Euler beam elements, only bending deformations in the vertical plane are considered, and shear deformations are ignored since the shear deformations are very small as compared to the bending deformations. The beam model contains a total of 12 DOFS, which includes five vertical DOFS and seven rotational DOFS.

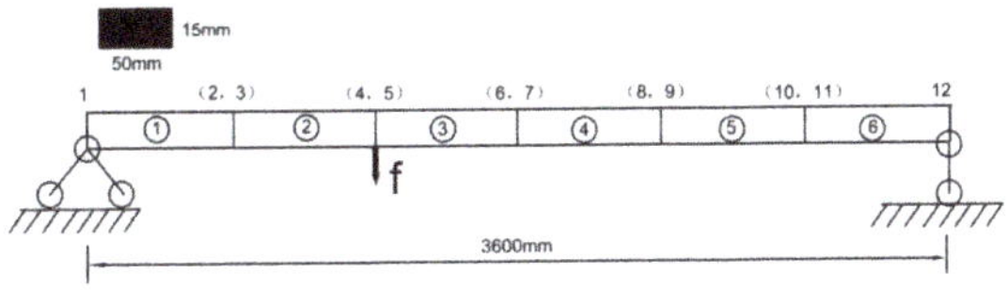

(a)

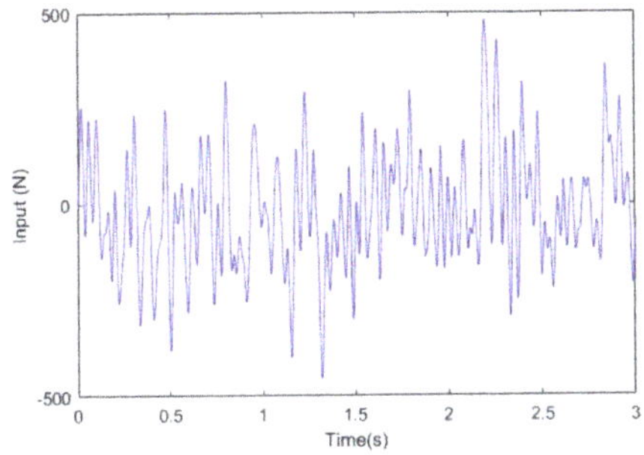

(b)

Figure 1. A simply supported beam subjected to white noise excitation. (**a**) A simply supported beam with unknown parameters; (**b**) broadband white noise excitation.

The beam elements adopt a consistent mass matrix and consistent stiffness matrix. Let $\mathbf{M}_i$ and $\mathbf{K}_i$ be the mass matrix and stiffness matrix of the i-th beam element in the element local coordinate, respectively. The local beam elemental consistent mass and consistent stiffness matrices are calculated using the cubic interpolation functions. The interpolation functions of the i-th element in its local coordinate can be obtained as follows:

$$\mathbf{H}_i = \left\{ 1 - 3\left(\frac{x}{l_i}\right)^2 + 2\left(\frac{x}{l_i}\right)^3, x - 2l_i\left(\frac{x}{l_i}\right)^2 + l_i\left(\frac{x}{l_i}\right)^3, 3\left(\frac{x}{l_i}\right)^2 - 2\left(\frac{x}{l_i}\right)^3, -l_i\left(\frac{x}{l_i}\right)^2 + l_i\left(\frac{x}{l_i}\right)^3 \right\} \tag{16}$$

where l_i is the length of the i-th beam element. The local mass and stiffness matrices for the i-th beam element are provided in Appendix A.

The equation of motion for the simply supported beam can be given by:

$$\mathbf{M}_b\ddot{\mathbf{z}} + \mathbf{C}_b\dot{\mathbf{z}} + \mathbf{K}_b\mathbf{z} = \boldsymbol{\eta}\mathbf{f} \tag{17}$$

where $\ddot{\mathbf{z}}$; $\dot{\mathbf{z}}$; $\mathbf{z}$ are the acceleration, velocity and displacement responses of the simply supported beam, respectively. $\mathbf{f}$ is a one-dimension external excitation vector and $\boldsymbol{\eta}$ is the influence matrix corresponding to $\mathbf{f}$.

$\mathbf{M}_b$ and $\mathbf{K}_b$ are the global mass and stiffness matrices of the simply supported beam, respectively, which can be obtained as the assembly of local element mass and local stiffness matrices. For the simply supported beam, the Rayleigh damping is assumed, and the global damping matrix $\mathbf{C}_b$ is expressed as:

$$\mathbf{C}_b = a_1\mathbf{M}_b + a_2\mathbf{K}_b \tag{18}$$

where a_1 and a_2 are the damping coefficients. The mass, stiffness and damping coefficients are considered unknown in this example. The state vector $\mathbf{X}_\mathbf{z}$ and parameter vector $\boldsymbol{\theta}_m$ of the structural system are combined to form the augmented state vector $\mathbf{X} = \left[\mathbf{z}^T, \dot{\mathbf{z}}^T, \overline{\mathbf{m}}^T, \mathbf{k}^T, a_1, a_2\right]^T$.

In this case, the length of the beam is 3.6 m, and the whole beam adopts a rectangular section with a uniform section size with a width of 50 mm and a height of 15 mm. The

Young's modulus and material density of the beam is chosen as 206 GPa and 7850 kg/m^3, respectively. The first two natural frequencies obtained from the beam model are 21.4 Hz and 133.7 Hz; a modal damping ratio of 5% is assumed for the first two modes.

Structural parameters are selected as follows: The cross-sectional area and moment of inertia of each beam element are $A_i = 7.5 \times 10^{-4}\,\text{m}^2$ and $I_{zi} = 1.406 \times 10^{-8}\,\text{m}^4$ $(i = 1,2\ldots 6)$, respectively. The linear stiffness of each beam element is defined as $k_i = \frac{EI_{zi}}{l_i} = 4.828\,\text{KN.m}$; the linear mass density of each element is defined as $\overline{m}_i = \rho_i A_i = 5.85\,\text{kg/m}\,(i = 1,2\ldots 6)$. The Rayleigh damping coefficients were calculated to be $a_1 = 1.356$ and $a_2 = 1.179 \times 10^{-3}$ according to the first two nature frequencies. The initial guess values for the unknown parameters are selected as: $\overline{m}_{i,0} = 4.68\,\text{kg/m}$, $k_{i,0} = 3.862\,\text{KN.m}\,(i = 1,2\ldots 6)$, $a_{1,0} = 1.084$, $a_{2,0} = 0.944 \times 10^{-3}$. The external excitation acting on a simply supported beam is assumed to be broadband white noise, which acts on the 4th DOF. From the finite element model built in matlab, the acceleration, velocity and displacement responses of the structure are obtained by solving differential Equation (17) using the 4th-order Ronge–Kutta integration method. Only five accelerometers are deployed at the 2nd, 4th, 6th, 8th and 10th DOFs, respectively, to measure the vertical accelerations. When considering the existence of measurement noise, a Gaussian white sequence with a 1% root-mean-square noise-to-signal ratio is added to the calculated response. The sampling frequency is 1000 Hz, and the sampling time is 3 s.

Figure 2 shows the comparisons of the identified and exact time histories of vertical displacement of nodal 2 (z_2), vertical velocity of nodal 2 $(\dot{z}_2)$, rotational displacement of nodal 2 (z_3) and rotational velocity of nodal 2 $(\dot{z}_3)$ obtained from the simulation case. It is shown that both structural displacement and velocity responses can be well tracked.

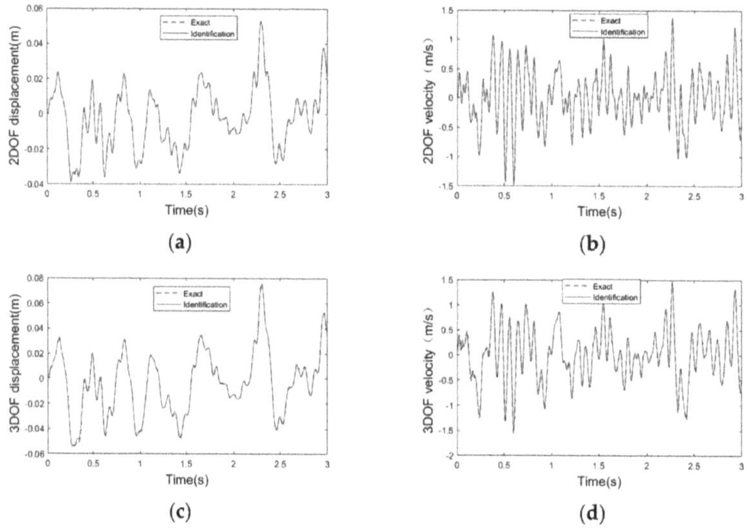

Figure 2. Comparisons of the exact and identified displacements and velocities. (**a**) Comparison of the exact and identified; (**b**) comparison of the exact and identified; (**c**) comparison of the exact and identified; (**d**) comparison of the exact and identified.

The convergence of six unknown parameters $(\overline{m}_1, \overline{m}_6, k_1, k_6, a_1, a_2)$ from the numerical case above was demonstrated in Figure 3. It can be noticed that these identified parameter values can converge to their exact values quite fast.

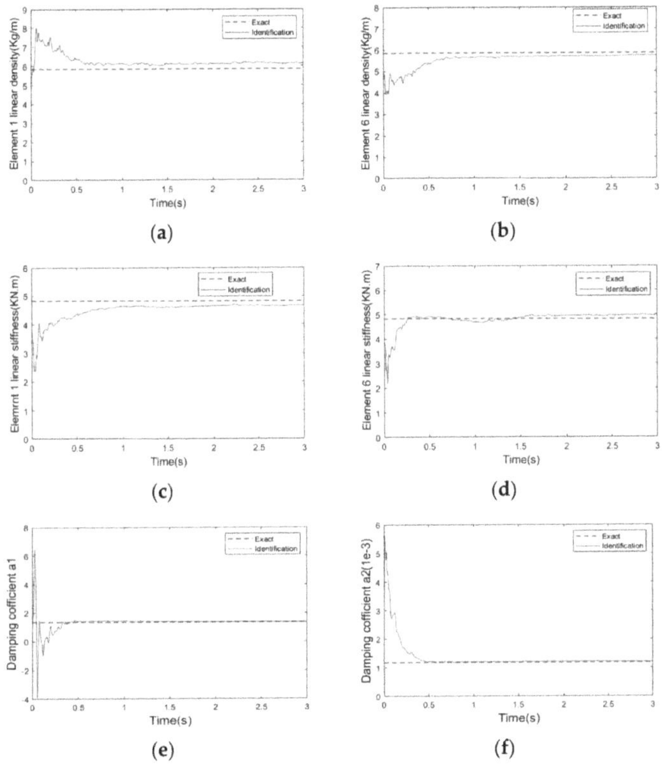

Figure 3. Parameter estimation results for the simply supported beam. (**a**) Convergence of identified linear density; (**b**) convergence of identified linear density; (**c**) convergence of identified linear stiffness k_1; (**d**) convergence of identified linear stiffness k_6; (**e**) convergence of identified damping coefficient a_1; (**f**) convergence of identified damping coefficient a_2.

The parameter identification values, true values and identified relative error of all elements are listed in Table 1. It is shown that all the elements' parameter identification results meet the accuracy requirements; the relative error of the identified mass, stiffness and damping coefficients relative to their true value is less than 5%.

Table 1. The parameters identification results of the simply supported beam.

Parameter	Identified	Actual	Relative Error (%)
k_1(KN.m)	4.675	4.828	−2.60
k_2(KN.m)	4.908	4.828	2.25
k_3(KN.m)	5.015	4.828	3.87
k_4(KN.m)	4.626	4.828	−4.18
k_5(KN.m)	4.628	4.828	−4.17
k_6(KN.m)	4.962	4.828	2.77
$\overline{m}_1$(kg/m)	6.140	5.850	4.95
$\overline{m}_2$(kg/m)	5.727	5.850	−2.10
$\overline{m}_3$(kg/m)	5.765	5.850	−1.45
$\overline{m}_4$(kg/m)	5.993	5.850	2.44
$\overline{m}_5$(kg/m)	5.724	5.850	−2.15
$\overline{m}_5$(kg/m)	6.075	5.850	3.67
a_1	1.380	1.356	1.77
$a_2(1 \times 10^{-3})$	1.206	1.179	2.29

3.2. Identification of a Plane Frame Structure Subjected to Stationary Filtered White Noise Excitation

In order to further verify the effectiveness of the proposed algorithm for the joint state-parameter identification of the structure with a non-diagonal mass matrix under unknown mass, a plane frame shown in Figure 4 is investigated in this case. A one-story two-span plane span was applied here for numerical simulation. All beams and columns adopt a consistent mass matrix and consistent stiffness matrix; the mass and stiffness matrix detailed representation are shown in Appendix B. Rayleigh damping is assumed in this numerical simulation case. The bases are assumed to be fixed. Therefore, the plane frame contains a total of 12 DOFS, and each node point has three degrees of freedom (horizontal, vertical and rotational).

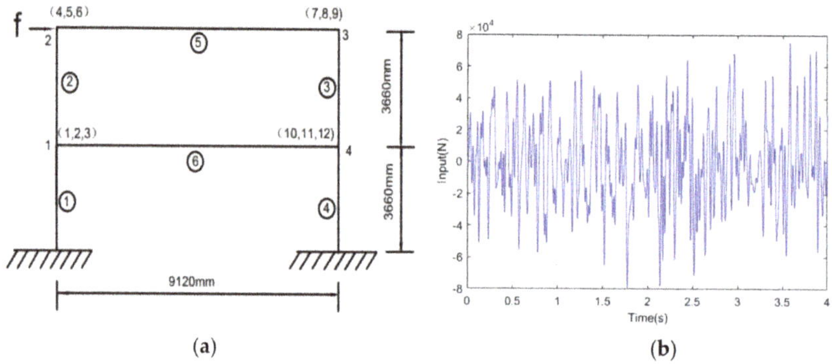

Figure 4. A plane frame subjected to stationary filtered white noise excitation. (**a**) A plane frame subjected to external excitation; (**b**) filtered white noise excitation generated according to the Kanai–Tajimi power spectrum.

In this numerical example, the height of the column is 3.66 m and the length of the beam is 9.12 m. The Young's modulus and material density are taken as 206 GPa and 7850 kg/m^3, respectively. The first two natural frequencies obtained from the plane frame model are 7.55 Hz and 27.3 Hz. A modal damping ratio of 5% is assumed for the first two modes.

Structural parameters are selected as follows: The cross-section area and moment of inertia of each element are $A_i = 2.68 \times 10^{-2}\,\text{m}^2$ and $I_{zi} = 4.870 \times 10^{-4}\,\text{m}^4\,(i = 1,2\ldots6)$. The linear stiffness of each column element is $k_i = \frac{EI_{zi}}{l_i} = 27{,}410\,\text{KN.m}\,(i = 1,2,3,4)$, linear stiffness of each beam element is $k_i = \frac{EI_{zi}}{l_i} = 11{,}000\,\text{KN.m}\,(i = 5,6)$. All the beams and columns adopt uniform linear density $\overline{m}_i = \rho_i A_i = 210.38\,\text{kg/m}\,(i = 1,2\ldots6)$. The Rayleigh damping coefficients were calculated to be $a_1 = 3.738$ and $a_2 = 4.538 \times 10^{-4}$ according to the first two nature frequencies. The initial guess values for the identified parameters are selected as: $k_{i,0} = 21{,}928\,\text{KN.m}\,(i = 1,2\ldots4)$, $k_{i,0} = 8800\,\text{KN.m}\,(i = 5,6)$, $\overline{m}_{i,0} = 166.3\,\text{Kg/m}\,(i = 1,2\ldots6), a_{1,0} = 2.990, a_{2,0} = 3.630$.

In order to verify the robustness of the algorithm to external excitation, this case adopts the filtered stationary white noise excitation generated according to the Kanai–Tajimi power spectrum as the external excitation [33].

$$S_{\ddot{x}_g}(\omega) = \frac{\omega_g^4 + 4\zeta_g^2\omega_g^2\omega^2}{(\omega^2 - \omega_g^2)^2 + 4\zeta_g^2\omega_g^2\omega^2} S_0 \tag{19}$$

where S_0 is the input white noise spectral density, ω_g and ξ_g are the characteristic frequency and characteristic damping ratio of the site, respectively. In this case, these parameters were taken as $\omega_g = 15.6(\text{rad/s})$, $\xi_g = 0.6$. In this study, the stationary filtered white noise time history generated according to the Kanai–Tajimi power spectrum was chosen as the external excitation acting at 4thDOF. From the finite element model built in matlab, the acceleration, velocity and displacement responses of the structure are obtained by solving differential Equation (17) using the 4th-order Ronge–Kutta integration method.

Suppose eight accelerometers are mounted on the plane frame to measure the horizontal accelerations at 1,2,3 and 4 nodal points, i.e., $(\ddot{x}_1, \ddot{x}_4, \ddot{x}_7, \ddot{x}_{10})$, two vertical accelerations at 1,4 nodal points, i.e., $(\ddot{x}_2, \ddot{x}_{11})$, and two rotational accelerations at 1,4 nodal points, i.e., $(\ddot{x}_3, \ddot{x}_{12})$. It is usually a good approximation to assume the rotational motion is related to horizontal motion through the static deflection relation; hence, we compute the rotational acceleration $\ddot{x}_3$ and $\ddot{x}_{12}$ from the horizontal acceleration $\ddot{x}_1$ at node 1. Consquently, our measured response vector is $y = [\ddot{x}_1, \ddot{x}_2, \ddot{x}_3, \ddot{x}_4, \ddot{x}_7, \ddot{x}_{10}, \ddot{x}_{11}, \ddot{x}_{12}]^T$.

When considering the existence of measurement noise, a Gaussian white sequence with a 1% root-mean-square noise-to-signal ratio was added to the calculated response. The sampling frequency is 1024 Hz, and the sampling time is 4 s.

Figure 5a,b show the comparisons of the identified and exact time histories of Horizontal displacement (z_1) of nodal 1 and rotational displacement (z_6) of nodal 4. The identified time histories and the exact time histories $\dot{z}_1, \dot{z}_6$, are depicted in Figure 5c,d, respectively. It is shown that both structural displacement and velocity responses can be identified effectively.

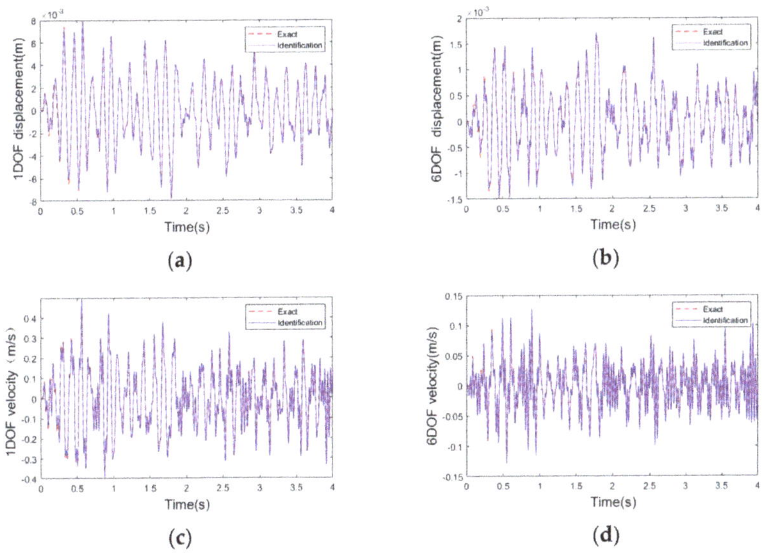

Figure 5. Comparisons of the exact and identified displacements and velocities. (**a**) Comparison of the exact and identified z_1; (**b**) comparison of the exact and identified z_6; (**c**) comparison of the exact and identified $\dot{z}_1$; (**d**) comparison of the exact and identified $\dot{z}_6$.

The convergence of six parameters $(\overline{m}_4, \overline{m}_5, k_4, k_5, a_1, a_2)$ is demonstrated in Figure 6. It shows that the identified linear density of element 4 $(\overline{m}_4)$ and element $5(\overline{m}_5)$, the identified linear stiffness of element $4(k_4)$ and element $5(k_5)$, and the dumping coefficients a_1, a_2 can be identified effectively. Table 2 provides all the elements' parameter identification values, true values and identified relative error of all elements. It is shown that all the elements' parameter identification results meet the accuracy requirements and the relative error of the identified mass, stiffness and damping coefficients relative to their true value is less than 5%.

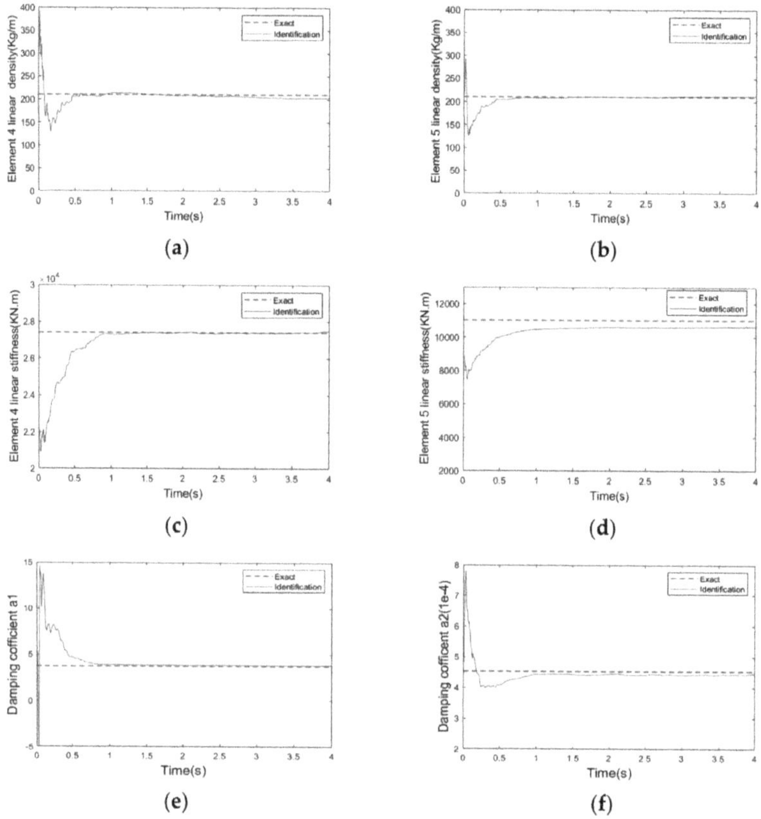

Figure 6. Parameter estimation results for the plane frame. (**a**) Convergence of identified linear density $\overline{m}_4$; (**b**) convergence of identified linear density $\overline{m}_5$; (**c**) convergence of identified linear stiffness k_4; (**d**) convergence of identified linear stiffness k_5; (**e**) convergence of identified dumping coefficient a_1; (**f**) convergence of identified dumping coefficient a_2.

Table 2. The parameters identification results of the plane frame.

Parameter	Identified	Actual	Relative Error (%)
k_1(KN.m)	27,800	27,410	1.42
k_2(KN.m)	27,906	27,410	1.81
k_3(KN.m)	28,434	27,410	3.73
k_4(KN.m)	27,484	27,410	0.27
k_5(KN.m)	10,635	11,000	−3.32
k_6(KN.m)	10,740	11,000	−2.36
$\overline{m}_1$(kg/m)	216.612	210.380	2.96
$\overline{m}_2$(kg/m)	204.300	210.380	−2.89
$\overline{m}_3$(kg/m)	211.701	210.380	0.63
$\overline{m}_4$(kg/m)	203.171	210.380	−3.42
$\overline{m}_5$(kg/m)	212.021	210.380	0.80
$\overline{m}_6$(kg/m)	206.901	210.380	−1.65
a_1	3.851	3.738	3.02
$a_2(1 \times 10^{-4})$	4.456	4.538	−1.80

4. Conclusions

In this paper, a joint state-parameter identification algorithm based on UKF was provided for a structure with unknown mass using partial acceleration measurements. Numerical verification was performed using a simply supported beam subjected to broadband white noise excitation and a one-span two-story plane frame subjected to filtered white noise excitation generated according to the Kanai–Tajimi power spectrum. The conclusions are as follows:

1. Numerical results indicate that the proposed approach can effectively identify the state and unknown parameters, including mass, stiffness, and damping coefficients of non-chain-like structures;
2. Unlike some existing methods, the proposed identification algorithm does not require iterative estimation at each time step, which makes the approach suitable for real-time identification;
3. The proposed algorithm for the identification of joint state-parameter is effective in a noisy environment. In this study, with reasonable noise included, the identification results for structural stiffness, damping and mass are robust to the measurement noises.

In summary, the proposed algorithm is suitable for the real-time identification of states and parameters of a structure with a non-diagonal mass matrix under unknown mass using partial acceleration measurement. Therefore, this paper provides a promising way for the joint state-parameter identification of non-chain-like structures with unknown mass information. However, this paper only demonstrates the algorithm for the identification of linear structures with a non-diagonal mass matrix and assumes that the external excitation is measurable. Extensions of such identification studies are conducted by the authors.

Author Contributions: S.W.: Writing—review, Investigation, Software, Data curation, Validation. Y.L.: Conceptualization, Methodology, Supervision. All authors have read and agreed to the published version of the manuscript.

Funding: The National Natural Science Foundation of China through the project No. 52178304.

Data Availability Statement: All data, models and code generated or used in this study are available from the corresponding author upon reasonable request.

Conflicts of Interest: The authors declare no conflict of interest.

Appendix A. The Local Mass and Stiffness Matrices

The local mass and stiffness matrices of the i-th element of the 2D simply supported beam can be expressed as:

$$\mathbf{M_i} = \frac{\overline{m}_i l_i}{420}\begin{bmatrix} 156 & 22l_i & 54 & -13l_i \\ 22l_i & 4l_i^2 & 13l_i & 13l_i \\ 54 & 13l_i & 156 & -22l_i \\ -13l_i & -3l_i^2 & -22l_i & 4l_i^2 \end{bmatrix} \tag{A1}$$

$$\mathbf{K}_i = k_i\begin{bmatrix} 12/l_i^2 & 6/l_i & -12/l_i^2 & 6/l_i \\ 6/l_i & 4 & -6/l_i & 2 \\ -12/l_i^2 & -6/l_i & 12/l_i^2 & -6/l_i \\ 6/l_i & 2 & -6/l_i & 4 \end{bmatrix} \tag{A2}$$

where mass is uniform along the length of the member, and its mass distribution along the length is defined by the linear density $\overline{m}_i$. The stiffness parameter of the i-th member is defined by the line stiffness $k_i = \frac{EI_{zi}}{l_i}$, in which E, I_{zi}, l_i represent Young's Modulus, inertia moment, and element length, respectively.

Appendix B. The Local Mass and Stiffness Matrices

The local mass and stiffness matrices of the i-th element of the 2D plane frame can be expressed as:

$$\mathbf{M_i} = \frac{\overline{m}_i l_i}{420}\begin{bmatrix} 140 & 0 & 0 & 70 & 0 & 0 \\ 0 & 156 & 22l_i & 0 & 54 & -13l_i \\ 0 & 22l_i & 4l_i^2 & 0 & 13l_i & -3l_i^2 \\ 70 & 0 & 0 & 140 & 0 & 0 \\ 0 & 54 & 13l_i & 0 & 156 & -22l_i \\ 0 & -13l_i & -3l_i^2 & 0 & -22l_i & 4l_i^2 \end{bmatrix} \quad \text{(A3)}$$

$$\mathbf{K}_i = k_i\begin{bmatrix} A_i/I_{zi} & 0 & 0 & -A_i/I_{zi} & 0 & 0 \\ 0 & 12/l_i^2 & 6/l_i & 0 & -12/l_i^2 & 6/l_i \\ 0 & 6/l_i & 4 & 0 & -6/l_i & 2 \\ -A_i/I_{zi} & 0 & 0 & A_i/I_{zi} & 0 & 0 \\ 0 & -12/l_i^2 & -6/l_i & 0 & 12/l_i^2 & -6/l_i \\ 0 & 6/l_i & 2 & 0 & -6/l_i & 4 \end{bmatrix} \quad \text{(A4)}$$

where mass is uniform along the length of the member, and its mass distribution along the length is defined by the linear density $\overline{m}_i$. The stiffness parameter of the i-th member is defined by the line stiffness $k_i = \frac{EI_{zi}}{l_i}$, in which E, I_{zi}, l_i represent Young's Modulus, inertia moment and element length, respectively. The cross-sectional area is represented by A_i.

References

1. Hwang, J.S.; Kwon, D.K.; Kareem, A. Frequency Domain State Space-Based Mode Decomposition Framework. *J. Eng. Mech.* **2019**, *145*, 04019051. [CrossRef]
2. Roy, K.; Ray-Chaudhuri, S. Fundamental mode shape and its derivatives in structural damage localization. *J. Sound Vib.* **2013**, *332*, 5584–5593. [CrossRef]
3. Santos, F.; Cismasiu, C.; Cismasiu, I.; Bedon, C. Dynamic Characterisation and Finite Element Updating of a RC Stadium Grandstand. *Buildings* **2018**, *8*, 141. [CrossRef]
4. Sivasuriyan, A.; Vijayan, D.S.; Górski, W.; Wodzyński, L.; Vaverková, M.D.; Koda, E. Practical Implementation of Structural Health Monitoring in Multi-Story Buildings. *Buildings* **2021**, *11*, 263. [CrossRef]
5. Chaudhary, P.K.; Anjneya, K.; Roy, K. Fundamental Mode Shape-Based Structural Damage Quantification Using Spectral Element Method. *J. Eng. Mech.* **2021**, *147*, 04021091. [CrossRef]
6. Tamuly, P.; Chakraborty, A.; Das, S. Simultaneous Input and Parameter Estimation of Hysteretic Structural Systems Using Quasi-Monte Carlo-Simulation-Based Minimum Variance Unbiased Estimator. *J. Bridge Eng.* **2021**, *26*, 04021081. [CrossRef]
7. Astroza, R.; Ebrahimian, H.; Li, Y.; Conte, J.P. Bayesian nonlinear structural FE model and seismic input identification for damage assessment of civil structures. *Mech. Syst. Signal Process.* **2017**, *93*, 661–687. [CrossRef]
8. Erazo, K.; Nagarajaiah, S. Bayesian structural identification of a hysteretic negative stiffness earthquake protection system using unscented Kalman filtering. *Struct. Control Health Monit.* **2018**, *25*, e2203. [CrossRef]
9. Ebrahimian, H.; Astroza, R.; Conte, J.P.; Papadimitriou, C. Bayesian optimal estimation for output-only nonlinear system and damage identification of civil structures. *Struct. Control Health Monit.* **2018**, *25*, e2128. [CrossRef]
10. Krishnan, M.; Bhowmik, B.; Hazra, B.; Pakrashi, V. Real time damage detection using recursive principal components and time varying auto-regressive modeling. *Mech. Syst. Signal Process.* **2018**, *101*, 549–574. [CrossRef]
11. Yuan, P.; Wu, Z.F.; Ma, X.R. Estimated mass and stiffness matrices of smear building from modal test data. *Earthq. Eng. Struct. Dyn.* **1998**, *27*, 415–421. [CrossRef]
12. Chakraverty, S. Identification of structural parameters of multistory shear buildings from modal data. *Earthq. Eng. Struct. Dyn.* **2005**, *34*, 543–554. [CrossRef]
13. Mukhopadhyay, S.; Lus, H.; Betti, R. Modal parameter based structural identification using input–output data: Minimal instrumentation and global identifiability issues. *Mech. Syst. Signal Process.* **2014**, *45*, 283–301. [CrossRef]
14. Mukhopadhyay, S.; Lus, H.; Betti, R. Structural identification with incomplete output-only data and independence of measured information for shear-type systems. *Earthq. Eng. Struct. Dyn.* **2016**, *45*, 273–296. [CrossRef]
15. Zhang, J.; Xu, J.C.; Guo, S.L.; Wu, Z.S. Flexibility-based structural damage detection with unknown mass for IASC-ASCE benchmark studies. *Eng. Struct.* **2013**, *48*, 486–496. [CrossRef]
16. Farshadi, M.; Esfandiari, A.; Vahedi, M. Structural model updating using incomplete transfer function and modal data. *Struct. Control. Health Monit.* **2016**, *24*, e1932. [CrossRef]
17. Mustafa, S.; Matsumoto, Y. Bayesian Model Updating and Its Limitations for Detecting Local Damage of an Existing Truss Bridge. *J. Bridge Eng.* **2017**, *22*, 04017019. [CrossRef]

18. Zeng, J.; Kim, Y.H. Identification of Structural Stiffness and Mass using Bayesian Model Updating Approach with Known Added Mass: Numerical Investigation. *Int. J. Struct. Stab. Dyn.* **2020**, *20*, 2050123. [CrossRef]
19. Mei, Q.P.; Gul, M. Novel Sensor Clustering-Based Approach for Simultaneous Detection of Stiffness and Mass Changes Using Output-Only Data. *J. Struct. Eng.* **2015**, *141*, 04014237. [CrossRef]
20. Do, N.T.; Gul, M. A time series based damage detection method for obtaining separate mass and stiffness damage features of shear-type structures. *Eng. Struct.* **2020**, *208*, 09914. [CrossRef]
21. Masri, S.F.; Bekey, G.A.; Sassti, H. Non-parametric identification of a class of nonlinear multi-degree dynamic systems. *Earthq. Eng. Struct. Dyn.* **1982**, *10*, 1–30. [CrossRef]
22. Nayeri, R.D.; Masri, S.F.; Ghanem, R.G.; Nigbo, R.L. A novel approach for the structural identification and monitoring of a full-scale 17-story building based on ambient vibration measurements. *Smart Mater. Struct.* **2008**, *17*, 025006. [CrossRef]
23. Zhan, C.; Lin, D.S.; Li, H.N. A local damage detection approach based on restoring force method. *J. Sound Vib.* **2014**, *333*, 4942–4959. [CrossRef]
24. Nayeri, R.D.; Tasbihgoo, F.; Wahbeh, M.; Caffrey, G.P.; Masri, S.F.; Conte, J.P.; Elgamal, A. Study of Time-Domain Techniques for Modal Parameter Identification of a Long Suspension Bridge with Dense Sensor Arrays. *J. Eng. Mech.* **2009**, *135*, 669–683. [CrossRef]
25. Xu, B.; He, J.; Dyke, S.J. Model-free nonlinear restoring force identification for SMA dampers with double Chebyshev polynomials: Approach and validation. *Nonlinear Dyn.* **2015**, *82*, 1507–1522. [CrossRef]
26. Huang, X.H.; Dyke, S.J.; Sun, Z.X.; Xu, Z.D. Simultaneous identification of stiffness, mass, and damping using an on-line model updating approach. *Struct. Control Health Monit.* **2017**, *24*, e1892. [CrossRef]
27. Reina, G.; Paiano, M.; Blanco-Claraco, J.L. Vehicle parameter estimation using a model-based estimator. *Mech. Syst. Signal Process.* **2017**, *87*, 227–241. [CrossRef]
28. Boada, B.L.; Boada, M.J.L.; Zhang, H. Sensor Fusion Based on a Dual Kalman Filter for Estimation of Road Irregularities and Vehicle Mass Under Static and Dynamic Conditions. *IEEE/ASME Trans. Mechatron.* **2019**, *24*, 1075–1086. [CrossRef]
29. Lei, Y.; Qiu, H.; Zhang, F.B. Identification of structural element mass and stiffness changes using partial acceleration responses of chain-like systems under ambient excitations. *J. Sound Vib.* **2020**, *488*, 115678. [CrossRef]
30. Zhang, D.Y.; Li, H. Loop substructure identification for shear structures of unknown structural mass using synthesized references. *Smart Mater. Struct.* **2017**, *26*, 085046. [CrossRef]
31. Zhang, D.Y.; Niu, Z.F.; Tian, J.D.; Li, H. Decentralized loop substructure identification for shear structures with virtual control system. *Struct. Control Health Monit.* **2021**, *29*, e2866. [CrossRef]
32. Xu, B.; Li, J.; Dyke, S.J.; Deng, B.C.; He, J. Nonparametric identification for hysteretic behavior modeled with a power series polynomial using EKF-WGI approach under limited acceleration and unknown mass. *Int. J. Non-Linear Mech.* **2020**, *119*, 103324. [CrossRef]
33. Song, W. Generalized minimum variance unbiased joint input-state estimation and its unscented scheme for dynamic systems with direct feedthrough. *Mech. Syst. Signal Process.* **2018**, *99*, 886–920. [CrossRef]

buildings

MDPI

Article

Deep Learning-Enriched Stress Level Identification of Pretensioned Rods via Guided Wave Approaches

Zi Zhang [1], Fujian Tang [2], Qi Cao [2], Hong Pan [1], Xingyu Wang [1] and Zhibin Lin [1,*]

[1] Department of Civil and Environmental Engineering, North Dakota State University, Fargo, ND 58018, USA
[2] School of Civil Engineering, Dalian University of Technology, Dalian 116024, China
* Correspondence: zhibin.lin@ndsu.edu; Tel.: +1-701-231-7204

Abstract: By introducing pre-compression/inverse moment through prestressing tendons or rods, prestressed concrete (PC) structures could overcome conventional concrete weakness in tension, and thus, these tendons or rods are widely accepted in a variety of large-scale, long-span structures. Unfortunately, prestressing tendons or rods embedded in concrete are vulnerable to degradation due to corrosion. These embedded members are mostly inaccessible for visual or direct destructive assessments, posing challenges in determining the prestressing level and any corrosion-induced damage. As such, ultrasonic guided waves, as one of the non-destructive examination methods, could provide a solution to monitor and assess the health state of embedded prestressing tendons or rods. The complexity of the guided wave propagation and scattering in nature, as well as high variances stemming from the structural uncertainty and noise interference PC structures may experience under complicated operational and harsh environmental conditions, often make traditional physics-based methods invalid. Alternatively, the emerging machine learning approaches have potential for processing the guided wave signals with better capability of decoding structural uncertainty and noise. Therefore, this study aimed to tackle stress level prediction and the rod embedded conditions of prestressed rods in PC structures through guided waves. A deep learning approach, convolutional neural network (CNN), was used to process the guided wave dataset. CNN-based prestress level prediction and embedding condition identification of rods were established by the ultrasonic guided wave technique. A total of fifteen scenarios were designed to address the effectiveness of the stress level prediction under different noise levels and grout materials. The results demonstrate that the deep learning approaches exhibited high accuracy for prestressing level prediction under structural uncertainty due to the varying surrounding grout materials. With different grout materials, accuracy could reach up to 100% under the noise level of 90 dB, and still maintain the acceptable range of 75% when the noise level was as high as 70 dB. Moreover, the t-distributed stochastic neighbor embedding technology was utilized to visualize the feature maps obtained by the CNN and illustrated the correlation among different categories. The results also revealed that the proposed CNN model exhibited robustness with high accuracy for processing the data even under high noise interference.

Keywords: guided wave; convolutional neural network; structural health monitoring; stress level prediction; t-distributed stochastic neighbor embedding

1. Introduction

Conventional reinforced concrete, due to the weakness of the concrete in tension, often shows cracking and corrosion at an early age [1,2]. PC has been proposed through prestressing/post-tensioning tendons/rods to compensate this drawback [3]. PC structures exhibit dramatically improved performance over conventional reinforced concrete ones, with the contribution of prestressing tendons or rods that enable them to cover a longer span, thereby providing an effective solution in large-scale buildings, bridges, dams, and nuclear power plant structures. It is known that PC structures often experience losses in prestress due to various factors, including shrinkage/creep of concrete and relaxation of tendons/rods, which in turn potentially lead to excessive deflection or cracking. As such, being able to quantify the stress level of these prestressing tendons/rods in service conditions is critical to ensure structural integrity and achieve successful performance. However, conventional visual or direct destructive examination tools are not valid, as the embedded prestressing tendons or rods with or without grout materials are often inaccessible. Therefore, non-destructive examination (NDE) methods and tools, including vibration-based sensors [4], distributed sensors [5], ultrasonic guided waves [6,7], or acoustic emission [8], could capture information of those far-reaching, inaccessible locations, while maintaining high-quality monitoring and assessing of structural health. For instance, Bartoli et al. [4] employed dynamic identification techniques to investigate the correlation between PC beam prestressing forces. Their results demonstrated that the vibration frequency could assist in identifying the prestress level. Chen et al. [5] used distributed sensor, long-gauge fiber Bragg grating to detect the damage of a bridge under stochastic traffic flow. In addition, the existence of anomalies is an important issue in monitoring data; hence, Zhang et al. [9] employed Bayesian dynamic regression to reconstruct missing data. Despite the merits of different sensing techniques, vibration-based methods are often limited in low frequency, while distributed sensors are often vulnerable to damage and anomalies. Alternatively, as stated in the literature [6,10–13], ultrasonic guided waves could be a better solution to tackle such situations, with the advantages of far-reaching, long-distance measurement and high accuracy in detecting small changes in material discontinuity. Additionally, as an active method, ultrasonic guided wave testing can judge the sensitivity and accuracy of a sensor by receiving the excitations, reducing the possibility of receiving abnormal data.

Guided waves are widely used to evaluate the health of beams, plates, and pipes, owing to the potential of long-distance propagation and sensitivity to mechanical damage. Three modes, namely longitudinal, flexural, and torsional modes, are generated when guided waves are propagated in a medium. Among them, longitudinal modes are more sensitive to tensile stress and easy to excite by piezoelectric actuators [14], and are used for the inspection of steel bars for corrosion, fracture, and stress reduction. Bread et al. [10] used the pulse-echo technique to detect the corrosion and fracture of grouted tendon anchors and rock bolts by ultrasonic guided waves. Lanza di Scalea et al. [15] applied guided waves through magnetostrictive transducers to monitor the stress in seven-wire strands. Their results demonstrated the feasibility of determining the prestress level using the guided wave method. Ervin et al. [16] created an embeddable ultrasonic sensor network to localize and monitor the corrosion of rebar embedded by mortar. They studied the characteristics of guided wave propagated in rebar and the effect forms for corrosion detection, and showed that the waves were sensitive to corrosion through scattering, mode conversions, and reflections. Chaki and Bourse [17] detected the stress level of the seven-wire steel strands by ultrasonic guided waves with L (0,1) mode. The typical calibration curves were plotted, which showed that the stress level corresponded to the phase velocity change in the guided waves. More recently, Treyssède and Laguerre [18] employed the semi-analytical finite element approach to study the guided wave propagation in multi-wire strands. In addition, high-order longitudinal modes were indicated to solve the leakage problem of fundamental mode L (0,1) by Dubuc et al. [19]. They used the acoustoelastic theory to propose an approximate theory for predicting the effect of stress on higher modes. Shoji Masanari [20] employed 60 kHz L (0,1) mode as the guided wave to inspect anchor rods embedded in soil, and unveiled the capability of the ultrasonic guided waves for stress identification in rods. While physics-based approaches have been used for the signal process of guided waves to identify stress changes in stressed rods, these methods still face challenges in handling the wave signals with a variety of structural uncertainties, signal attenuation, and environmental noises during testing.

In this way, recently emerging machine learning, particularly deep learning, could provide potential solutions to improve the signal process of guided waves [6,7,21,22]. Deep learning algorithms have been employed in time series [21,23–26] and image processing [27–30]. Guo et al. [31] utilized a sparse coding-based deep learning algorithm to process wireless sensory data of a three-span bridge. The features of the dataset were learned by sparse coding and then trained by the network. Cha et al. [27] proposed a vision-based method by a deep learning network to detect concrete cracks without calculating the features. The comparative study indicated that the deep learning-based technique had better performance than the conventional physics-based methods. Furthermore, Cha et al. [28] investigated the fast region-based convolutional neural network to detect five types of damage in real time. Zhang et al. [32] proposed a CNN framework with some convolutional kernels to identify vibration signals. Harsh et al. [33] applied high-frequency guided waves to detect damages in railheads, generated the data by experiment and simulation study, and then set up a framework to detect damage of railheads by a machine learning method. The error rate was from 2% to 16.67%. Tabian et al. [34] used guided waves to detect impact energy, localization, and characterization of complex composite structures. The waves transferred into 2D images and were identified by a CNN algorithm. The results showed that the accuracy was above 95%. Zargar and Yuan [35] used a unified CNN-RNN network to extract the information of aluminum plates. This research focused on the wave propagation in both spatial and temporal domains. While deep learning approaches have been successfully used in many aspects of structural health monitoring, less research is involved in deep learning-based ultrasonic guided wave diagnoses.

Therefore, we aimed to develop and implement the deep learning-enriched guided wave technique to quantify the stress level of prestressed rods used in PC structures. The CNN framework was utilized for processing guided wave signals to predict the stress level of the rods with varying grout materials. t-distributed stochastic neighbor embedding (t-SNE) was employed to visualize the features extracted by the CNN model. Moreover, different noise levels were considered to examine the robustness of the CNN classifier.

2. Guided Waves as NDE Approach for Prestressed Rods

2.1. Governing Equations and Simulation of Guided Waves along a Rod

Guided waves were first introduced in cylinder structures in the 19th century [36]. The governing equation of the wave propagating in isotropic cylinders is expressed as [37]

$$(\lambda + 2\mu)\nabla(\nabla * \boldsymbol{u}) + \mu\nabla^2\boldsymbol{u} + \boldsymbol{f} = \boldsymbol{\rho}\left(\frac{\partial^2 u}{\partial x^2}\right) \tag{1}$$

where $\boldsymbol{u}$ represents the displacement vector, x is the time, ∇^2 is the three-dimensional Laplace operator, λ and μ indicate Lame's constants, $\boldsymbol{\rho}$ is the mass density, and the body force $\boldsymbol{f}$ is equal to zero. Then, the Helmholtz decomposition is used in Equation (1) to simplify the problem as

$$\boldsymbol{u} = \nabla\varphi + \nabla * \mathbf{H} \tag{2}$$

$$\nabla * \mathbf{H} = 0 \tag{3}$$

where φ and $\mathbf{H}$ represent the scalar and vector potentials.

Three types of guided waves, namely longitudinal mode (L (0, m)), torsional mode (T (0, m)), and flexural mode (F (n, m)), were generated to propagate through a cylindrical structure. In the modes, n and m denote the circumferential order and modulus, respectively. When $n = 0$, the waves are symmetrical, such as L (0, m) and T (0, m). Otherwise, the waves are asymmetrical.

Stress affects the phase velocity of the guided wave. The change in the phase velocity ΔC is expressed as

$$\Delta C = -\left[\frac{(C^0)^2}{l}\right]\Delta t \tag{4}$$

where C^0 is the unstressed velocity, l represents the length of the wave propagation in the stress area, and Δt is the time change.

Figure 1 shows the phase velocity and group velocity derived by MATLAB PCDisp [38,39]. The lower frequency of the excitation waves, less than 50 kHz, was used to reduce the dispersion of the guided waves.

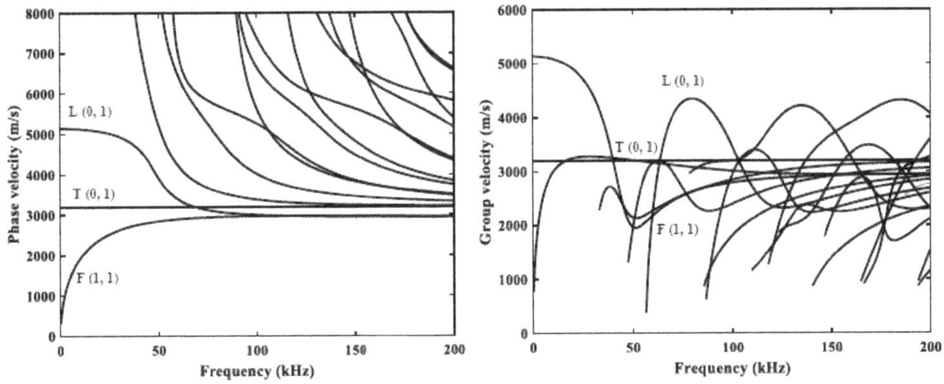

Figure 1. Phase velocities and group velocities.

As such, the ultrasonic guided waves used in this study were numerically simulated by Multiphysics Finite Element (FE) software COMSOL, and their propagation characteristics along the prestressed rod under different conditions, including under varying grout materials and different stress levels, were then modeled and extracted using machine learning to assist in data classification, as discussed in Section 3.

2.2. Calibration of the FE Analysis of the Ultrasonic Guided Waves through the Rod

We sought to ensure that proper parameters were used for the rod simulation and calibrate the effectiveness of FE-based simulation for capturing the characteristics of the wave propagation along rods. One case of the characterization of ultrasonic guided waves along an anchor rod was selected from the literature [40], in which the rod had a diameter of 21 mm and was 2.3 m in length, and was embedded in a concrete block with a cross-section of 1.0 m by 1.0 m and a depth of 2.0 m, as shown in the FE meshed model in Figure 2a. The excitation of the ultrasonic guided waves was a six-cycle tone burst with a frequency of 35 kHz. A pulse-echo test was set up on the rod where the actuators and receivers were on the same side. Excitations were generated by Wavemaker 16 equipment, which is used for long-range inspection of pipes. The rod was embedded in the concrete block along 2 m and the remaining length of the rod from both ends of the block [40]. The comparison results are shown in Figure 2b, where signals in the literature are marked with red lines and the black ones were generated from this study, and the three circled wave packets denote the excitation signals, the first right boundary reflection, and the second right boundary reflection. As shown in Figure 2b, the simulated guided waves in this study matched well with the experimental data collected from the literature in most cases, where three wave packets were captured well. The first boundary wave reflections from simulated signals occurred at 0.001 s, identical to the experimental data with comparable amplitudes. Note that some deviations occurred after the excitation, and a potential reason could partially result from the attenuation of the concrete block where the simulation did not capture well. However, the entire trend and amplitudes in most cases were in agreement with the literature, suggesting that the simulation used in this study was appropriate to ensure capturing of the characteristics of the ultrasonic guided waves through the rods.

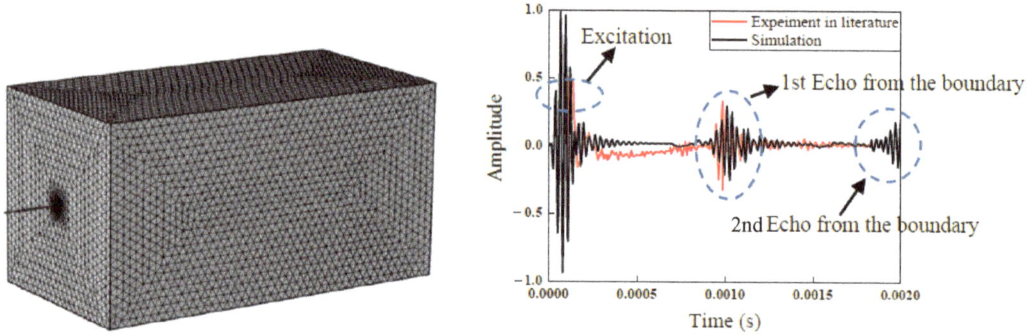

(**a**) (**b**)

Figure 2. Calibration of the FE model. (**a**) Meshing of the rod embedded in concrete. (**b**) Comparison of signals with the literature.

2.3. Design of Scenarios

Followed by the calibration in Section 2.2, this section details the design of different scenarios to generate datasets that helped to elucidate the critical factors affecting the characteristics of ultrasonic guided wave propagation along stressed rods, thus paving the way for stress level prediction using machine learning in Section 3. As such, the prototype of the stressed rods was derived from the literature [41]. The rod was 31.75 mm in diameter with a length of 3657.6 mm. The material properties of the rod were density of 7800 kg/m^3, Young's modulus of 2×10^5 MPa, and Poisson ratio of 0.3. A clamp served as the anchorage of the rod, which was located 50.8 mm away from the free end. Besides the reference where the rod had no grout, two grout materials, namely grease and cement, were selected to unveil their effects on the effectiveness of the proposed methods. The density of the cement was 1440 kg/m^3, the Young's modulus was 2.5×10^4 MPa, and the Poisson ratio was 0.25. The density of the grease was 2600 kg/m^3. The detailed information of the rod models is shown in Figure 3. The entire rod with the clamp is illustrated in Figure 3a (note that the meshing figure is a schematic diagram; the actual meshing unit is much smaller), and the embedded rod is shown in Figure 3b. Specifically, the rod passed through the plastic pipe and then added grease and cement to fill the gap between the pipe and the rod. The outer diameter of the plastic pipe was 52.07 mm, and the thickness was 2.54 mm. In finite element studies, meshing is one of the critical parts in simulation. Free triangular element was selected in this model. Guided wave simulation requires a high-quality meshing system to minimize the propagation error of guided waves. Thus, a wavelength needs to contain at least eight elements. In this study, the maximum element size in the model was 2 mm and time steps were 5×10^{-6} s. Guided waves could be excited in the rod by adding displacement loads in all nodes of the left boundary in the model. The excitation waves were 35 kHz five-cycle sinusoidal waves modulated by the Hanning window. Four points, as received points, distributed the circumferential surface of the rod. Positions of received nodes were 5 mm away from the left side. Received signals were time series data, which intercept the first 1000 data points as results for further study.

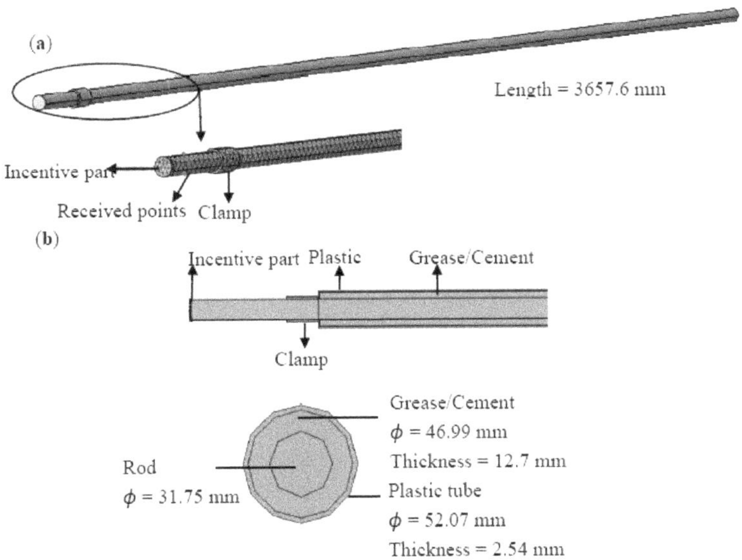

Figure 3. Rod models.

To simulate the stress reduction in prestressed components, five different pressure levels were loaded into each rod: no prestress (State #1), 20% ultimate tensile strength (UTS) (State #2), 40% UTS (State #3), 60% UTS (State #4), and 80% UTS (State #5). In total, 15 cases were designed in this section, which are shown in Table 1. Noise was added into the data to increase the uncertainty of the dataset.

Table 1. Test matrix for computation modeling.

Case	State	Prestressing Level (UTS)	Grout Material	Noise Level
1 (no grout)	# 1	zero	\	100 dB–60 dB
	# 2	20%	\	
	# 3	40%	\	
	# 4	60%	\	
	# 5	80%	\	
2 (grease)	# 6	zero	Grease	
	# 7	20%	Grease	
	# 8	40%	Grease	
	# 9	60%	Grease	
	# 10	80%	Grease	
3 (cement)	# 11	zero	Cement	
	# 12	20%	Cement	
	# 13	40%	Cement	
	# 14	60%	Cement	
	# 15	80%	Cement	

2.4. Data Collection and Augmentation

Figure 4 shows the time records with five different prestress levels derived from the finite element simulation in Case 1 from Table 1. The received point was located 5 mm away from the left side of the rod. To ensure the input wave had similar energy, all the received signals were normalized, and the maximum amplitude of the first packet was equal to 1. The time series were from 0 s to 0.005 s; the guided wave can propagate and reflect at least twice throughout the rod. Figure 4a–e illustrates the received signals of the steel with prestress levels equal to 0% UTS, 20% UTS, 40% UTS, 60% UTS, and 80% UTS. Specifically, at low prestress level states (0–40% UTS), the results clearly exhibit three main wave packets that represent the initial excitation and the first and second reflections from the boundary. Following the first packet, fluctuations with small amplitude were echoes from the clamp. With the stress level increased, the amplitude of this part was smaller, and it was hard to detect at 80% UTS. In addition, the velocity of guided waves reduced with increasing the stress of the rod. The first reflection from the right boundary was around 0.0018 s in the base state. The value was raised to 0.0019 s and 0.002 s when stresses were 40% UTS and 80% UTS. At 60% and 80% UTS (shown in Figure 4d,e), only two main packets existed in the signals, where one was initial input waves, and the other was the boundary reflection. The energy of guided waves was dissipated when waves propagated in the second cycle. Thus, it was difficult to detect the second echo from the boundary.

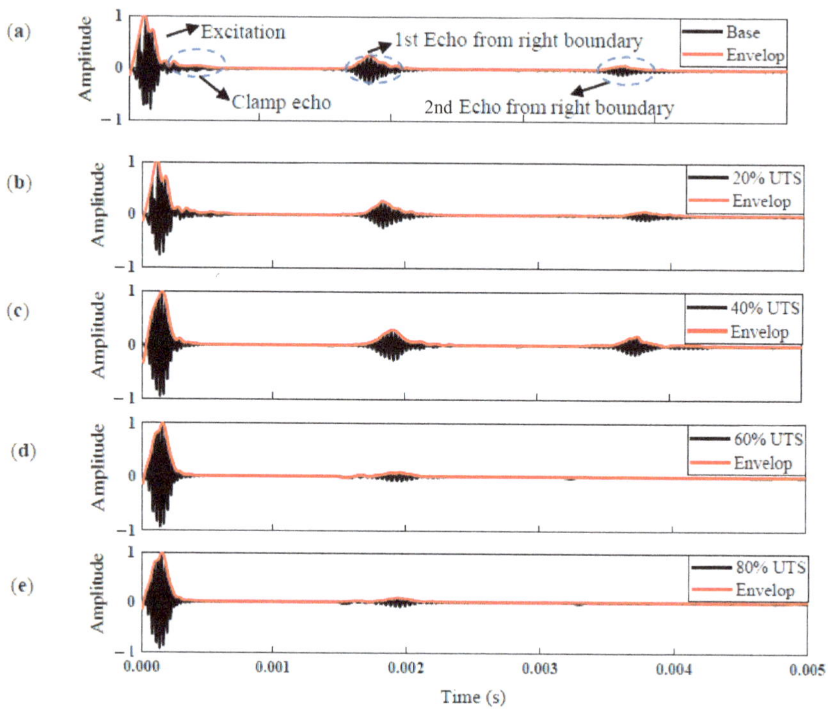

Figure 4. Received signals at different stress levels: (**a**) zero; (**b**) 20% UST; (**c**) 40% UTS; (**d**) 60% UTS; (**e**) 80% UTS.

As illustrated in Figure 5, the collected guided waves of the rod exhibited different patterns from different grout materials. Figure 5 illustrates the signals collected from States #1, #6, and #11 (without stress) in the time domain and frequency domain. It was observed from the time domain that with long distance propagation, the energy of the reflection waves was reduced progressively. However, comparing these three states, the rod embedded in cement had the highest attenuation, where the peak value of reflections was reduced from 0.2588 to 0.1075. After propagating to the second cycle, the peak value of the second boundary echo was reduced to 0.0897 in the unembedded state, and the value of the rod in cement was the lowest, 0.0113. On the contrary, grease had less of an effect on guided waves. With 1463.04 mm of propagation, the peak value of the reflected waves was 0.0455, which was close to the unembedded state. In the frequency domain, it was clear that the main frequency of waves was 35 kHz. Some weak peaks occurred at the low frequency and the high frequency due to reflections from the clamp and the boundary.

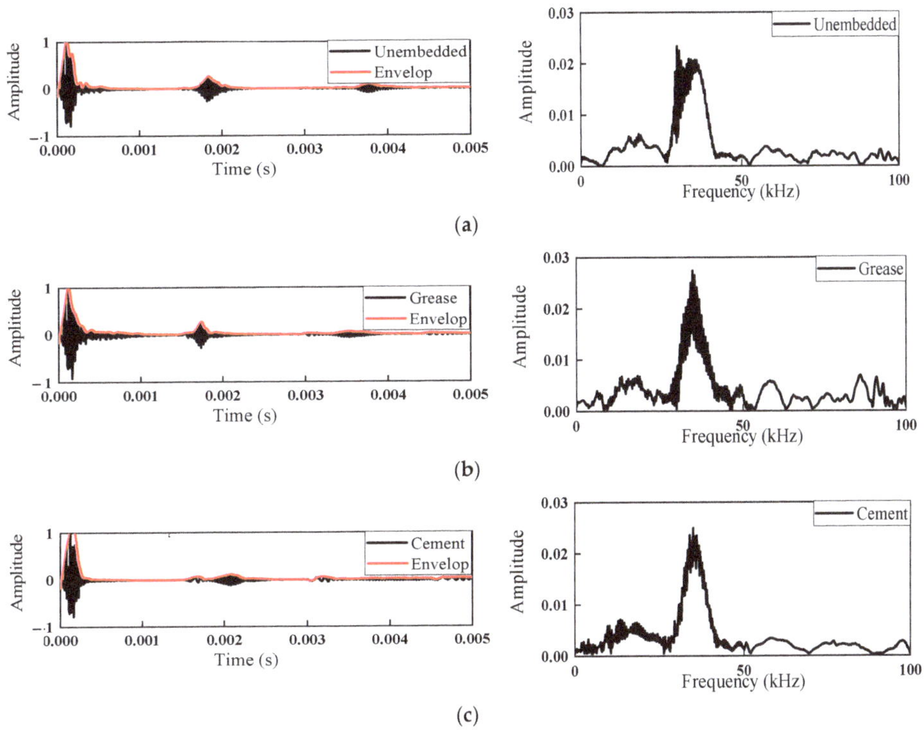

Figure 5. Received signals for rods with different grout methods: (**a**) no grout; (**b**) grease; (**c**) cement.

A total of 15 states were designed to simulate the actual situation of the rod. In each state, four received nodes were distributed around the circumference and were located 5 mm away from the left side. Since the received waves were easily contaminated by noise, five noise levels based on the signal to noise ratio (SNR) were added into the received signals. In addition, noise involved in the signals could increase the uncertainties of the data, so we attempted to investigate the robustness of the deep learning methods. Figure 6 illustrates the collected signals at five different noise levels. When SNR exceeded 80 Db, the interference from noise was obvious and covered some original features of the initial signals. Especially, at 60 dB, the amplitude of the noise was greater than the signal amplitude, which was not conductive for further study.

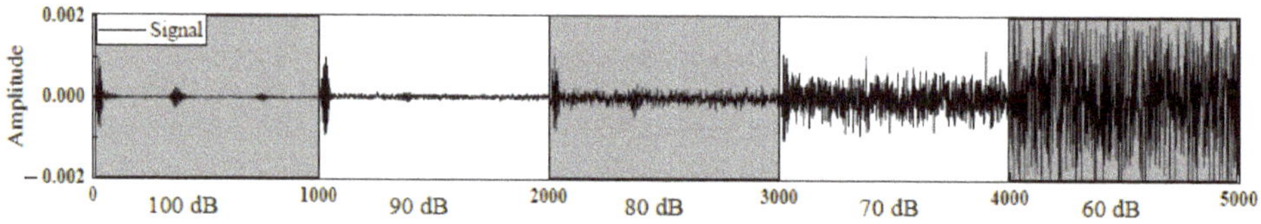

Figure 6. Received signals at different noise levels.

3. Deep Learning Framework

CNN has been widely adopted in many application domains, such as image classification and segmentation, speech recognition, and computer vision tasks. The CNN framework contains multiple layers, including a convolutional layer, pooling layer, fully connected layer, and ReLU layer. These layers help to decode the input data into high-dimensional slices, extract the intricate features, and then encode them into the target values. In this study, CNN was used to identify the complicated guided wave signals. The architecture of the CNN trained by guided wave signals is described. The model of the CNN was changed from LeNet-5.

3.1. Introductions of CNN

The input data consisted of a series of signals with m detection points and n time steps. Thus, the size of the input layer was n × m.

The convolutional layer is one of the most crucial layers in a CNN. In this layer, each element from the kernel is multiplied with the data in the previous layer. The size of the kernel determines the operation area, and the number of kernels decides the third dimension of the output. The kernel size is much smaller than the input layer, so the kernel moves step by step to implement. The stride defines the length of each step, which causes the output data reduction. The size of the stride is an essential value which affects the efficiency and performance of the layer. A bigger size may cause the loss of some important features, and a small size may cause an increase in calculation. The initial kernels are generated randomly, and they update by learning from each iterator. A bias is added after summing all the multiplication results in the operation area. When all the kernels finish the multiplication with the input data and summarize, the result is the output in this layer.

The pooling layer is used to reduce the size of the previous layer. Two pooling layers can be selected, namely max pooling and mean pooling. In max pooling, the maximum values in the operation area are taken as the result. In mean pooling, the average values are the result. The operation area is also moved by setting the values of stride. After adding a pooling layer to a CNN framework, the output of the convolutional bands has a dramatic decrease. In this CNN, all the pooling layers were set as max pooling layers.

The activation layer adds nonlinearity into the CNN. In this model, ReLU was chosen as the activation layer in the CNN trained by guided wave signals. ReLU changes some of the neurons to zero, which will thin the network, reduce the interdependence between parameters, and avoid overfitting to some extent. In addition, it also saves computation and improves the efficiency of deep learning models, compared with other activation functions.

After the cooperation of several layers, the initial data are changed into a series of feature maps, and the size is deformed. To transfer these feature maps into their own category, a fully connected layer is necessary. The result of this layer is the probability that the data belong to each label. The entire process is shown in Figure 7.

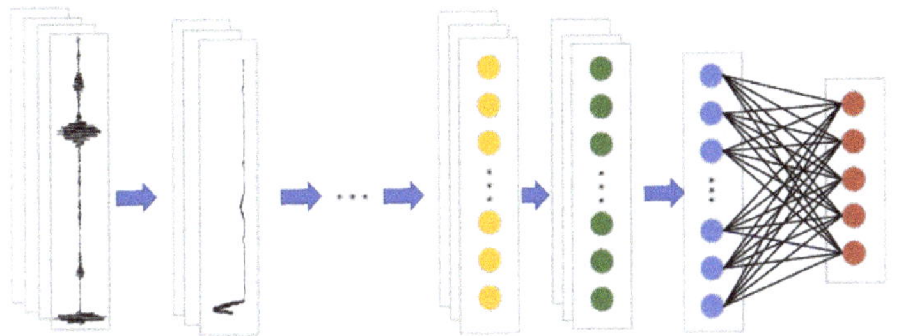

Figure 7. Flowchart of CNN.

3.2. CNN Architecture

The proposed CNN in this study was an eight-layer network, including three convolutional layers, two max pooling layers, a ReLU layer, a fully connected layer, and a softmax layer. The selection of hyperparameters affects the performance of the neural network. Different methods [27,42,43] have been proposed to select these parameters. For instance, the learning rate is to adjust the gradient update, the kernel number size is to adjust the receptive filed; in addition, stride step and batch size are critical in parameter design [26,41]. In this study, the hyperparameter selection stems from previous studies, maintaining the basic network architecture of LeNet-5. Note that several studies revealed that the introduction of Bayesian optimization [44,45] in determining the hyperparameters could further enhance the architecture of the CNN with less trial and error, and thus improve the accuracy. As part of the ongoing investigation of the applications of deep learning in civil structures, we consider advances in hyperparameter design, including using Bayesian optimization, for optimizing the CNN architecture.

The detailed information of each CNN layer is given in Table 2. The input data are a matrix sized 1000×4, representing four collected signals in a rod sample. A total of 1000 data points were intercepted from received signals, which included the excitation and the reflection from the right boundary. Four is the number of received signals. In the first convolutional layer, 20 filters sized 25×2 were generated and operated the input data into $976 \times 3 \times 20$. The following was a max pooling layer with the stride equal to 5. The layer captured the maximum value in the response field and significantly cut down the size of the input. After that, the output of the data was $195 \times 3 \times 20$. Then, the data experienced cooperation from the convolutional layer and the max pooling layer, involving 40 filters, and the pooling size was 5×1. At this moment, the output was 34×1, a dramatic decline compared with the initial input. The third convolutional layer had a small size, 5×1, and a ReLU was implemented to increase the nonlinearity. Finally, the fully connected layer and softmax layer transferred the data into several probabilities in each label.

Table 2. Details of the proposed CNN.

Name	Filters	Filter Size	Stride	Bias	Output Layer Size
Input layer	–	–	–	–	1000×4
Convolutional layer (C_1)	20	25×2	1	20	976×3
Max pooling layer (P_1)	20	5×1	5	–	195×3
Convolutional layer (C_2)	40	25×3	1	40	171×1
Max pooling layer (P_1)	40	5×1	5	–	34×1
Convolutional layer (C_3)	20	5×1	1	20	30×1
ReLU	–	–	–	–	30×1
Fully connected layer (F_1)	5	30×1	1	5	5
Softmax	–	–	–	–	5

3.3. Feature Visualization with t-SNE

The CNN classifier achieves better performance since it automatically extracts features from the training data. It expands the data into high-dimensional matrices by multiple filters. Thus, these features are usually high-dimensional, which is not conductive to understanding. However, stochastic neighbor embedding (SNE) was introduced to reduce the dimensions of the features, making it possible to visualize the feature. SNE aims to convert the high-dimensional Euclidean distance between data samples into conditional probabilities. The t-distributed stochastic neighbor embedding (*t*-SNE) [46] proposed by Maaten and Hinton transforms a high-dimensional dataset into a pairwise similarity matrix and minimizes the gap between the distribution in two spaces. This method is popular for feature visualization in machine learning algorithms.

4. Results and Discussion

4.1. Feature Visualization

In this study, 500 data points in each state emerged by adding white Gaussian noise, including 60% data for training, 20% for validation, and 20% for testing. The CNN model was well trained after studying the features from the training data. The feature maps can estimate the efficiency of the proposed method. The following figures show the high-dimensional feature maps in two-dimensional space by *t*-SNE.

Figure 8 depicts the features in Case 1, where the prestress level of the unembedded rod is 80 dB. States #1–5 represent the rod with the prestress level equal to 0% (base state), 20% UTS, 40% UTS, 60% UTS, and 80% UTS. In total, 2500 samples comprised the dataset, where 1500 were used to train the model, 500 for validation, and the remaining 500 for testing. Figure 8a displays the feature maps of the test set after the first convolutional layer. Through the distribution of features, data labeled as 40% UTS (green upper triangle) were isolated from the entire dataset. This indicates that the features in this label extracted from the first layer of the CNN model were much easier to separate than others because the Euclidean distance is larger. However, the clusters in the red diamond (base state) and yellow circle (20% UTS) located on the right side overlapped. At least one-quarter of the data were mixed and difficult to separate, which means errors will occur. In addition, the rod samples prestressed in 60% UTS (blue lower triangle) and 80% UTS (purple star) were tangled together. Figure 8b represents the feature maps from the last layer of the CNN. After eight layers' processing, most of the samples were separated, except one outlier in the base class clustered in 20% UTS and a small overlap appeared between the samples in 60% UTS and 80% UTS. The results demonstrate that the features became more sensitive after operating the whole CNN process.

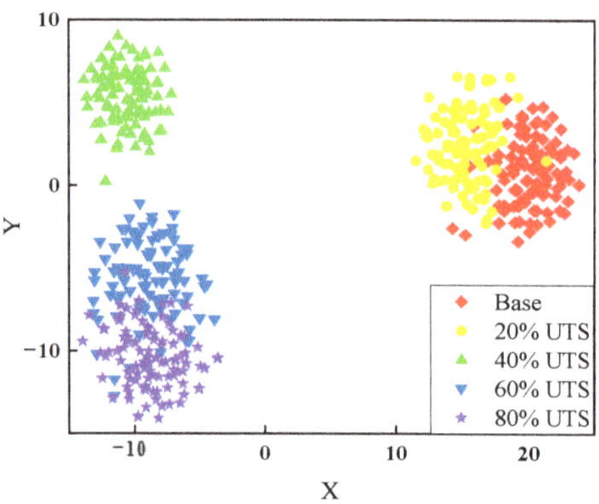

(a)

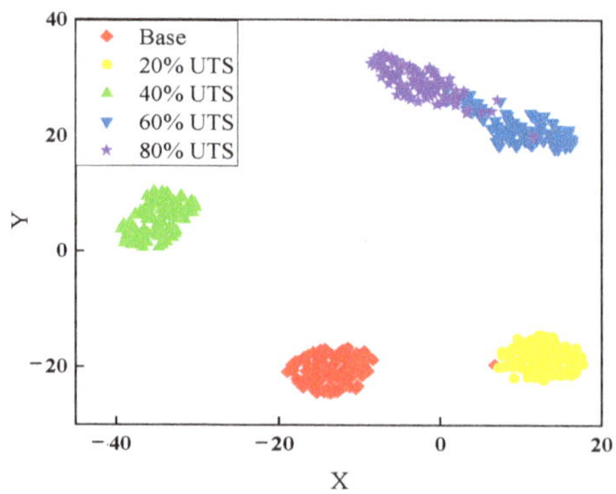

(b)

Figure 8. Feature visualization by *t*-SNE. (**a**) Feature maps in the first convolutional layer; (**b**) feature maps in the last layer.

The CNN model sliced input signals into several small pieces, which enlarged the sensitive features and cut off the excess. Figure 9 plots the feature maps from five different prestress levels after three convolutional layers. The data are signals from five different prestress levels in Case 1 when SNR is equal to 100 dB. The differences between the five figures are clear. The shape and peak values of lines characterize signals from different groups. However, with the increasing noise level, extracting features became harder.

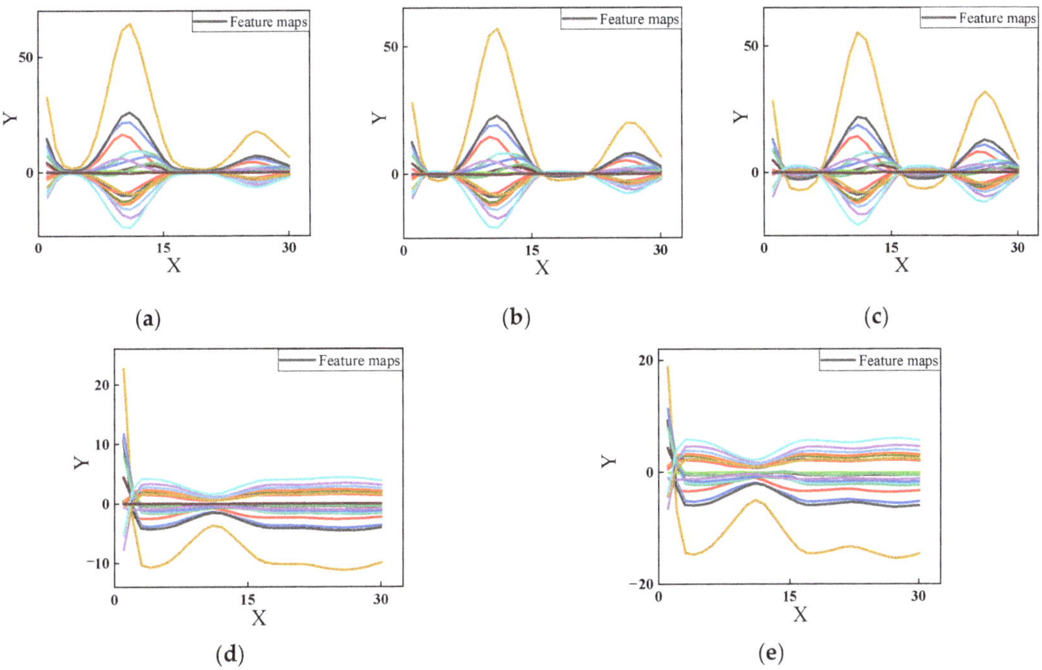

Figure 9. Feature maps: (**a**) 0% UTS; (**b**) 20% UTS; (**c**) 40% UTS; (**d**) 60% UTS; (**e**) 80% UTS.

Figure 10 illustrates the feature visualization in Case 1 when SNRs were equal to 100 dB, 70 dB, and 60 dB. At 100 dB (in Figure 10a), the five clusters were far from each other, and each cluster of data is relatively concentrated, with an average standard deviation close to 2.53. With the noise level increased to 80 dB (shown in Figure 10b), a small overlap appeared between the 60% UTS and 80% UTS data, but the other three groups were clearly separated. The standard deviation in this situation was large, 2.83. However, when SNR was lower than 80 dB, the distributions of features changed dramatically. At 70 dB, features in either the base state and 20% UTS or 60% UTS and 80% UTS were blended into each other, and only the data in the green upper triangles were independently located below the graph, demonstrated in Figure 10b. In Figure 10c, feature maps at 60 dB had a worse situation, as all the data interwove together entirely. It is hard to ascertain the boundaries of each group. Therefore, this proved that noise had an adverse effect on the feature extraction of the CNN.

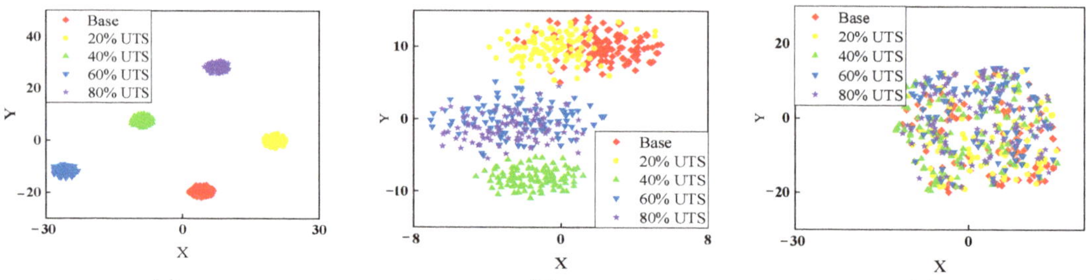

(**a**) (**b**) (**c**)

Figure 10. Feature visualization by *t*-SNE. (**a**) SNR = 100 dB; (**b**) SNR = 70 dB; (**c**) SNR = 60 dB.

4.2. Classification for Prestress Levels of the Rod by CNN

The performance of CNN trained by the data under various noise levels is shown in Figure 11, which classified the prestress levels of the rod in Case 1. Of the 2500 data points, 1500 were used for training the CNN model, and the training curves are illustrated in Figure 11a. Specifically, these training curves started near 20%, and then converged to 100%. When the noise level was 100 dB, the model only spent seven epochs to improve the accuracy to 100%. That epoch number was enlarged to 25 at 90 dB. With SNR increased, more epochs were spent on converging, and the error reduced to 0 with 34 epochs at 80 dB. The slope of training curves from 100 dB to 80 dB were much larger than the curves at 70 dB and 60 dB. In detail, the accuracy of classification was raised from 20% to 31% by 15 epochs at 60 dB and then close to 100% after the 30th epoch. The validation set included 500 data points for modifying the parameters in CNN. The validation curves for 40 epochs showed that the accuracy of CNN started at 20% and increased with the epochs. At 100 dB and 90 dB, the accuracies reached 100% after training 6 and 20 epochs, respectively. When SNR was 80 dB, the curve was close to 0.98 after the 28th epoch and would not improve as the epoch increased. However, the accuracies were lower at 70 dB and 60 dB, where the curve converged to 0.78 and 0.34 after training 40 epochs. Approximately 32% of the data were misjudged as the incorrect category at 70 dB. The situation at 60 dB was much worse, as most of the data could not be classified into the right category.

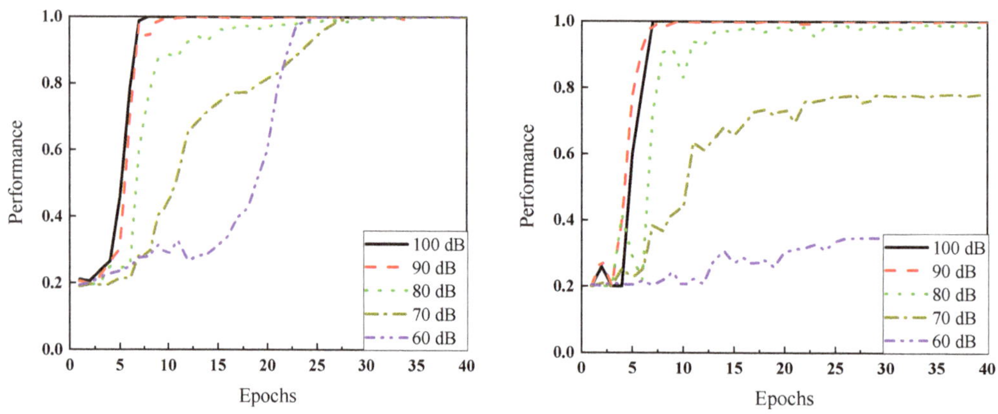

(**a**) (**b**)

Figure 11. Learning results of CNN at various noise levels. (**a**) Training curve; (**b**) test curve.

The test results at various noise levels are shown in Table 3. When SNR was no less than 90 dB, the CNN model classified the test data correctly into the corresponding categories because the training curves and the validation curves reached 100% after training. At 80 dB, the accuracy of test data was 98%, and some misjudgments appeared in the base state and the 80% UTS state. The results of the feature map visualization show that some data in the base state were dropped into the 20% UTS group, and the 60% UTS and 80% UTS clusters overlapped (shown in Figure 11b). The mixed features caused the misjudgments in the test. At 70 dB, only 74% of the data were identified correctly, and 25% of the data in the first category were misclassified into the second category. On the contrary, 27% of the data that belonged to the second category were placed into the first category. The accuracies of the fourth and fifth categories were both equal to 63%. The conclusion is consistent with the previous analysis in feature visualization and accurate curves. At 60 dB, the CNN model was not suitable in this situation because the noise covered the signals entirely and all the analysis focused on the noise. Thus, all results had low accuracy.

Table 3. Confusion matrices at various noise levels.

			90 dB (100%)					**80 dB (98%)**		
	Base	**20%**	**40%**	**60%**	**80%**	**Base**	**20%**	**40%**	**60%**	**80%**
Base	100.0%	0.0%	0.0%	0.0%	0.0%	99.0%	0.0%	0.0%	0.0%	0.0%
20%	0.0%	100.0%	0.0%	0.0%	0.0%	1.0%	100%	0.0%	0.0%	0.0%
40%	0.0%	0.0%	100.0%	0.0%	0.0%	0.0%	0.0%	100%	0.0%	0.0%
60%	0.0%	0.0%	0.0%	100.0%	0.0%	0.0%	0.0%	0.0%	98%	7.0%
80%	0.0%	0.0%	0.0%	0.0%	100.0%	0.0%	0.0%	0.0%	2.0%	93.0%
			70 dB (74%)					**60 dB (26.8%)**		
Base	75%	27.0%	0.0%	0.0%	0.0%	28.0%	27.0%	20.0%	12.0%	15.0%
20%	25.0%	71.0%	1.0%	0.0%	0.0%	28.0%	22.0%	17.0%	15.0%	12.0%
40%	0.0%	0.0%	98%	0.0%	0.0%	20.0%	16.0%	34.0%	17.0%	22.0%
60%	0.0%	1.0%	1.0%	63.0%	37.0%	14.0%	14.0%	12.0%	23.0%	15.0%
80%	0.0%	1.0%	0.0%	37.0%	63.0%	10.0%	21.0%	17.0%	33.0%	36.0%

The performance of the CNN classifier in Cases 2 and 3 is shown in Figure 12. In Case 2, rods were embedded in cement. The training and validation results at five noise levels are illustrated in Figure 12a. The training and validation curves had better results when SNRs were higher than 70 dB. All the curves fluctuated during the first five epochs and then quickly converged to 0. Nearly 10 epochs were spent to increase the accuracies to 100%. At 70 dB, although the training curve took about 20 epochs to converge to 100%, the validation curve only reached 0.78. However, the performance of the CNN dropped sharply at 60 dB. In Case 3, the error occurred at 80 dB, where the training accuracy and validation accuracy were close to 0.99 and 0.94, respectively. The error rate increased at 70 dB, where the validation curve reached 0.77 at the 20th epoch and then flattened out. The results in Case 3 were similar to those of Case 1 because the grease had less of an effect on guided wave propagation.

The test results of Cases 2 and 3 are shown in Table 4 when SNRs were from 90 dB to 60 dB. In Case 2, all the test data were identified correctly at 90 dB and 80 dB. At 70 dB, only the base state had a classification rate of 98%, and the other four groups had lower rates. Among them, 12 of 100 data samples in the second category were misclassified into the third and fourth categories. The accuracy of the third category was 74% with 26% misjudgment. In addition, the error rates in the fourth and fifth categories were 66% and 64%, respectively. When the noise level increased to 60 dB, the accuracy of prestress identification was the lowest (46.6%). Most of the data could not be identified, except the data in the base condition (at 99%). Compared with Case 2, the approach exhibited a slightly lower accuracy for Case 3, where 100% of data in 90 dB were classified, 94.4% were identified at 80 dB, 76% at 70 dB, and 41.6% at 60 dB. At 80 dB and 70 dB, most of the errors occurred between the fourth and fifth categories. Specifically, the predictions of the fourth and fifth categories reached 78% and 94% at 80 dB. At 70 dB, the proportions were reduced to 52% and 51%, respectively. At 60 dB, only the data under the base condition could be identified, and it was hard to identify the data in other conditions.

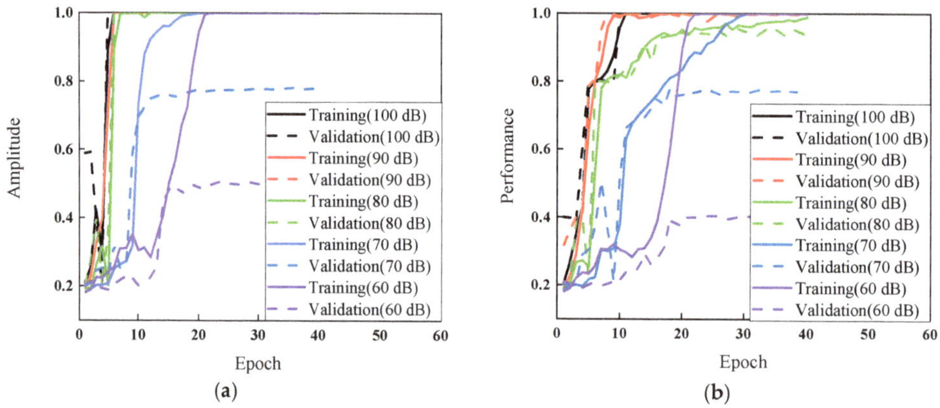

Figure 12. Results of CNN at various noise levels. (**a**) Case 2; (**b**) Case 3.

Table 4. Confusion matrices in Case 2 and 3.

			Case 2 (Grease as the Grout Material)							
			90 dB (100%)					**80 dB (100%)**		
	Base	**20%**	**40%**	**60%**	**80%**	**Base**	**20%**	**40%**	**60%**	**80%**
Base	100.0%	0.0%	0.0%	0.0%	0.0%	100.0%	0.0%	0.0%	0.0%	0.0%
20%	0.0%	100.0%	0.0%	0.0%	0.0%	0.0%	100.0%	2.0%	5.0%	4.0%
40%	0.0%	0.0%	100.0%	0.0%	0.0%	0.0%	8.0%	100.0%	15.0%	18.0%
60%	0.0%	0.0%	0.0%	100.0%	0.0%	0.0%	4.0%	14.0%	100.0%	14.0%
80%	0.0%	0.0%	0.0%	0.0%	100.0%	0.0%	0.0%	10.0%	14.0%	100.0%
			70 dB (78%)					**60 dB (46.6%)**		
Base	98%	0.0%	0.0%	0.0%	0.0%	99%	1%	0%	1%	0%
20%	2.0%	88%	2.0%	5.0%	4.0%	0%	50%	16%	29%	8%
40%	0.0%	8.0%	74%	15.0%	18.0%	0%	13%	24%	16%	25%
60%	0.0%	4.0%	14.0%	66%	14.0%	1%	26%	31%	26%	33%
80%	0.0%	0.0%	10.0%	14.0%	64.0%	0%	10%	29%	28%	34%
			Case 3 (Cement as the grout material)							
			90 dB (100%)					**80 dB (94.4%)**		
Base	100.0%	0.0%	0.0%	0.0%	0.0%	100.0%	0.0%	0.0%	0.0%	0.0%
20%	0.0%	100.0%	0.0%	0.0%	0.0%	0.0%	100.0%	0.0%	0.0%	0.0%
40%	0.0%	0.0%	100.0%	0.0%	0.0%	0.0%	0.0%	100.0%	0.0%	0.0%
60%	0.0%	0.0%	0.0%	100.0%	0.0%	0.0%	0.0%	0.0%	78.0%	6.0%
80%	0.0%	0.0%	0.0%	0.0%	100.0%	0.0%	0.0%	0.0%	22.0%	94.0%
			70 dB (76%)					**60 dB (41.6%)**		
Base	99.0%	1.0%	0.0%	0.0%	0.0%	79.0%	18.0%	2.0%	5.0%	3.0%
20%	0.0%	99.0%	0.0%	0.0%	0.0%	18.0%	38.0%	20.0%	9.0%	11.0%
40%	1.0%	0.0%	79.0%	13.0%	12.0%	1.0%	19.0%	29.0%	29.0%	23.0%
60%	0.0%	0.0%	11.0%	52.0%	37.0%	1.0%	15.0%	25.0%	27.0%	28.0%
80%	0.0%	0.0%	10.0%	35.0%	51.0%	1.0%	10.0%	24.0%	30.0%	35.0%

4.3. Classification for Embedding Situation by CNN

The embedding situation of the rod could also be predicted by the CNN classifier. A total of 1500 data points were used for training the CNN, including unembedded rods, rods embedded with cement, and rods embedded with grease. The training and validation curves are shown in Figure 13. At a lower noise level, the classifier entirely predicted the state of the rod. Until SNR was 70 dB, the validation curve was 0.956, training 40 epochs.

However, the accuracies at 60 dB dropped sharply. The test results are given in Table 5. Predictions were equal to 100% in each category when SNRs were from 100 dB to 80 dB. Several errors occurred at 70 dB and the accuracy was 92%. Specifically, the rod samples labeled as unembedded were much easier to confuse as rods embedded in grease, as 15% of the data in that category were misclassified into the grease state. On the other hand, 7% of the data were misclassified into the unembedded state. Only 1% of data in cement were predicted incorrectly. The worst results were the classification at 60 dB, where the total accuracy was only 35%, which means the model cannot identify the signals.

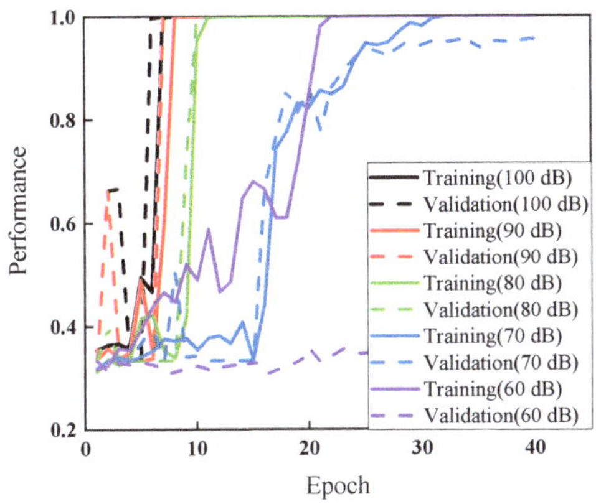

Figure 13. Results of CNN at various noise levels.

Table 5. Confusion matrices in varying embedding conditions.

	90 dB (100%)			80 dB (100%)		
	Unembedded	**Cement**	**Grease**	**Unembedded**	**Cement**	**Grease**
Unembedded	100.0%	0.0%	0.0%	100.0%	0.0%	0.0%
Cement	0.0%	100.0%	0.0%	0.0%	100.0%	0.0%
Grease	0.0%	0.0%	100.0%	0.0%	0.0%	100.0%
	70 dB (92%)			**60 dB (35%)**		
Unembedded	84.0%	0.0%	7.0%	37.0%	24.0%	39.0%
Cement	1.0%	99.0%	0.0%	28.0%	43.0%	36.0%
Grease	15.0%	1.0%	93.0%	35.0%	33.0%	25.0%

5. Conclusions

We investigated the deep learning-based guided wave process for stress level prediction of prestressed rods. The CNN model was established for automatically encoding the hidden information from complex signals that accounted for the impacts of different noise levels and embedded grout materials. Some conclusions can be listed as follows:

(a) The deep learning method effectively encoded the guided waves under complex uncertainties and assisted in stress level prediction and the embedded conditions of the rods, thereby showing potential for signal processing of NDE methods in structural health monitoring of PC structures.

(b) The feature visualization method, *t*-SNE, provided an effective window that the different feature patterns could be clearly identified from visual two-dimensional plots. The distances between each feature point indicated the correlation among data. In addition, the impacts of noise interference on the data were observed with the use of this approach.

(c) The deep learning approach also exhibited high accuracy and robustness for data with high noise interference. The CNN classification for most cases could reach up to 100% when the noise levels were lower (80 dB–100 dB). However, with the energy of the noise (SNR = 70 dB) close to the signals, data classification exhibited a certain level of reduction, and the error rates were close to 80%. Particularly, when the noise increased to a much higher level (60 dB), all the signals were contaminated, and the effectiveness of the classification dropped.

(d) The proposed method can also identify the embedding conditions. The identification is no less than 92% when the noise level is lower than 60 dB. However, the accuracy drops to 35% at 60 dB, which means it is difficult to distinguish the embedding conditions of rods.

This study simulated the different levels of PC structure's prestress loss by a deep learning method. To accommodate engineering concerns, noise interference was added. In future work, the methods will be explored on more large-scale structures in laboratory and field conditions. As high levels of noise prevent the model from achieving high accuracy, future perspective will focus on this issue to improve the accuracy of identification under higher noise levels (60 dB). In addition, different kinds of damage will occur at same time; thus, future study will investigate more complex situations. Furthermore, as the selection of hyperparameters is also critical, future research will also focus on deep learning with optimization methods (Bayesian optimization).

Author Contributions: Z.Z. designed, conducted research, and wrote the paper under the supervision of Z.L.; F.T. and Q.C. edited the paper; H.P. and X.W. assisted with the experiments and edited the paper. All authors have read and agreed to the published version of the manuscript.

Funding: This research received no external funding.

Conflicts of Interest: The authors declare no conflict of interest.

References

1. Gui, G.; Pan, H.; Lin, Z.; Li, Y.; Yuan, Z. Data-driven support vector machine with optimization techniques for structural health monitoring and damage detection. *KSCE J. Civ. Eng.* **2017**, *21*, 523–534. [CrossRef]
2. Tang, F.; Lin, Z.; Chen, G.; Yi, W. Three-dimensional corrosion pit measurement and statistical mechanical degradation analysis of deformed steel bars subjected to accelerated corrosion. *Constr. Build. Mater.* **2014**, *70*, 104–117. [CrossRef]
3. Lin, Z.; Mostafa, F.; Wu, C.H.; Chen, G.; Bevans, W.J.; Gunasekaran, A.V.K.; Sedigh, S. Design, Construction, Laboratory and Field Testing of the Bridge on the Arnault Branch, Washington County, Missouri. In *Center for Transportation Infrastructure and Safety/NUTC Program*; Missouri University of Science and Technolog: Rolla, MO, USA, 2013.
4. Bartoli, I.; Salamone, S.; Phillips, R.; Lanza di Scalea, F.; Sikorsky, C.S. Use of Interwire Ultrasonic Leakage to Quantify Loss of Prestress in Multiwire Tendons. *J. Eng. Mech.* **2011**, *137*, 324–333. [CrossRef]
5. Chen, S.Z.; Wu, G.; Feng, D.C.J.M.S.; Processing, S. Damage detection of highway bridges based on long-gauge strain response under stochastic traffic flow. *Mech. Syst. Signal Process.* **2019**, *127*, 551–572. [CrossRef]
6. Zhang, Z.; Pan, H.; Wang, X.; Tang, F.; Lin, Z. Ultrasonic guided wave approaches for pipeline damage diagnosis based on deep leaning. In Proceedings of the ASCE Pipelines 2022 Conference, Indianapolis, IN, USA, 31 July–3 August 2022.
7. Zhang, Z.; Pan, H.; Wang, X.; Lin, Z. Machine learning-enabled Lamb wave approaches for damage detection. In Proceedings of the 2021 10th International Conference on Structural Health Monitoring of Intelligent Infrastructure, Porto, Portugal, 30 June–2 July 2021.
8. Tang, F.; Lin, Z.; Qu, H. Investigation into corrosion-induced bond degradation between concrete and steel rebar with acoustic emission and 3D laser scan techniques. *J. Infrastruct. Preserv. Resil.* **2022**, *3*, 1–18. [CrossRef]
9. Zhang, Y.M.; Wang, H.; Bai, Y.; Mao, J.X.; Xu, Y.C. Bayesian dynamic regression for reconstructing missing data in structural health monitoring. *Struct. Health Monit.* **2022**, *21*, 2097–2115. [CrossRef]
10. Beard, M.; Lowe, M.; Cawley, P. Ultrasonic guided waves for inspection of grouted tendons and bolts. *J. Mater. Civ. Eng.* **2003**, *15*, 212–218. [CrossRef]
11. Cobb, A.C.; Kwun, H.; Caseres, L.; Janega, G. Torsional guided wave attenuation in piping from coating, temperature, and large-area corrosion. *NDT E Int.* **2012**, *47*, 163–170. [CrossRef]
12. Zhang, Z.; Pan, H.; Wang, X.; Lin, Z. Machine Learning-Enriched Lamb Wave Approaches for Automated Damage Detection. *Sensors* **2020**, *20*, 1790. [CrossRef]
13. Zhang, Z.; Pan, H.; Wang, X.; Lin, Z. Deep Learning Empowered Structural Health Monitoring and Damage Diagnostics for Structures with Weldment via Decoding Ultrasonic Guided Wave. *Sensors* **2022**, *22*, 5390. [CrossRef]
14. Sun, H.; Zhu, J. Nondestructive evaluation of steel-concrete composite structure using high-frequency ultrasonic guided wave. *Ultrasonics* **2020**, *103*, 106096. [CrossRef]

15. Lanza di Scalea, F.; Rizzo, P.; Seible, F. Stress measurement and defect detection in steel strands by guided stress waves. *J. Mater. Civ. Eng.* **2003**, *15*, 219–227. [CrossRef]
16. Ervin, B.L.; Kuchma, D.A.; Bernhard, J.T.; Reis, H. Monitoring corrosion of rebar embedded in mortar using high-frequency guided ultrasonic waves. *J. Eng. Mech.* **2009**, *135*, 9–19. [CrossRef]
17. Chaki, S.; Bourse, G. Guided ultrasonic waves for non-destructive monitoring of the stress levels in prestressed steel strands. *Ultrasonics* **2009**, *49*, 162–171. [CrossRef]
18. Treyssede, F.; Laguerre, L. Investigation of elastic modes propagating in multi-wire helical waveguides. *J. Sound Vib.* **2010**, *329*, 1702–1716. [CrossRef]
19. Dubuc, B.; Ebrahimkhanlou, A.; Salamone, S. Higher order longitudinal guided wave modes in axially stressed seven-wire strands. *Ultrasonics* **2018**, *84*, 382–391. [CrossRef]
20. Shoji, M. Ultrasonic Guided Wave Inspection of Anchor Rods Embedded in Soil. *J. Nondestruct. Eval.* **2019**, *38*, 96. [CrossRef]
21. Pan, H.; Gui, G.; Lin, Z.; Yan, C. Deep BBN learning for health assessment toward decision-making on structures under uncertainties. *KSCE J. Civ. Eng.* **2018**, *22*, 928–940. [CrossRef]
22. Lin, Z.; Pan, H.; Wang, X.; Li, M. Data-driven structural diagnosis and conditional assessment: From shallow to deep learning. In Proceedings of the Sensors and Smart Structures Technologies for Civil, Mechanical, and Aerospace Systems 2018, Denver, CO, USA, 4–8 March 2018; International Society for Optics and Photonics: Bellingham, WA, USA, 2018; p. 1059814.
23. Abdeljaber, O.; Avci, O.; Kiranyaz, S.; Gabbouj, M.; Inman, D.J. Real-time vibration-based structural damage detection using one-dimensional convolutional neural networks. *J. Sound Vib.* **2017**, *388*, 154–170. [CrossRef]
24. Fawaz, H.I.; Forestier, G.; Weber, J.; Idoumghar, L.; Muller, P.-A. Deep learning for time series classification: A review. *Data Min. Knowl. Discov.* **2019**, *33*, 917–963. [CrossRef]
25. Pan, H.; Azimi, M.; Yan, F.; Lin, Z. Time-frequency-based data-driven structural diagnosis and damage detection for cable-stayed bridges. *J. Bridge Eng.* **2018**, *23*, 04018033. [CrossRef]
26. Zhang, Z.; Wang, X.; Pan, H.; Lin, Z. Corrosion-induced damage identification in metallic structures using machine learning approaches. In Proceedings of the 2019 Defense TechConnect Innovation Summit, National Harbor, MD, USA, 7–10 October 2019.
27. Cha, Y.J.; Choi, W.; Büyüköztürk, O. Deep learning-based crack damage detection using convolutional neural networks. *Comput. Aided Civ. Infrastruct. Eng.* **2017**, *32*, 361–378. [CrossRef]
28. Cha, Y.J.; Choi, W.; Suh, G.; Mahmoudkhani, S.; Büyüköztürk, O. Autonomous structural visual inspection using region-based deep learning for detecting multiple damage types. *Comput. Aided Civ. Infrastruct. Eng.* **2018**, *33*, 731–747. [CrossRef]
29. An, Y.-K.; Jang, K.; Kim, B.; Cho, S. Deep learning-based concrete crack detection using hybrid images. In Proceedings of the Sensors and Smart Structures Technologies for Civil, Mechanical, and Aerospace Systems 2018, Denver, CO, USA, 4–8 March 2018; International Society for Optics and Photonics: Bellingham, WA, USA, 2018; p. 1059812.
30. Pan, H.; Zhang, Z.; Wang, X.; Lin, Z. Image-based damage conditional assessment of large-scale infrastructure systems using remote sensing and deep learning approaches. In Proceedings of the 2019 TechConnect World Innovation Conference, Boston, MA, USA, 17–19 June 2019.
31. Guo, J.; Xie, X.; Bie, R.; Sun, L. Structural health monitoring by using a sparse coding-based deep learning algorithm with wireless sensor networks. *Pers. Ubiquitous Comput.* **2014**, *18*, 1977–1987. [CrossRef]
32. Zhang, W.; Peng, G.; Li, C.; Chen, Y.; Zhang, Z. A new deep learning model for fault diagnosis with good anti-noise and domain adaptation ability on raw vibration signals. *Sensors* **2017**, *17*, 425. [CrossRef]
33. Mahajan, H.; Banerjee, S. A machine learning framework for guided wave-based damage detection of rail head using surface-bonded piezo-electric wafer transducers. *Mach. Learn. Appl.* **2021**, *7*, 100216. [CrossRef]
34. Tabian, I.; Fu, H.; Sharif Khodaei, Z. A Convolutional Neural Network for Impact Detection and Characterization of Complex Composite Structures. *Sensors* **2019**, *19*, 4933. [CrossRef]
35. Zargar, S.A.; Yuan, F.-G. Impact diagnosis in stiffened structural panels using a deep learning approach. *Struct. Health Monit.* **2021**, *20*, 681–691. [CrossRef]
36. Chree, C. The equations of an isotropic elastic solid in polar and cylindrical co-ordinates their solution and application. *TCaPS* **1889**, *14*, 250.
37. Raghavan, A.; Cesnik, C.E. 3-D elasticity-based modeling of anisotropic piezocomposite transducers for guided wave structural health monitoring. *J. Vib. Acoust.* **2007**, *129*, 739–751. [CrossRef]
38. Seco, F.; Martín, J.M.; Jiménez, A.; Pons, J.L.; Calderón, L.; Ceres, R. PCDISP: A tool for the simulation of wave propagation in cylindrical waveguides. In Proceedings of the 9th International Congress on Sound and Vibration, Orlando, FL, USA, 8–11 July 2002.
39. Seco, F.; Jiménez, A.R. Modelling the generation and propagation of ultrasonic signals in cylindrical waveguides. *Ultrason. Waves* **2012**, 1–28. [CrossRef]
40. Beard, M.; Lowe, M. Non-destructive testing of rock bolts using guided ultrasonic waves. *Int. J. Rock Mech. Min. Sci.* **2003**, *40*, 527–536. [CrossRef]
41. Haskins, R.W. Detection of Microcracks in Trunnion Rods Using Ultrasonic Guided Waves. 2015. Available online: https://apps.dtic.mil/sti/pdfs/ADA619008.pdf (accessed on 23 September 2022).
42. Zhang, B.; Hong, X.; Liu, Y. Multi-task deep transfer learning method for guided wave-based integrated health monitoring using piezoelectric transducers. *IEEE Sens. J.* **2020**, *20*, 14391–14400. [CrossRef]

43. Liang, X. Image-based post-disaster inspection of reinforced concrete bridge systems using deep learning with Bayesian optimization. *Comput. Aided Civ. Infrastruct. Eng.* **2019**, *34*, 415–430. [CrossRef]
44. Zhang, Y.-M.; Wang, H.; Mao, J.-X.; Xu, Z.-D.; Zhang, Y.-F. Probabilistic framework with bayesian optimization for predicting typhoon-induced dynamic responses of a long-span bridge. *J. Struct. Eng.* **2021**, *147*, 04020297. [CrossRef]
45. Gulgec, N.S.; Takáč, M.; Pakzad, S.N. Structural damage detection using convolutional neural networks. In *Model Validation and Uncertainty Quantification, Volume 3*; Springer: Cham, Switzerland, 2017; pp. 331–337.
46. Maaten, L.v.d.; Hinton, G. Visualizing data using t-SNE. *J. Mach. Learn. Res.* **2008**, *9*, 2579–2605.

buildings

MDPI

Article

Numerical Study on Elastic Parameter Identification of Large-Span Steel Truss Structures Based on Strain Test Data

Yuxin Zhang [1,*], Ao Zhou [1], Helong Xu [1] and Hexin Zhang [2]

[1] School of Civil Engineering, Shanghai Normal University, Shanghai 201400, China
[2] School of Computing, Engineering and the Built Environment, Edinburgh Napier University, Edinburgh EH10 5DT, UK
* Correspondence: zyx@shnu.edu.cn

Abstract: Large-span steel trusses are widely used in public buildings such as large-span factory buildings, exhibition halls, gymnasiums, and bridges because of their fast construction speed and easy industrial manufacturing. Due to construction errors and environmental factors, the material properties may change during their service life, and it is an important prerequisite for the structural safety assessment to identify the true material parameters of the structure. Among the many parameters, the elastic modulus is one that has the greatest impact on the accuracy of structural safety analysis. In this paper, a mathematical analysis model of elastic modulus identification was constructed, based on the strain test data and the improved gradient regularization method. The relationship between the strain test data and elastic moduli was established. A common finite element program based on the method was developed to identify the elastic modulus. A series of numerical simulations was carried out on a 53-element steel truss model to study the availability and numerical stability of the method. The effects of different initial values, numbers of strain tests, and locations of the strain test as well as the number of unknown parameters on the identification results were studied. The results showed that the proposed method had very high accuracy and computational efficiency. For the case of 53 unknown parameters without considering the test error, the identification accuracy could reach a 1×10^{-10} order of magnitude after only several iterations. This paper provides an effective solution to obtain the actual values of the elastic modulus of steel truss structures in practical engineering.

Keywords: parameter identification; regularization; gradient matrix; elastic modulus; strain

1. Introduction

Large-span steel truss structures are widely used in public buildings such as large-span factory buildings, exhibition halls, gymnasiums, and bridges because of their fast construction speed and easy industrial manufacturing [1,2]. Due to construction errors and environmental factors, the structural material properties may change during the service, and it is an important prerequisite for structural safety assessment to fully understand the real material parameters of the structure. Therefore, the parameter and damage identification of steel truss structures have been of significant concern.

Chang [3] presented the preliminary results of modal-parameter identification and vibration-based damage detection of a damaged steel truss bridge. Zhuo [4] studied the damage identification of bolt connections in steel truss structures by using sound signals. Luong [5] proposed a methodology to identify multiple axial forces in members of a truss structure based on the modal parameters. Luong [6] investigated the inverse identification of the stress state in axially loaded slender members of steel truss structures using measured dynamic data. Liu [7] adopted inverse sensitivity analysis to estimate the unknown system parameter perturbation from the difference between the observed output data and the corresponding analytical output data calculated from the original system model. Cho [8] performed system identification on the swing span of a steel truss bridge dating from 1896 using acceleration data collected from a wireless sensor network (WSN). Terlaje [9] used displacement measurements resulting from applied static point loads as constraints in an optimization algorithm that employed optimality criterion methods to extract the cross-sectional properties of elements within a mathematical model of a structure. Chakraborty [10] presented a methodology to diagnose and quantify the damage at the element level in a truss structure with the measured static strain properties of the truss.

In the above works, we found that most of the parameter identification work for steel truss structures was based on the dynamic response test, and mainly identified the dynamic characteristics of the structure. Works based on the static test and identifying the elastic parameters of the structure are limited. This is mainly because the dynamic test method can achieve real-time monitoring without artificially applying loads and blocking traffic. However, because it is related to modal identification, there are higher requirements for the accuracy of the test instruments and identification methods. A disadvantage of the static test method is the need to apply a load to excite static response, which blocks traffic, but it has the advantage of a good identification effect and easy measurement of the required data. Moreover, the static equilibrium equation is only related to the nature of the structural stiffness, and it is easy to calculate the structural stiffness according to the measured static data. Additionally, the static test equipment is cheaper, the test technology is more advanced, and the deformation of the structure can be measured more accurately, so it is beneficial to study the elastic parameter identification method based on the static test. In fact, the damage detection method based on the static test has also received extensive attention in the field of civil engineering. Song [11] studied the problem of the optimal strain sensor placement in the damage detection of truss elements. Wang [12] identified moving train load parameters including the train speed, axle spacing, gross train weight, and axle weights based on the strain-monitoring data. Compared with other static response tests, the strain test has unique advantages because of the strain gauges' small mass, high accuracy, easy installation and fixation, and low comprehensive cost, so it is widely used in engineering [13–15].

The elastic modulus is also one of the most important parameters that affect the structural safety assessment because it directly affects the composition of the structural stiffness matrix. Although work on the elastic parameter identification of steel truss structures is limited, this problem has been widely considered in the field of mechanical inverse problems, and many research methods have been proposed. In these methods, the common approach is to reflect the local parameter variation onto the actual response value based on the relationship between the structural parameter variation and the actual measured data of a certain response. Then, the problem is transformed into the minimization of the objective function with the unknown parameters as the unknown variables and the minimum difference between the theoretical response value and the measured response value, which is also a typical engineering inverse problem. According to different solution methods, the problem can be subdivided into the neural network method [16,17], Levenberg–Marquardt method [18,19], Tikhonov regularization method [20–22], Gauss–Newton method [23], genetic algorithm [24], and so on.

The gradient regularization method (hereafter GRM) is a method for solving the inverse problem. It was first proposed in [25] and its applicability to one-dimensional hyperbolic equations (one-dimensional wave equations) was verified. The applicability of the GRM in identifying the parameters of two-dimensional elliptic operators was demonstrated in [26]. The nonlinear inverse problem is transformed into the problem of solving linear equations by expanding the unknown parameters in series with the supplementary conditions of the inverse problem. Then, the GRM is used to solve the ill-posed linear equations. This method starts from the generality of the inverse problem, without any special constraints, and is not limited by the space dimension when solving the inverse problem, so it is a very general method for solving the inverse problem. The GRM solves the difficulties of ill-posedness, strong nonlinearity, and large calculation requirement in the process of solving inverse problems, and its advantages of a high accuracy of calculation results and short calculation time make it applicable in related fields. The elastic modulus of concrete dam was identified by using the GRM and the displacement monitoring data in Liu [27]. Zhang [28] realized the elastic modulus identification of bar structures based on the displacement test data and the GRM.

The choice of the regularization parameter α during the solution process of the regularization method is very important. If α is too large, the stability of the solution is guaranteed, but the accuracy is reduced; if α is too small, the stability of the solution is difficult to guarantee. Based on the idea of homotopy mapping, Cui et al. [29] extended the solution path, effectively expanded the convergence domain, and reduced the dependence on the initial value of iteration. Reichel [30] et al. selected the appropriate regularization parameter when the truncated singular value decomposition method and LSQR iterative Krylov subspace could not accurately estimate the data error. Hua [31] et al. studied the selection of regularization parameters in model updates and proposed that the selection of adaptive regularization parameters was more effective than that of the fixed regularization parameters. Hansen [32] proposed a more efficient regularization parameter selection method based on the L curve. Bucataru et al. [33] studied the numerical reconstruction of thermal boundary data on a part of the boundary occupied by an anisotropic solid, and used gradient regularization to solve the inverse problem.

In this paper, the problem was transformed into identifying the elastic modulus of the structure by measuring the strain data at several points of the structure, which is a typical inverse problem of operator identification [34]. Based on the GRM, the diagonal elements of the Jacobi matrix in the solution process were normalized through linear transformation, which improved the solution speed and accuracy. A problem solution model was derived and constructed based on the strain test data and the improved gradient regularization–finite element method for the first time. A general finite element calculation program was developed. A series of numerical simulation tests were carried out on a 53-element steel truss model to study the availability and numerical stability of the method. The effects of different initial values, different numbers of strain test, different locations of the strain test, and the number of unknown parameters on the identification results were studied. The results showed that the proposed method had very high accuracy and computational efficiency. Without considering the test error, only a few iterations were needed, and the identification accuracy could reach the order of 1×10^{-10}. For large-scale calculation, the advantages of this work will be more prominent compared with the traditional optimal solution method, and the identification accuracy does not depend on the selection of the initial value, so it has strong practicability. The proposed method provides an effective solution for obtaining accurate design values of the elastic parameters of steel truss structures in practical engineering.

2. Elastic Parameter Identification Model Based on Strain Test Data

2.1. Mathematical Solution Model of the Problem

For any structural members, when we artificially add an external load, P, it will cause additional displacement and stress–strain changes, and the strain is actually a function of the displacement. The static equation of a steel truss structure solved by the structural finite element is:

$$[K(E)]\{U\} = \{P\} \tag{1}$$

$$\{\varepsilon\} = [B]\{U\} \tag{2}$$

where $\{U\} = \begin{pmatrix} u_1 & u_2 & \ldots\ldots & u_n \end{pmatrix}^T$ is a column vector composed of unknown node displacements and $\{P\} = (q_1, q_2, \ldots\ldots, q_n)^T$ is a column vector of known nodal loads; $[K(E)] = \begin{pmatrix} k_{11}(E) & \ldots & k_{1n}(E) \\ \vdots & \ddots & \vdots \\ k_{n1}(E) & \cdots & k_{nn}(E) \end{pmatrix}$ is the global stiffness matrix of the structure; and E is the elastic modulus of the structure.

B is the transfer matrix between the strain and displacement, which is determined according to the specific problem.

Formula (1) is the solution equation of the verse problem, that is, to solve the response with the given load action and design parameter information. The correlation between the elastic modulus and strain, ε, is established by Equation (1). However, if the design parameters are unknown and the load and test data of the local response are known, can the information of design parameters be obtained by reverse solving? This is a typical inverse problem of operator identification. If the structure is regarded as a continuum in a certain spatial domain, and E, ε, and P are functions about x, then the following mathematical equations for solving the inverse problem can be established:

$$K(E(x))U(x) = P(x)\ x \in \partial R^r \tag{3}$$

$$\varepsilon(x) = [B(x)]U(x) \tag{4}$$

$$B_1(\varepsilon(x)) = \varepsilon_c(x) \quad x \in \partial R1^r \tag{5}$$

$$B_2(\varepsilon(x))_{x=xs} = \varepsilon_s(x) \quad x \in \partial R2^r \tag{6}$$

where K is the operator of $E(x)$; B_1, B_2 is the boundary condition operator and the additional condition operator, respectively; $s = 1,\ m$, and m is the number of known strain test data. R is the spatial domain of the problem, equal to 2 or 3 for the plane and spatial domain problems, respectively; x is the coordinate defined on R^r; R_1^r is the domain of given boundary conditions; R_2^r is the domain of the given supplementary conditions.

2.2. Establishing the Objective Function

Since only the strain data at the positions of the representative elements can be obtained, the solution of the problem described in Equations (3)–(6) is not unique, and only a set of optimal solutions satisfying the additional strain test data can be found. Considering the existence of the test error and numerical error, this paper used the strain relative value to establish the following constraint objective function.

Seeking E:

$$\text{Let } f(E) \le err \tag{7}$$

$$f(E) = \sum_{i=1}^{s} |\tilde{\varepsilon}_i(E)| \tag{8}$$

$$\tilde{\varepsilon}_i(E) = \frac{\varepsilon_{si} - \varepsilon_i}{\varepsilon_{si}} \tag{9}$$

where err is a very small given value; ε_{si} is strain test data for the ith element; ε_i is the calculated strain value for the ith element.

2.3. Solution of Gradient Regularization Method [25] Based on Strain Test Data

Assuming the strain test data, $\varepsilon^*(x)$, is an exact solution satisfying Equations (3)–(6), $\varepsilon^*(x)$ and the supplementary condition $B_2(\varepsilon^*(x))$ are expanded by first-order Taylor expansion in the vicinity of $E_0(x)$, then there will be:

$$\varepsilon^*(x) = \varepsilon_0(x) + \int_R \frac{\partial \varepsilon(x)}{\partial E(x')} \Delta E(x') dx' \tag{10}$$

$$B_2(\varepsilon^*(x)) = B_2(\varepsilon_0(x)) + \int_R \frac{\partial B_2(\varepsilon(x))}{\partial E(x')} \Delta E(x') dx' \tag{11}$$

Since $\varepsilon^*(x)$ is an exact solution, we have

$$B_2(\varepsilon^*(x)) = \varepsilon_s(x) \tag{12}$$

If we make:

$$D(x) = \varepsilon_s(x) - B_2(\varepsilon_0(x)) \tag{13}$$

$$G(E(x'), x) = \frac{\partial B_2(\varepsilon(x))}{\partial E(x')} \tag{14}$$

then we substitute Equations (12)–(14) into Equation (11) to obtain:

$$\int_R G(E(x'), x) \cdot \Delta E(x') dx' = D(x) \tag{15}$$

Here, $G(E(x'), x)$ is the gradient operator of the supplementary condition $\partial B_2(\varepsilon(x))$ to $E(x')$ at $E(x') = E_0(x')$.

This is an ill-posed problem, hence, we construct the regularization functional:

$$J(\Delta E(x), \alpha) = \rho^2\left(\int_W G(k(x'), x)\Delta E(x')dx', D(x)\right) + \alpha\Theta(\Delta E(x')) \quad x' \in W, x \in \partial W_2 \tag{16}$$

$$\rho^2\left(\int_W G(E(x'), x)\Delta E(x')dx', D(x)\right) = \int_{\partial W_2}\left(\int_W G(E(x'), x)\Delta E(x')dx' - D(x)\right)^2 dx \tag{17}$$

Find $\Delta E(x')$ from Formula (18):

$$\min J(\Delta E(x'), \alpha) = \rho^2\left(\int_R G(E(x'), x)\Delta E(x')dx', D(x)\right) + \alpha\Theta(\Delta E(x')) \tag{18}$$

where α is the regularization parameter and $\Theta(\Delta E(x'))$ is the regularization functional. Equation (18) is discretized to give:

$$\min J(\Delta\tilde{E}, \alpha) = (\tilde{G} \cdot \Delta\tilde{E} - \tilde{D})^T(\tilde{G}\Delta\tilde{E} - \tilde{D}) + \alpha(\tilde{D}\Delta\tilde{E})^T(\tilde{R}\Delta\tilde{E}) \tag{19}$$

Let the first-order partial derivative of Equation (19) be equal to 0, and the extreme value of Equation (19) can be obtained as follows:

$$\left(\tilde{G}^T\tilde{G} + \alpha\tilde{H}\right)\Delta\tilde{E} = \tilde{G}^T\tilde{D} \tag{20}$$

where $\tilde{H}$ is the derived matrix of $\Theta(E(x'))$.

For a linear problem, $\Delta\tilde{E}$ can be solved from Equation (20). For a nonlinear problem, it needs to be solved by many iterations, thus:

$$\tilde{E}_{n+1} = \tilde{E}_n + \Delta\tilde{E}_n \tag{21}$$

When the convergence condition of Equation (21) is satisfied, the value E_{n+1} is taken as the real elastic modulus.

3. Improvement of Gradient Regularization Method [28]

The regularization parameter, α, plays a key role in the process of solving the inverse problem, and it will affect the stability and accuracy of the solution. When it becomes large, the stability of the solution is improved, but the accuracy is reduced, and vice versa when it becomes small. Therefore, the choice of α is key to the balance between the accuracy and stability.

In the original GRM, when looking for the parameters, the values are completely different in each iteration, even by several orders of magnitude. The search for α is difficult. Therefore, based on the original GRM, the following linear transformation was adopted to normalize the diagonal elements of the Jacobi matrix.

After discretization, we have:

$$G = (G_1 \ldots G_j \ldots G_n) \tag{22}$$

$$\Delta E = (\Delta E_1 \ldots \Delta E_j \ldots \Delta E_n)^T \tag{23}$$

$$G_j = (G_{1j} \ldots G_{ij} \ldots G_{mj})^T = \left(\frac{\partial B_1}{\partial E_j} \cdots \frac{\partial B_i}{\partial E_j} \cdots \frac{\partial B_m}{\partial E_j} \right)^T \tag{24}$$

where n is the number of unknown elastic moduli and m is the number of supplementary test data. Let:

$$\overline{G} = \left(\frac{G_1}{\|G_1\|_2} \cdots \frac{G_j}{\|G_j\|_2} \cdots \frac{G_n}{\|G_n\|_2} \right) \tag{25}$$

$$\Delta\overline{E} = (\|G_1\|_2 \Delta E_1 \ldots \|G_j\|_2 \Delta E_j \ldots \|G_n\|_2 \Delta E_n)^T \tag{26}$$

Equation (15) is discretized as:

$$\tilde{G}\Delta\tilde{E} = \tilde{D} \tag{27}$$

Substituting $\overline{G}, \Delta\overline{E}$ into Equation (20):

$$\left(\overline{G}^T\overline{G} + \alpha\tilde{R}^T\tilde{R}\right)\Delta\overline{E} = \overline{G}^T\tilde{D} \tag{28}$$

where

$$\overline{G}^T\overline{G} = \left(\frac{G_i^T}{\|G_i\|_2} \cdot \frac{G_j}{\|G_j\|_2} \right)_{n \times n} \tag{29}$$

When $i = j$,

$$\frac{G_i^T}{\|G_i\|_2} \cdot \frac{G_j}{\|G_j\|_2} = \frac{G_i^T}{\|G_i\|_2} \cdot \frac{G_i}{\|G_i\|_2} = 1 \tag{30}$$

$$\Delta\tilde{E} = \left\{ \frac{\Delta\tilde{E}_1}{\|G_1\|_2} \cdots \frac{\Delta\tilde{E}_j}{\|G_j\|_2} \cdots \frac{\Delta\tilde{E}_n}{\|G_n\|_2} \right\}^T \tag{31}$$

So far, the Jacobi matrix is normalized, which will not only improve the search speed, but also increase the accuracy of its solution.

4. Numerical Experiments and Analysis

To verify the method in this paper, a typical steel truss bridge model was selected, and a series of numerical experiments were carried out with the common finite element analysis program developed with Fortran Language based on the GRM method.

4.1. Prototype for Numerical Experiments

As shown in Figure 1, a plane steel truss model was selected for the numerical simulation analysis [35]. In this structure, each bar had a pipe cross section with an outer diameter of 1.71 cm, and a wall thickness of 0.2 cm. The total length of the truss was 5.6 m, with 0.4 m in each bay, and the height of the truss was 0.4 m. The model has 53 bar elements, 28 nodes, and 81 degrees of freedom. Among them, Nos. 1–14 elements were lower chord bars; No. 15–26 elements were upper chord bars; Nos. 27–53 elements were vertical bars, and the material parameters of each bar was the same (see Table 1 for specific parameter information). The bars were connected at pinned joints. There were two supports in this truss structure: a pin support at the left end and a roller support at the right end of the lower chord. The roller support at the right end was constrained in the vertical direction.

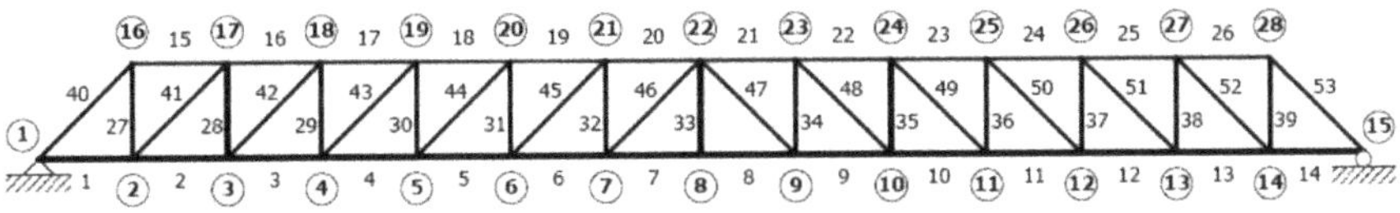

Figure 1. Schematic diagram of the plane hinged truss model.

Table 1. Basic material parameters of the truss bridge.

Parameter	Value
Modulus of elasticity, E	2.0×10^8 Pa
Moment of inertia, I	3.556×10^{-4} m^4
Section area of each bar, A	1.38×10^{-4} m^2
Density, *r*	5.69×10^7 kg/m^3

It was assumed that the external load acts on node 21, F = −10 kN (vertically downward), and the strain value of each element can be calculated through the verse problem calculation with the known design values of material parameters, as shown in Table 2.

Table 2. The calculated strain value of each element.

Element	Strain ($\times 10^{-3}$)	Element	Strain ($\times 10^{-3}$)
1	9.54	28	8.13
2	9.17	29	7.23
3	8.46	30	5.96
4	7.40	31	4.34
5	5.98	32	2.35
6	4.25	33	3.54
7	2.45	34	2.35
8	2.45	35	4.34
9	4.25	36	5.96
10	5.98	37	7.23
11	7.40	38	8.13
12	8.46	39	8.67
13	9.17	40	9.11
14	9.54	41	8.74
15	9.17	42	8.02
16	8.45	43	6.94
17	7.38	44	5.49
18	5.96	45	3.69
19	4.21	46	1.53
20	2.21	47	1.53
21	2.21	48	3.69
22	4.21	49	5.49
23	5.96	50	6.94
24	7.38	51	8.02
25	8.45	52	8.74
26	9.17	53	9.11
27	8.67		

4.2. Study on the Availability of the Method

To verify the availability of the method, it was assumed that the material parameters are unknown. To simplify the problem of the regularity study of the method, the initial elastic modulus values of the bottom chord, the middle web member, and top chord were assumed to be the same and set as E1, E2 and E3, respectively, according to the position of the member; these values were estimated according to experience. The strain values of elements 2, 21, 36, and 44 were obtained by the verse problem calculation as the simulated strain test data. The termination criterion was set to 1×10^{-10}. If the method is correct, the design parameter values will be identified. The iteration process of the numerical experiment is shown in Table 3.

Table 3. Iteration process (unit: 10^8 N/m^2).

Step	E1	E2	E3	Objective Function
E0	1.00	1.50	1.80	/
1	1.49	1.85	2.02	7.55×10^{-1}
2	1.87	1.97	2.00	1.51×10^{-1}
3	1.99	2.00	2.00	9.01×10^{-3}
4	2.00	2.00	2.00	2.68×10^{-5}
5	2.00	2.00	2.00	6.68×10^{-9}
6	2.00	2.00	2.00	2.16×10^{-13}
True value	2.00	2.00	2.00	

The data in Table 3 show that the calculated elastic modulus value finally converged to the design value of the model and satisfied the requirement of the objective function in the sixth iteration step, which proves the availability of the method and reflects its high efficiency.

4.3. Elastic Parameter Identification under Different Initial Elastic Moduli Values

The strains of elements 2, 21, 36, and 44 were still selected as supplementary conditions for the inversion calculation. Three groups of different elastic moduli values were taken as the initial elastic moduli values, and the calculation results are shown in Table 4 (the iteration convergence progress is shown in Figure 2). The identification results in Table 4 show that the selection of the initial value of the elastic moduli had little effect on the identification results as long as the supplementary test information was accurate, but it would have a certain impact on the identification speed. However, because of the high efficiency of the GRM, the impact on the computational efficiency was almost negligible. Still, for large-scale engineering calculations in practical application, the initial value should be estimated according to the engineering information as far as possible to improve the calculation efficiency as much as possible. The information can be the initial design value of the elastic modulus in the design files, or some test data at the beginning of construction, etc.

4.4. Effect of Amounts of Strain Test Data on the Identification Results

The initial elastic moduli values (1.3, 1.6, 1.9) $\times$ 10^8 N/m^2 were selected, along with different amounts of strain data from Table 2 as supplementary conditions to conduct the numerical simulation. The calculation results are shown in Table 5 (the convergence process is shown in Figure 3).

Table 4. Iteration results with different initial values.

Set No.	Initial Value ($\times 10^8$ N/m^2)	Total Iteration Steps	Solution ($\times 10^8$ N/m^2)	Error (%)
1	E1 = 1.0	6	2.0	2.10×10^{-11}
	E2 = 1.3		2.0	5.80×10^{-11}
	E2 = 1.5		2.0	7.85×10^{-11}
2	E1 = 1.2	5	2.0	7.65×10^{-10}
	E2 = 1.5		2.0	6.20×10^{-10}
	E3 = 1.7		2.0	5.60×10^{-11}
3	E1 = 1.3	5	2.0	5.14×10^{-10}
	E2 = 1.6		2.0	2.86×10^{-10}
	E3 = 1.9		2.0	3.00×10^{-12}

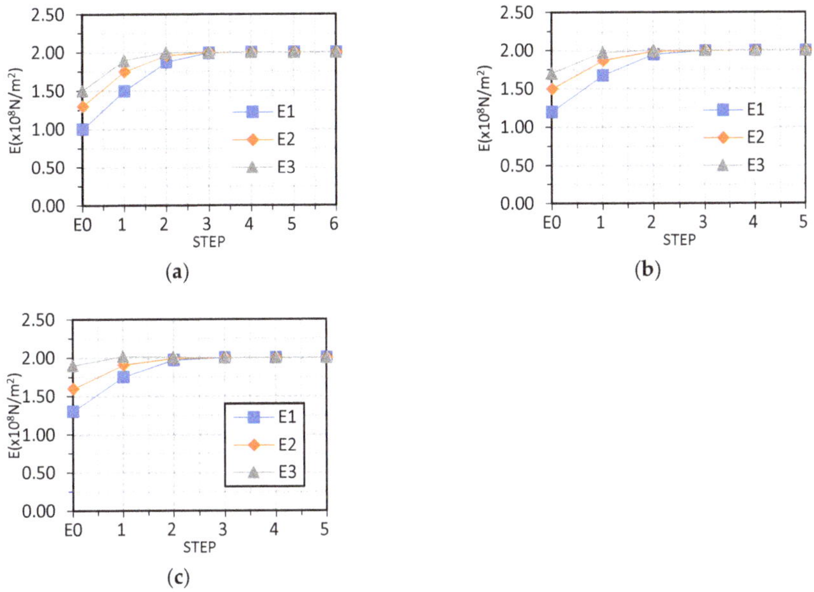

Figure 2. Results under different sets of initial values. (**a**) Results under the first set of initial values. (**b**) Results under the second set of initial values. (**c**) Result under the third set of initial values.

Table 5. Results with different numbers of strain test data ($\times 10^8$ N/m^2).

Number of Strain Test Data	Total Iteration Steps	E1, E2, E3	Error (%)
2	5	1.69 1.71 3.57	−15.5% −14.5% 157.0%
3	5	2.0 2.0 2.0	4.77×10^{-10} 1.74×10^{-10} 1.00×10^{-11}
5	5	2.0 2.0 2.0	5.43×10^{-10} 1.38×10^{-10} 2.50×10^{-12}
10	5	2.0 2.0 2.0	8.61×10^{-10} 9.85×10^{-11} 1.15×10^{-11}
15	5	2.0 2.0 2.0	9.72×10^{-10} 2.10×10^{-10} 5.00×10^{-13}
53	6	2.0 2.0 2.0	1.15×10^{-11} 1.20×10^{-11} 1.10×10^{-11}

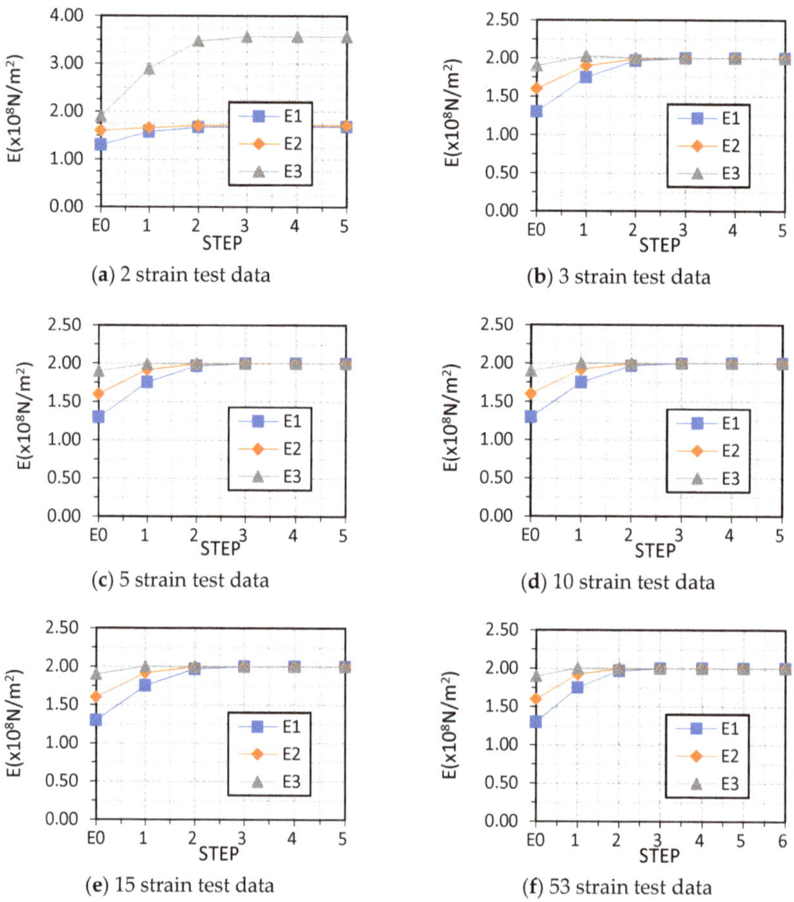

Figure 3. Calculation results of different numbers of strain test data.

It can be seen from the above numerical experiment results that when the number of supplementary conditions is less than the number of unknown elastic parameters, the calculation cannot converge to the true elastic moduli value. Thus, the number of supplementary conditions in practical projects is at least greater than or equal to the number of unknown parameters. However, the calculated results are still convergent to the elastic parameters of the model regardless of whether the least three elements or all 53 elements of the strain data are selected as the supplementary conditions. Therefore, in an actual project, it is enough to select a certain number of measured values as additional information, more is not necessarily better. The number of measuring points can be selected according to the actual situation of the site, which not only reduces the construction cost, but also reduces the time cost and improves the operation efficiency.

4.5. Identification of Elastic Parameters with Strain Test Data at Different Locations

We still took (1.3, 1.6, 1.9) $\times$ 10^8 N/m^2 as the initial elastic moduli values. When the locations were all scattered at the lower chord (the red part as shown in Figure 4a) and the locations were all concentrated at the support position (the red part as shown in Figure 4b), the iteration process is shown in Tables 6 and 7, respectively.

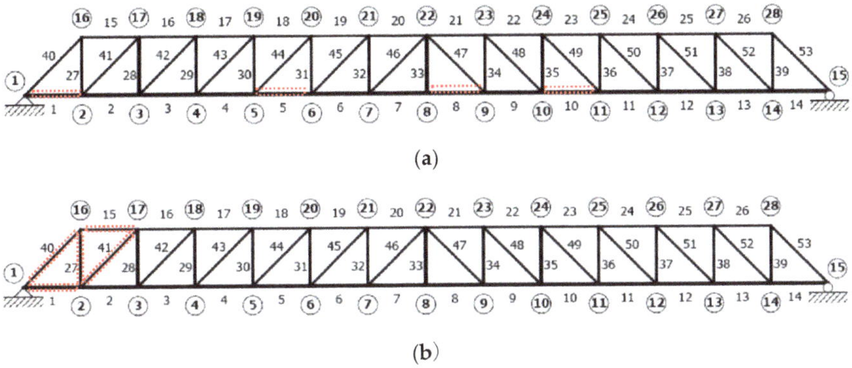

Figure 4. (a) Selection of scattered measuring points of the lower chord. (b) Selection of centralized measuring points at the support.

Table 6. Calculation results with centralized measuring points of the lower chord.

Number of Iterations	E1	E2	E3	Objective Function
E_0	1.30	1.60	1.90	/
0	1.52	2.78	2.43	3.86×10^{-1}
1	1.81	2.32	2.09	1.26×10^{-1}
2	1.96	2.03	2.02	2.71×10^{-2}
3	2.00	2.00	2.00	1.20×10^{-3}
4	2.00	2.00	2.00	5.66×10^{-8}
5	2.00	2.00	2.00	1.52×10^{-12}

Table 7. Calculation results with centralized measuring points at the support.

Number of Iterations	E1	E2	E3	Objective Function
E0	1.30	1.60	1.90	/
0	1.64	1.96	2.32	3.23×10^{-1}
1	1.91	2.01	2.01	1.15×10^{-1}
2	2.00	2.00	2.00	5.20×10^{-3}
3	2.00	2.00	2.00	8.13×10^{-6}
4	2.00	2.00	2.00	1.11×10^{-9}
5	2.00	2.00	2.00	1.09×10^{-12}

The above numerical experimental results show that there was little difference in the accuracy of the identification results across different locations of the selected measuring points. The true elastic moduli values could be accurately identified. This is of great significance for practical engineering applications. In practical engineering, according to the actual construction conditions, the time-saving and labor-saving points can be preferentially considered for strain measurement to improve the construction efficiency. Of course, the measuring points with obvious strain change should be selected wherever possible, which is conducive to ensuring the accuracy of the solution.

4.6. Studies on Numerical Stability

In practical engineering, test errors are often caused by other factors such as the accuracy of the instrument or the non-standard test operation. Random errors of +5% and +10% were artificially applied to the strain values of elements 2, 21, 36, and 44 in Table 2, and the calculation results are shown in Table 8. It can be seen from the identification results in Table 8 that after considering the model error, the iteration process is stable and the solution changes regularly and stably with the change in the error, which proves the good numerical stability of the method in this paper.

Table 8. Calculation results under different error conditions.

Step	0%	+5%	+10%
E0	1.3 1.6 1.9	1.3 1.6 1.9	1.3 1.6 1.9
1	1.75 1.91 2.01	1.71 1.84 1.92	1.67 1.77 1.83
2	1.97 1.99 2.01	1.88 1.90 1.91	1.81 1.82 1.82
3	2.00 2.00 2.00	1.90 1.90 1.90	1.82 1.82 1.82
4	2.00 2.00 2.00	1.90 1.90 1.90	1.82 1.82 1.82
err(%)	0 0 0	4.76 4.76 4.76	9.09 9.09 9.09

4.7. Identification under Different Numbers of Unknown Parameters

In practical engineering, the elastic modulus of each region may not be equal, so it should be assumed that the elastic parameters of each measured region are unknown and must be identified. Since the initial value of the initial elastic modulus had no effect on the parameter identification, it was assumed that the initial elastic modulus of all members was 1.5e8 N/m^2. The identification results under five unknowns (six simulated test strains were randomly selected), 10 unknowns (11 simulated test strains were randomly selected), 15 unknowns (16 simulated test strains were randomly selected), and the elastic moduli of 53 elements were all unknown (53 test strains are all selected) were studied, respectively. The calculation results of the objective function are shown in Table 9, and the convergence process is shown in Figure 5.

Table 9. Convergence process of the objective function under different unknowns.

Number of Iterations	Number of Unknowns			
	5	10	15	53
0	4.00×10^{-1}	7.335×10^{-1}	1.071	3.536
1	2.35×10^{-2}	4.249×10^{-2}	6.250×10^{-1}	2.048×10^{-1}
2	7.34×10^{-5}	1.194×10^{-4}	2.968×10^{-3}	1.245×10^{-3}
3	3.07×10^{-8}	6.589×10^{-8}	2.278×10^{-3}	3.475×10^{-4}
4	4.79×10^{-13}	6.917×10^{-13}	2.078×10^{-5}	1.969×10^{-6}
5			2.482×10^{-9}	1.347×10^{-10}
6			2.090×10^{-12}	1.277×10^{-12}

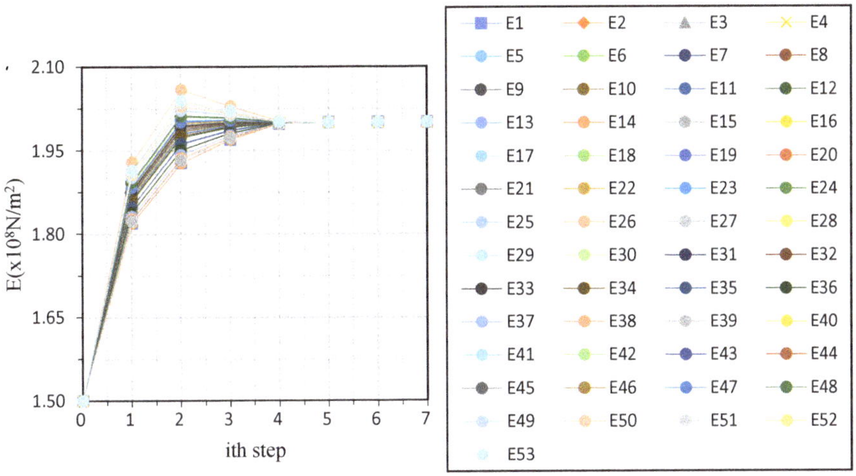

Figure 5. Calculation results under 53 unknowns.

When the number of unknowns increased to 53, the calculation results still converged to the true elastic moduli values, which proves the effectiveness of the method when applied to large-scale operations. However, Table 8 shows that when the number of unknowns increased to 15 and 53, more iterations were needed to reach the objective function value than when the number of unknowns was 5 or 10. Thus, when the number increases to a certain extent, the solution process will become longer. In practical engineering, according to the position of the members and their properties in the structure, the members with the same elastic modulus value should be assessed and set as the same unknown parameters as much as possible in combination with prior experience to improve the efficiency of the solution and reduce unnecessary test work.

5. Conclusions

In this paper, a mathematical analysis model of elastic modulus identification was constructed, based on strain test data and the improved gradient regularization–finite element method. The numerical analysis showed that the proposed method had very high accuracy and computational efficiency. Finally, the related problems in this paper can be explained and discussed as follows:

(1) The elastic parameter identification method based on the strain test and gradient regularization–finite element method in this paper is available and efficient. However, in practical application, the accuracy of the test data must be ensured. At present, the strain measurement technology for civil structure health monitoring has been very well-developed, the most commonly used being resistance type strain gauges [36], vibrating wire type gauges [37], and fiber optic sensors [38]. At present, the accuracy of the commonly used strain sensors can reach 0.1~0.5 $\mu\varepsilon$. The rapid development of modern testing technology provides a technical guarantee for the proposed method.

(2) In the elastic stage, the proposed method only needs to determine the additional external load and the geometric dimensions of the structure at the moment of the test and has nothing to do with the initial stress of the structure, so it needs less input information and improves the identification accuracy.

(3) In fact, the identification method in this paper is also applicable to the identification of elastic parameters of other structures, but the solution model used in the verse analysis process is different according to the characteristics of various structures.

Author Contributions: Conceptualization, Y.Z.; methodology, Y.Z.; software, Y.Z. and A.Z. and H.X.; validation, Y.Z.; formal analysis, Y.Z. and A.Z.; investigation, Y.Z. and A.Z.; data curation, Y.Z. and A.Z.; writing—original draft preparation, Y.Z. and A.Z.; writing—review and editing, Y.Z. and H.Z.; supervision, Y.Z.; project administration, Y.Z.; funding acquisition, Y.Z. All authors have read and agreed to the published version of the manuscript.

Funding: This research was funded by [National Natural Science Foundation of China] grant number [50808125].

Conflicts of Interest: The authors declare no conflict of interest.

References

1. Yin, Y.; Wang, S.; Fan, Z.L. Verification of pushover analysis for a long-span steel truss structure. *J. Vibroeng.* **2019**, *21*, 420–430. [CrossRef]
2. Quaranta, G.; Demartino, C.; Xiao, Y. Experimental dynamic characterization of a new composite glubam-steel truss structure. *J. Build. Eng.* **2019**, *25*, 100773. [CrossRef]
3. Chang, K.C.; Kim, C.W. Modal-parameter identification and vibration-based damage detection of a damaged steel truss bridge. *Eng. Struct.* **2016**, *122*, 156–173. [CrossRef]
4. Zhuo, D.B.; Cao, H. Damage identification of bolt connection in steel truss structures by using sound signals. *Struct. Health Monit.* **2022**, *21*, 501–517. [CrossRef]
5. Luong, H.T.M.; Zabel, V.; Lorenz, W.; Rohrmann, R.G. Vibration-based Model Updating and Identification of Multiple Axial Forces in Truss Structures. *Procedia Eng.* **2017**, *188*, 385–392. [CrossRef]
6. Luong, H.T.M.; Zabel, V.; Lorenz, W.; Rohrmann, R.G. Non-destructive Assessment of the Axial Stress State in Iron and Steel Truss Structures by Dynamic Measurements. *Procedia Eng.* **2017**, *199*, 3380–3385. [CrossRef]
7. Liu, K.; Law, S.S.; Zhu, X.Q. System parameter identification from projection of inverse analysis. *J. Sound Vib.* **2017**, *396*, 83–107. [CrossRef]
8. Cho, S.; Ciles, R.K.; Spencer, B.F. System identification of a historic swing truss bridge using a wireless sensor network employing orientation correction. *Struct. Control. Health Monit.* **2015**, *22*, 255–272. [CrossRef]
9. Terlaje, A.S.; Trurnan, K.Z. Parameter Identification and Damage Detection Using Structural Optimization and Static Response Data. *Adv. Struct. Eng.* **2007**, *10*, 607–621. [CrossRef]
10. Chakraborty, S.; Islam, A.A. Damage Identification of Steel Trusses from Static Strain Data. *Indian Eng.* **2007**, *83*, 229–232.
11. Song, J.H.; Lee, E.T.; Eun, H.C. Optimal sensor placement through expansion of static strain measurements to static displacements. *Int. J. Distrib. Sens. Netw.* **2021**, *17*, 1550147721991712. [CrossRef]
12. Wang, H.; Zhu, Q.X.; Li, J.; Mao, J.X.; Hu, S.T.; Zhao, X.X. Identification of moving train loads on railway bridge based on strain monitoring. *Smart Struct. Syst.* **2019**, *23*, 263–278.
13. Meoni, A.; D'Alessandro, A.; Mancinelli, M.; Ubertini, F. A Multichannel Strain Measurement Technique for Nanomodified Smart Cement-Based Sensors in Reinforced Concrete Structures. *Sensors* **2021**, *21*, 5633. [CrossRef]
14. Matta, F.; Bastianini, F.; Galati, N.; Casadei, P.; Nanni, A. Distributed Strain Measurement in Steel Bridge with Fiber Optic Sensors: Validation through Diagnostic Load Test. *J. Perform. Constr. Facil.* **2008**, *22*, 264–273. [CrossRef]
15. Goldfeld, Y.; Klar, A. Damage Identification in Reinforced Concrete Beams Using Spatially Distributed Strain Measurements. *J. Struct. Eng.* **2012**, *139*, 1. [CrossRef]
16. Nick, H.; Aziminejad, A. Vibration-Based Damage Identification in Steel Girder Bridges Using Artificial Neural Network Under Noisy Conditions. *J. Nondestruct. Eval.* **2021**, *40*, 15. [CrossRef]
17. Weinstein, J.C.; Sanayei, M.; Brenner, B.R. Bridge Damage Identification Using Artificial Neural Networks. *J. Bridge Eng.* **2018**, *23*, 04018084. [CrossRef]
18. Kuzina, V.; Koshev, A. Modification of the Levenberg–Marquardt Algorithm for Solving Complex Computational Construction Problems. *IOP Conf. Ser. Mater. Sci. Eng.* **2020**, *960*, 032039. [CrossRef]
19. Izmailov, A.F.; Solodov, M.V.; Uskov, E.I. A globally convergent Levenberg–Marquardt method for equality-constrained optimization. *Comput. Optim. Appl.* **2019**, *72*, 215–239. [CrossRef]
20. Naseralavi, S.S.; Shojaee, S.; Ahmadi, M. Modified gradient methods hybridized with Tikhonov regularization for damage identification of spatial structure. *Smart Struct. Syst.* **2016**, *18*, 839–864. [CrossRef]
21. Kadaj, R. The method of detection and localization of configuration defects in geodetic networks by means of Tikhonov regularization. *Rep. Geod. Geoinform.* **2021**, *112*, 19–25. [CrossRef]
22. Farzaneh, S.; Safari, A.; Parvazi, K. Improving Dam Deformation Analysis Using Least-Squares Variance Component Estimation and Tikhonov Regularization. *J. Surv. Eng.* **2021**, *147*, 04020024. [CrossRef]
23. Smirnova, A.; Tuncer, N. Inverse problem of groundwater modeling by iteratively regularized Gauss–Newton method with a nonlinear regularization term. *Nonlinear Anal.* **2011**, *74*, 5987–5998. [CrossRef]
24. Artar, M.; Daloğlu, A. Damage detection in simulated space frames using genetic algorithms. *Sigma J. Eng. Nat. Sci.* **2015**, *33*, 166–187.

25. Tang, L.M.; Zhang, W.F.; Liu, Y.X. Gradient-Regularization Method to Solve the Inverse Problems of Differential Equation. *Chin. J. Comput. Mech.* **1991**, *8*, 123–129.
26. Liu, Y.X.; Wang, D.G.; Zhang, J.L.; Li, J.S.; Lu, Z.G.; Yu, Y.J. Identification of material parameters with gradient regularization method. *Chin. J. Comput. Mech.* **2000**, *17*, 69–75.
27. Giraud, J.; Lindsay, M.; Ogarko, V.; Jessell, M.; Martin, R.; Pakyuz-Charrier, E. Integration of geoscientific uncertainty into geophysical inversion by means of local gradient regXularization. *Solid Earth* **2019**, *10*, 193–210. [CrossRef]
28. Zhang, Y.X.; Li, G.Q.; Zhang, J.L. Structural elastic modulus identification based on finite element regularization method. *Eng. Mech.* **2007**, *10*, 6–10.
29. Cui, K.; Li, X.S.; Li, B.Y.; Yang, G.W. Global convergence gradient regularization algorithm for solving nonlinear inverse problems. *Chin. J. Comput. Mech.* **2005**, *4*, 415–419.
30. Reichel, L.; Rodriguez, G. Old and new parameter choice rules for discrete ill-posed problems. *Numer. Algorithms* **2013**, *63*, 65–87. [CrossRef]
31. Hua, X.G.; Ni, Y.Q.; Ko, J.M. Adaptive regularization parameter optimization in output-error-based finite element model updating. *Mech. Syst. Signal Process.* **2009**, *23*, 563–579. [CrossRef]
32. Hansen, P.C.; Leary, D.P. The Use of the L-Curve in the Regularization of Discrete Ill-Posed Problems. *SIAM J. Sci. Comput.* **2006**, *14*, 1487–1503. [CrossRef]
33. Bucataru, M.; Cimpean, I.; Marin, L. A gradient-based regularization algorithm for the Cauchy problem in steady-state anisotropic heat conduction. *Comput. Math. Appl.* **2022**, *119*, 220–240. [CrossRef]
34. Morozov, V.A. *Methods for Solving Incorrectly Posed Problems*; Springer: New York, NY, USA, 1984.
35. Gao, Y. *Structural Health Monitoring Strategies for Smart Sensor Networks*; Newmark Structural Engineering Laboratory (NSEL) Report Series, No. 13; University of Illinois: Urbana, IL, USA, 2008; Available online: http://hdl.handle.net/2142/8802 (accessed on 10 October 2021).
36. Copertaro, E. Assessment of resistive strain gauges measurement performances in experimental modal analysis and their application to the diagnostics of abrasive waterjet cutting machinery. *Measurement* **2022**, *188*, 110626. [CrossRef]
37. Peng, L.; Jing, G.Q.; Wang, Y.X.; Miao, N. Research on Calibration Model of Bridge Health Monitoring Based on Vibrating Wire Strain Sensor. *Earth Environ. Sci.* **2021**, *692*, 042099. [CrossRef]
38. Yan, M.; Tan, X.; Mahjoubi, S.; Bao, Y. Strain transfer effect on measurements with distributed fiber optic sensors. *Autom. Constr.* **2022**, *139*, 104262. [CrossRef]

buildings

MDPI

Article

Estimation of Soil–Structure Model Parameters for the Millikan Library Building Using a Sequential Bayesian Finite Element Model Updating Technique

Hamed Ebrahimian [1,*], Abdelrahman Taha [1], Farid Ghahari [2], Domniki Asimaki [3] and Ertugrul Taciroglu [2]

1 Department of Civil and Environmental Engineering, University of Nevada, Reno, NV 89577, USA
2 Department of Civil and Environmental Engineering, University of California, Los Angeles, CA 90095, USA
3 Department of Mechanical and Civil Engineering, California Institute of Technology, Pasadena, CA 91125, USA
* Correspondence: hebrahimian@unr.edu

Abstract: We present a finite element model updating technique for soil–structure system identification of the Millikan Library building using the seismic data recorded during the 2002 Yorba Linda earthquake. A detailed finite element (FE) model of the Millikan Library building is developed in OpenSees and updated using a sequential Bayesian estimation approach for joint parameter and input identification. A two-step system identification approach is devised. First, the fixed-base structural model is updated to estimate the structural model parameters (including effective elastic modulus of structural components, distributed floor mass, and Rayleigh damping parameters) and some uncertain components of the foundation-level motion. Then, the identified structural model is used for soil–structure model updating wherein the Rayleigh damping parameters, the stiffness and viscosity of the soil subsystem (modeled using a substructure approach), and the foundation input motions (FIMs) are estimated. The identified model parameters are compared with state-of-practice recommendations. While a specific application is made for the Millikan Library, the present work offers a framework for integrating large-scale FE models with measurement data for model inversion. By utilizing this framework for different civil structures and earthquake records, key structural model parameters can be estimated from the real-world recorded data, which can subsequently be used for assessing and improving, as necessary, state-of-the-art seismic analysis and structural modeling techniques. This paper presents an effort towards using real-world measurements for large-scale FE model updating in the soil and structure, uniform soil time domain for joint parameter and input estimation, and thus paves the way for future applications in system identification, health monitoring, and diagnosis of civil structures.

Keywords: finite element model updating; soil–structure interaction; system identification; joint system and input identification; Bayesian estimation; Millikan Library

1. Introduction

The dynamic response of a building structure to an earthquake excitation is occasionally the result of a complex interaction between the structural system and the underlying and surrounding geology. Since modeling the physics of a coupled soil–structure system in detail is an intricate undertaking, the state of practice has adopted simplified modeling and analysis procedures, such as the substructure approach (e.g., [1,2]). In this approach, the soil flexibility and energy dissipation are modeled using distributed spring and dashpot elements to which the foundation input motions (FIMs) are applied as uniform base excitations [3]. The stiffness and viscosity of these elements are derived using simplified analytical methods, which are nonetheless based on idealized and restrictive assumptions—e.g., linear elastic response behavior of soil and structure, uniform soil half-space (or soil profiles with stiffness gradually varying with depth [4]), etc. These simplifying assumptions and the empirical nature of the mechanical analogs (e.g., soil springs and dashpots) could potentially lead to unquantified errors in predicting the seismic responses of real-world building structures, even though the simplified models have demonstrated acceptable accuracy in idealized case studies (e.g., [5]). Furthermore, integrating the substructure method with the Rayleigh damping model—which itself has a highly empirical nature—introduces another source of uncertainty in the seismic analysis of building structures. Clearly, there is an inconsistency between the relatively sophisticated mechanics-based techniques available for structural system modeling and the crude simplicity of the underlying assumptions that guide the damping and soil–structure interaction (SSI) models. It appears clear that the efforts on damping and SSI characterization should be guided by observations distilled from real-life data.

Soil–structure interaction (SSI) analysis has been an active research subject for more than 40 years (e.g., [6]). SSI effects can be classified into two categories: kinematic and inertial interaction effects [7]. Referred to as the FIMs, the earthquake excitations experienced by a structure–foundation system are affected by the stiffness contrast between the foundation stiffness and the surrounding geology. Therefore, FIMs are generally different from the Free-Field Motions (FFMs) that would have been recorded in the absence of the foundation system. The effects of foundation stiffness and geometry, which result in discrepancies between FIMs and FFMs, are collectively referred to as the kinematic interaction effects. Inertial interaction effects are a result of the foundation–superstructure mass, which imparts inertial forces onto the surrounding soil and causes the foundation to experience a response different from the FIMs. Due to the inertial interaction effects, the vibrating structure operates as a wave source and alters the wave field around the foundation system. In the substructure modeling approach, these two effects are treated separately (e.g., [8]). That is, FIMs are analytically or numerically calculated based on the soil and foundation properties (e.g., [9]). Then, inertial effects are represented by frequency-dependent impedance functions that are estimated through analytical, numerical, or experimental studies (e.g., [10]). Finally, the superstructure on top of the impedance functions is analyzed under the estimated FIMs.

In this study, we develop a model inversion framework that can be used to back-calculate the soil–structure model parameters from the seismic responses of real-world buildings. By repeating this effort for different buildings and earthquake records, the estimation results can be summarized and compared with the state-of-practice recommendations to improve the present seismic analysis capabilities and predictive models for building structures. We apply the model inversion framework that is devised in the present study to the data recorded from the Millikan Library building during the 2002 Yorba Linda earthquake, which was a low-amplitude seismic event. The response of the building during this small event was nearly linear elastic. As such, it provides an opportunity to compare the estimation results with linear elastic seismic analysis practices.

The utilized model inversion framework is based on a finite element (FE) model updating approach formulated in the time domain, using a sequential Bayesian estimation method [11]. The FE model updating approach has been developed and verified previously using numerically simulated data (e.g., [12,13]). This paper presents an effort towards using real-world measurements for large-scale FE model updating in the time domain, and thus paves the way for future applications. Although the presented model updating approach is used along with a linear FE model in this study, the approach can be readily extended to nonlinear FE models of civil structures.

In what follows, the Bayesian FE model updating approach and its algorithmic details are briefly discussed in Section 2. Section 3 provides the modeling details of the Millikan Library building. A two-step system identification approach is introduced in Section 4, which is applied to the Millikan Library data in Sections 5 and 6. An identifiability analysis accompanies each system identification step to select the estimation parameters. Finally, the key estimation results are summarized and compared with available literature and the state of practice in Section 7.

2. Data Assimilation through FE Model Updating in the Time Domain

The finite element model of a building structure depends on various "model parameters". Model parameters include, but are not limited to, inertial properties, damping parameters, parameters that characterize the stiffness properties of the structural and nonstructural components, and the time histories of the unmeasured components of base excitation. If the soil–structure interaction effects are included in the FE model, then the parameters used to model the soil subsystem and the time history of FIMs are also unmeasurable and unknown. Using the recorded acceleration response time histories of the building during an earthquake, our objective is to identify the best estimates of the unknown model parameters along with the input excitation time history, in order to minimize the discrepancy between the FE-predicted and measured structural responses.

The estimation problem is tackled by updating iteratively and sequentially the joint probability distribution function (PDF) of the unknown model parameters and the discrete values of the input motion time histories using a Bayesian inference method. The basics of the sequential FE model updating method for joint system and input identification are presented in [11]. Here, we briefly review the technical background of this method and highlight the new improvements introduced for this study.

2.1. Bayesian FE Model Updating in Time Domain for Joint System and Input Identification

The predicted response of a FE model at time step i to an earthquake excitation time history can be expressed as [14]

$$\hat{\mathbf{y}}_i = \mathbf{h}_i\left(\boldsymbol{\theta}, \ddot{\mathbf{u}}^g_{1:i}\right) \quad (1)$$

where $\hat{\mathbf{y}}_i \in \mathbb{R}^{n_y \times 1}$ denotes the vector of structural response quantities predicted by the FE model at time step i (i = 1, 2, …, k), $\boldsymbol{\theta} \in \mathbb{R}^{n_\theta \times 1}$ denotes the FE model parameter vector, $\ddot{\mathbf{u}}^g_i \in \mathbb{R}^{n_{\ddot{u}} \times 1}$ denotes the vector of unknown or unmeasured input ground motion at time step i, $n_{\ddot{u}}$ denotes the number of unknown or unmeasured input ground motion components, and $\ddot{\mathbf{u}}^g_{1:i} = \left[\left(\ddot{\mathbf{u}}^g_1\right)^T \; \left(\ddot{\mathbf{u}}^g_2\right)^T \; \ldots \; \left(\ddot{\mathbf{u}}^g_i\right)^T\right]^T$ is the vector of the unknown or unmeasured input ground motion time history. $\mathbf{h}_i(\ldots)$ is the response function of the FE model at time step i, which, in general, is a nonlinear function of $\boldsymbol{\theta}$ and $\ddot{\mathbf{u}}^g_{1:i}$. The measured response vector of the structure, $\mathbf{y}_i$, is related to the FE-predicted response, $\hat{\mathbf{y}}_i$, as ([12–14])

$$\mathbf{v}_i\left(\boldsymbol{\theta}, \ddot{\mathbf{u}}^g_{1:i}\right) = \mathbf{y}_i - \hat{\mathbf{y}}_i \quad (2)$$

where $\mathbf{v}_i \in \mathbb{R}^{n_y \times 1}$ is the simulation error vector, which accounts for the misfit between the measured and FE-predicted responses. This misfit stems from the output measurement noise, parameter uncertainties, and model errors. Model errors stem from simplifying idealizations adopted when devising the model as well as deviations in its geometry and configuration from the real-life structure it represents, which result in an inherent discrepancy between the FE model prediction and the measured structural responses [15].

Given an unbiased estimate of $\boldsymbol{\theta}$ and $\ddot{\mathbf{u}}^g_{1:k}$, it is assumed here that the simulation error can be modeled as a stationary, zero-mean, and independent Gaussian white noise process [16]. Therefore, the likelihood function can be derived as

$$\begin{aligned} p\left(\mathbf{y}_i \mid \boldsymbol{\theta}, \ddot{\mathbf{u}}^g_{1:i}\right) &= p(\mathbf{v}_i) = \frac{1}{(2\pi)^{n_y/2}|\mathbf{R}|^{1/2}} e^{-\frac{1}{2}\mathbf{v}_i^T \mathbf{R}^{-1} \mathbf{v}_i} \\ &= \frac{1}{(2\pi)^{n_y/2}|\mathbf{R}|^{1/2}} e^{-\frac{1}{2}\left(\mathbf{y}_i - \mathbf{h}_i(\boldsymbol{\theta}, \ddot{\mathbf{u}}^g_{1:i})\right)^T \mathbf{R}^{-1} \left(\mathbf{y}_i - \mathbf{h}_i(\boldsymbol{\theta}, \ddot{\mathbf{u}}^g_{1:i})\right)^T} \end{aligned} \quad (3)$$

where $|\mathbf{R}|$ denotes the determinant of the diagonal matrix $\mathbf{R} \in \mathbb{R}^{n_y \times n_y}$, which is the time-invariant covariance matrix of the simulation error vector, i.e., $\mathbf{R} = E\left(\mathbf{v}_i \mathbf{v}_i^T\right), \forall i$. The unknown model parameter vector and time histories of the unknown input ground motion can be stacked into an extended unknown parameter vector defined as $\boldsymbol{\psi} = \left[\boldsymbol{\theta}^T, \left(\ddot{\mathbf{u}}^g_{1:k}\right)^T\right]^T$. In a stochastic model inversion, the uncertainty in the unknown parameter vector is characterized by a joint probability distribution function (PDF). The objective in FE model updating is to find an estimate of the unknown parameters such that their joint posterior PDF given the measured output response of the structure is maximized. This is referred to as the maximum *a posteriori* (MAP) estimate. Using Bayes' rule, it can be observed that

$$\left(\hat{\boldsymbol{\psi}}\right)_{\text{MAP}} = \underset{\boldsymbol{\psi}}{\operatorname{argmax}}\ p(\boldsymbol{\psi} \mid \mathbf{y}_{1:k}) = \underset{\boldsymbol{\psi}}{\operatorname{argmax}}\ p(\mathbf{y}_{1:k} \mid \boldsymbol{\psi}) \times p(\boldsymbol{\psi}) \tag{4}$$

where $p(\mathbf{y}_{1:k} \mid \boldsymbol{\psi}) = \prod_{i=1}^{k} p(\mathbf{v}_i)$, and $p(\boldsymbol{\psi})$ is the joint prior PDF of the extended unknown parameters.

2.2. Sequential Finite Element Model Updating Using Model Linearization

Solving the MAP estimation problem shown in Equation (4) using a batch optimization algorithm can be computationally demanding [11]. Therefore, a sequential estimation approach is introduced in [11] to improve the computational efficiency and convergence rate. In this approach, the estimation time domain is divided into successive overlapping time windows, referred to as the estimation windows. The estimation problem is solved iteratively at each estimation window to estimate the posterior mean vector and covariance matrix of the unknown parameters. At each iteration, the FE model is linearized to propagate the uncertainties in the prior PDF into the FE model response. The posterior mean vector and covariance matrix of the unknown parameter vector are then derived using Equation (4), and used as prior information for the next iteration. Once the iterations converge for an estimation window, the estimated mean vector and covariance matrix are transferred to the next estimation window and are used as prior information. In what follows, the mathematical process of the sequential Bayesian estimation approach is outlined.

Assume that the mth estimation window with length t_l spans from time step t_1^m to time step t_2^m, where $t_l = t_2^m - t_1^m$. Considering that $p\left(\mathbf{y}_{t_1^m:t_2^m} \,\middle|\, \boldsymbol{\psi}_m\right) = \prod_{i=t_1^m}^{t_2^m} p(\mathbf{v}_i)$, and assuming a Gaussian distribution for the joint prior PDF, the natural logarithm of the posterior joint PDF of the extended unknown parameter vector, i.e., $\boldsymbol{\psi}_m = \left[\boldsymbol{\theta}^T, \left(\ddot{\mathbf{u}}^g_{t_1^m:t_2^m}\right)^T\right]^T$, can be derived as [11]

$$\begin{aligned} log\left(p\left(\boldsymbol{\psi}_m \mid \mathbf{y}_{t_1^m:t_2^m}\right)\right) = k_0 &- \tfrac{1}{2}\left(\mathbf{y}_{t_1^m:t_2^m} - \mathbf{h}_{t_1^m:t_2^m}\left(\boldsymbol{\psi}_m, \hat{\ddot{\mathbf{u}}}^g_{1:t_1^m-1}\right)\right)^T \widetilde{\mathbf{R}}^{-1}\left(\mathbf{y}_{t_1^m:t_2^m} - \mathbf{h}_{t_1^m:t_2^m}\left(\boldsymbol{\psi}_m, \hat{\ddot{\mathbf{u}}}^g_{1:t_1^m-1}\right)\right) \\ &- \tfrac{1}{2}\left(\boldsymbol{\psi}_m - \hat{\boldsymbol{\psi}}_m^-\right)^T\left(\hat{\mathbf{P}}_{\boldsymbol{\psi},m}^-\right)^{-1}\left(\boldsymbol{\psi}_m - \hat{\boldsymbol{\psi}}_m^-\right) \end{aligned} \tag{5}$$

where k_0 is a constant, and $\hat{\boldsymbol{\psi}}_m^-$ and $\hat{\mathbf{P}}_{\boldsymbol{\psi},m}^-$ are the prior mean vector and the covariance matrix of the extended parameter vector at the mth estimation window. $\widetilde{\mathbf{R}} \in \mathbb{R}^{(t_l \times n_y) \times (t_l \times n_y)}$ is a block diagonal matrix, in which the diagonals denote the simulation error covariance matrix $\mathbf{R}$, and $\hat{\ddot{\mathbf{u}}}^g_{1:t_1^m-1}$ is the estimated input motion time history from the previous estimation windows, which is treated as a known vector. To determine the MAP estimate of $\boldsymbol{\psi}_m$, we solve $\frac{\partial log\left(p\left(\boldsymbol{\psi}_m \mid \mathbf{y}_{t_1^m:t_2^m}\right)\right)}{\partial \boldsymbol{\psi}_m} = 0$, which results in a nonlinear algebraic equation in $\boldsymbol{\psi}_m$, as shown in [11]. This equation can be solved using an iterative first-order approximation of the FE response function at $\hat{\boldsymbol{\psi}}_m^-$, which requires the computation of FE response sensitivities with respect to the extended unknown parameter vector, evaluated at the prior mean values.

The FE response sensitivity matrix is denoted by $\mathbf{C}$ hereafter. As a result, the following (first-order approximate) equation for the MAP estimate of ψ_m can be obtained [11] as

$$\hat{\psi}^{+}_{m,j} = \hat{\psi}^{-}_{m,j} + \mathbf{K}\left(\mathbf{y}_{t_1^m:t_2^m} - \mathbf{h}_{t_1^m:t_2^m}\left(\hat{\psi}^{-}_{m,j}, \hat{\ddot{\mathbf{u}}}^{g}_{1:t_1^m-1}\right)\right) \tag{6}$$

where $\hat{\psi}^{+}_m$ is the updated (or the posterior) mean estimate of ψ_m. The term $\mathbf{K} = \hat{\mathbf{P}}_{\psi\mathbf{y}}(\hat{\mathbf{P}}_{\mathbf{yy}})^{-1}$ is referred to as the Kalman gain matrix [17]. The matrix $\hat{\mathbf{P}}_{\psi\mathbf{y}} = \hat{\mathbf{P}}^{-}_{\psi,m}\mathbf{C}^T$ is the cross-covariance matrix of the unknown parameter vector and measured data, and $\hat{\mathbf{P}}_{yy} = \mathbf{C}\hat{\mathbf{P}}^{-}_{\psi,m}\mathbf{C}^T + \tilde{\mathbf{R}}$ is the covariance matrix of the measured data (the reader is referred to [11,18] for further derivation details). The subscript j is appended to the estimated mean vector and covariance matrix to denote the iteration number. The updated $\hat{\psi}^{+}_{m,j}$ from Equation (6) is iteratively used as the new point for the linearization of the nonlinear FE model to improve the estimation. This iterative prediction–correction procedure at each estimation window is equivalent to an iterative extended Kalman filtering (EKF) method for parameter-only estimation [17]. Following the EKF procedure, the prior covariance matrix $\hat{\mathbf{P}}^{-}_{\psi,m,j}$ is updated to the posterior covariance matrix $\hat{\mathbf{P}}^{+}_{\psi,m,j}$ at each prediction–correction iteration as in

$$\hat{\mathbf{P}}^{+}_{\psi,m,j} = (\mathbf{I} - \mathbf{KC})\hat{\mathbf{P}}^{-}_{\psi,m,j}(\mathbf{I} - \mathbf{KC})^T + \mathbf{K}\tilde{\mathbf{R}}\mathbf{K}^T. \tag{7}$$

Furthermore, to improve the convergence characteristics of the iterative prediction–correction procedure, a disturbance matrix is added to the posterior covariance matrix at each iteration to provide the prior covariance matrix for the next iteration, i.e., $\hat{\mathbf{P}}^{-}_{\psi,m,j+1} = \hat{\mathbf{P}}^{+}_{\psi,m,j} + \mathbf{Q}$, in which $\mathbf{Q}$ is a diagonal matrix with small positive diagonal entries (relative to the diagonal entries of the matrix $\hat{\mathbf{P}}^{+}_{\psi,m,j}$). The matrix $\mathbf{Q}$ is referred to as the process noise covariance matrix in the Kalman filtering world [17].

Accurately computing the FE response sensitivities is an integral part of the presented sequential FE model updating. In previous studies (e.g., [11,12]), the response sensitivities have been computed using a direct differentiation method (DDM) (e.g., [19]), which requires extending the FE numerical scheme and implementing new codes in the FE software to calculate the response sensitivities. Although available in OpenSees [20], this implementation is not readily available as part of other FE simulation platforms. Therefore, to extend the applicability of the sequential FE model updating procedure to other FE simulation platforms, here we use the finite difference method (FDM) to compute the response sensitivities. The FDM method is implemented by perturbing the estimation parameters one at a time based on a central difference method with unequal spacing. We use parallel computing to enhance the computational efficiency of the FDM implementation. The Bayesian FE model updating algorithms are implemented in MATLAB [21], which calls several instances of OpenSees in parallel for FE response and response sensitivity computations at each prediction–correction iteration. Table 1 summarizes the sequential FE model updating algorithm for the joint estimation of model parameters and input motion time histories that is described above. It should be noted that the presented Bayesian FE model updating framework is applicable to both linear and nonlinear FE models. In case the FE model is linear, the model response function will still be a nonlinear function of the model parameters and input motion; however, the model updating process will become more computationally efficient.

Table 1. Sequential FE model updating algorithm for joint estimation of model parameters and input motion time histories.

1. Determine the start and end points of the estimation windows—i.e., t_1^m, t_2^m, where m = 1, 2, …, N and N is the number of estimation windows.
2. Set the initial mean vector and covariance matrix of the model parameter vector $(\hat{\theta}_0, \hat{\mathbf{P}}_{\theta_0})$ and input ground motion time history vector across the first estimation window $(\hat{\ddot{\mathbf{u}}}^g_{0,t_1^1:t_2^1}, \hat{\mathbf{P}}_{\ddot{\mathbf{u}}_0^g})$. The initial mean values of the ground motion time history are usually set as zero, i.e., $\hat{\ddot{\mathbf{u}}}^g_{0,t_1^1:t_2^1} = \mathbf{0}$.
3. Use the inverse of the initial values of the model parameters as the scaling factors, i.e., $a_n = \frac{1}{\theta_{0,n}}$. Set up $\hat{\psi}^+_{0,0} = \begin{bmatrix} \mathbf{A}\hat{\theta}_0 & \hat{\ddot{\mathbf{u}}}^g_{0,t_1^1:t_2^1} \end{bmatrix}^T$ and $\hat{\mathbf{P}}^+_{\psi,0,0} = \begin{bmatrix} \mathbf{A}\hat{\mathbf{P}}_{\theta_0}\mathbf{A}^T & 0 \\ 0 & \hat{\mathbf{P}}_{\ddot{\mathbf{u}}_0^g} \end{bmatrix}$.
4. For the mth estimation window:

4.1. Setup the initial mean vector and covariance matrix of the unknown parameters (i.e., $\hat{\psi}^+_{m,0}$ and $\hat{\mathbf{P}}^+_{\psi,m,0}$) based on the posterior mean vector and covariance matrix obtained from the last estimation window (i.e., $\hat{\psi}^+_{m-1}$ and $\hat{\mathbf{P}}^+_{\psi,m-1}$) (refer to [11] for details).

4.2. Define the process noise covariance matrix $\mathbf{Q}$ and the simulation error covariance matrix $\mathbf{R}$. Set up $\tilde{\mathbf{R}}$.

4.3. Iterate (j = 1, 2, …):

4.3.1. Set $\hat{\psi}^-_{m,j} = \hat{\psi}^+_{m,j-1}, \hat{\mathbf{P}}^-_{\psi,m,j} = \hat{\mathbf{P}}^+_{\psi,m,j-1} + \mathbf{Q}$.

4.3.2. Obtain the FE responses using $\hat{\psi}^-_{m,j}$, i.e., $\hat{\mathbf{y}}_{t_1^m:t_2^m} = \mathbf{h}_{t_1^m:t_2^m}(\hat{\psi}^-_{m,j}, \hat{\ddot{\mathbf{u}}}^g_{1:t_1^m-1})$. Obtain the FE response sensitivities using a finite difference method. Set up the sensitivity matrix with respect to the scaled model parameters, $\mathbf{C}^s$.

4.3.3. Compute the Kalman gain matrix: $\mathbf{K} = (\hat{\mathbf{P}}^-_{\psi,m,j}(\mathbf{C}^s)^T)(\mathbf{C}^s\hat{\mathbf{P}}^-_{\psi,m,j}(\mathbf{C}^s)^T + \tilde{\mathbf{R}})^{-1}$.

4.3.4. Find the corrected estimates of the mean vector and covariance matrix of the extended parameter vector: $\hat{\psi}^+_{m,j} = \hat{\psi}^-_{m,j} + \mathbf{K}\left(\mathbf{y}_{t_1^m:t_2^m} - \mathbf{h}_{t_1^m:t_2^m}\left(\hat{\psi}^-_{m,j}, \hat{\ddot{\mathbf{u}}}^g_{1:t_1^m-1}\right)\right), \ldots$ $\hat{\mathbf{P}}^+_{\psi,m,j} = (\mathbf{I} - \mathbf{K}\mathbf{C}^s)\hat{\mathbf{P}}^-_{\psi,m,j}(\mathbf{I} - \mathbf{K}\mathbf{C}^s)^T + \mathbf{K}\tilde{\mathbf{R}}\mathbf{K}^T$.

4.3.5. Correct for constraints.

4.3.6. Update the model parameter scaling factors using Equation (8). Update $\hat{\psi}^+_{m,j}$ and $\hat{\mathbf{P}}^+_{\psi,m,j}$ based on the new model parameter scale factors.

4.3.7. Check for convergence: if $|\hat{\psi}^+_{m,j} - \hat{\psi}^+_{m,j-1}| < 0.01 \times |\hat{\psi}^+_{m,j-1}|$ or $j > 15$, then move to the next estimation window ($m = m + 1$, go to step 4); otherwise, iterate again at the current estimation window ($j = j + 1$, go to step 4.3).

2.3. Correction for Constraints

The estimation process described earlier is not constrained and may result in the estimation of non-physical values for the model parameters. This issue is resolved by correcting the posterior estimation results for a set of predefined lower- and upper-bound constraints. Once any of the posterior mean values exceed their designated lower or upper limit, the posterior Gaussian PDF (characterized by the updated mean vector and covariance matrix) is truncated at the constraint edges. The mean vector and the covariance matrix of the truncated PDF are then calculated. The constraint correction approach is borrowed from [17,22]. The algorithmic details are provided in [23] and are not repeated here for brevity.

2.4. Adaptive Scaling of the Unknown Model Parameters

In order to improve the performance of the estimation algorithm, the FE model parameters are scaled adaptively at each iteration. The scaling allows the FE model responses to have relatively similar sensitivities with respect to different estimation parameters at each estimation point. The scaled FE model parameter vector is defined as $\boldsymbol{\theta}^s = \mathbf{A}\boldsymbol{\theta}$, in which $\mathbf{A}$ is an $n_\theta \times n_\theta$ diagonal scaling matrix. It should be noted that only the FE model parameters are scaled, and the vector of unknown input ground motion remains unscaled. The parameter scaling factors (i.e., the diagonal entries of matrix $\mathbf{A}$ denoted by a_n, $n = 1, 2, \ldots, n_\theta$) are calculated to result in equal corresponding diagonal entries in the Fisher Information Matrix of the scaled parameters, which is approximated as $\mathbf{I}^s = \left(\mathbf{C}^s_\theta\right)^T\left(\mathbf{C}^s_\theta\right)$ [14], wherein $\mathbf{C}^s_\theta$ denotes the FE response sensitivity matrix with respect to the scaled FE model parameter vector. The nth diagonal entry of $\mathbf{I}^s$ is $I^s_n = \frac{1}{a_n{}^2}\left(\frac{\partial \mathbf{h}}{\partial \theta_n}\right)^T\left(\frac{\partial \mathbf{h}}{\partial \theta_n}\right)$. The scaling factor a_n is calculated so that I^s_n is equal to the mean of diagonal entries of the Fisher Information Matrix corresponding to the vector of unknown input ground motion, i.e., for the mth estimation window:

$$\begin{aligned} \frac{1}{a_n{}^2}\left(\frac{\partial \mathbf{h}}{\partial \theta_n}\right)^T\left(\frac{\partial \mathbf{h}}{\partial \theta_n}\right) &= mean\left[diag\left(\left(\frac{\partial \mathbf{h}}{\partial \ddot{\mathbf{u}}^g_{t_1^m:t_2^m}}\right)^T\left(\frac{\partial \mathbf{h}}{\partial \ddot{\mathbf{u}}^g_{t_1^m:t_2^m}}\right)\right)\right] \Rightarrow a_n \\ &= \sqrt{\frac{\left(\frac{\partial \mathbf{h}}{\partial \theta_n}\right)^T\left(\frac{\partial \mathbf{h}}{\partial \theta_n}\right)}{mean\left[diag\left(\left(\frac{\partial \mathbf{h}}{\partial \ddot{\mathbf{u}}^g_{t_1^m:t_2^m}}\right)^T\left(\frac{\partial \mathbf{h}}{\partial \ddot{\mathbf{u}}^g_{t_1^m:t_2^m}}\right)\right)\right]}} \end{aligned} \quad (8)$$

3. The Millikan Library Building

The Millikan Library is a reinforced concrete shear wall building, with a basement and nine stories above the ground, located on the California Institute of Technology (Caltech) campus (Figure 1). The Millikan Library has been the subject of many studies, especially in the fields of soil–structure system identification, due to its unique structural and soil properties, and its relatively old and dense instrumentation, which recorded its seismic responses during numerous earthquakes [24–30]. The Millikan Library building is 43.9 m tall above ground, including the roof level. It has a 4.3 m deep basement below the ground level. The basement is encased by surrounding shear walls. Precast concrete claddings are installed on the north and south faces of the building from the second floor up to the roof. The lateral force-resisting system comprises shear walls in the north–south direction on the east and west sides of the building and core walls around the elevator shaft in the north–south and east–west directions. The floor system consists of concrete slabs supported by reinforced concrete beams. Lightweight aggregate concrete is used for slabs and beams at all floors, while regular aggregate concrete is used for foundations, columns, and walls. The foundation system consists of a 1.2 m deep central pad, 6.0 m below the ground level, and two foundation strips on the north and south sides of the building 5.0 m below the ground level. Four stepped beams at the four corners of the foundation connect the central foundation pad to the north and south foundation strips.

The building is instrumented using a total of 36 uniaxial accelerometers, measuring six acceleration responses at the foundation (or basement floor) level, and three translational acceleration responses on each floor. The instrumentation details are obtained from [31] and are provided in Figure A1 and Table A1 in Appendix A. Multiple earthquake records are available for the Millikan Library; a relatively recent survey of earthquake data recorded at the building can be found in [28]. For the purpose of this study, we use the 2002 Yorba Linda earthquake record, which is a low-amplitude earthquake (PGA of ~0.6% g). Two notes should be made about this earthquake record of the Millikan Library. First, the location of sensors on the foundation (or basement floor) could not be determined precisely. Second, recordings of the east sensor on the second and eighth floors (measuring NS direction) were not available during the 2002 Yorba Linda Earthquake. Therefore, only 34 measurement channels are available for this record. Since the level of considered earthquakes is low, the effects of material and geometric nonlinearities in the dynamic response of the structure are expected to be minor, and therefore, the use of a linear model is justified.

Figure 1. Millikan Library building.

FE Model Development

Using the available structural drawings, a detailed FE model of the structural system was developed. We used the graphical user interface of SAP2000 software [32] to develop the initial geometry of the model. The SAP2000 model was then transferred to OpenSees [20] using a custom-developed MATLAB script based on the SAP2000 Application Programming Interface (API). For the purpose of this study, we prepared a linear FE model of the Millikan Library building. The model comprises linear elastic beam–column elements for the buildings' beams and columns, and quadrilateral shell (DKGQ) elements with linear elastic (elastic membrane plate) sections are used for shear walls and slabs. The kinematic interaction of precast claddings (installed on the north and south faces of the building) with the structural system is modeled using diagonal brace elements, which are included in the north and south frame bays from the second story to the roof. The brace elements are assumed to be linear elastic with square cross-sections of 0.1 m $\times$ 0.1 m. By adjusting the elastic modulus of the diagonal braces, the stiffness and force contributions of the precast panels to the lateral response of the structure can be characterized. The kinematic interaction between the structure and other nonstructural components and systems (e.g., partition walls and stairs) are not modeled explicitly; they are expected to be lumped into the stiffness properties of the structural members and precast claddings.

The inertial effects of nonstructural components (i.e., stairs, precast panels, and elevators) are accounted for by modeling their corresponding mass and weight contributions. The weight of each precast panel is estimated to be about 50 kN, which conforms with [24]. A uniformly distributed load is applied on the floor slabs to account for live loads and weight of partition walls, later treated as an unknown parameter to be estimated. A uniformly distributed load of 2.5 kN/m^2 is applied on the foundation to account for the live load of library shelves in the basement. Moreover, the mass of soil filling between the foundation surface and basement floor is accounted for by applying the corresponding mass on the central foundation pad and the two foundation strips. The mass of structural components is calculated based on the nominal concrete densities specified in the structural drawings: $\rho_{c-LW} = 1760\frac{\text{kg}}{\text{m}^3}$ for lightweight aggregate concrete material, $\rho_{RC-LW} = 1900\frac{\text{kg}}{\text{m}^3}$ for lightweight aggregate reinforced concrete, and $\rho_{RC} = 2500\frac{\text{kg}}{\text{m}^3}$ for regular aggregate reinforced concrete. Figure 2 shows the geometry of the model. Different colors in this figure present different material property groups used to define the shell (shear wall and slab) and beam–column elements. More details on material parameters are provided in Section 5. A sensitivity analysis is performed to ensure convergence of analysis results with respect to the mesh size. The model in the presented configuration consists of 1885 frame elements and 4043 shell elements, corresponding to a total of 27,526 degrees of freedom.

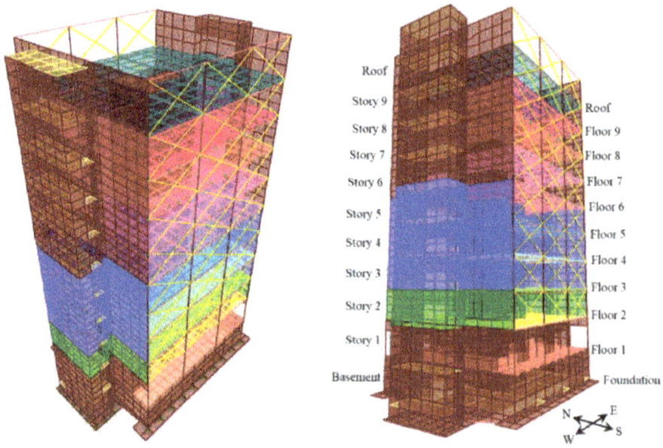

Figure 2. Developed FE model of the Millikan Library structure.

A gravity analysis was performed before dynamic time history analyses by applying constant g acceleration in the vertical direction. The damping energy dissipation for the time history analysis is defined using mass- and stiffness-proportional Rayleigh damping. The Rayleigh damping parameters are treated as unknowns to be estimated. For the purpose of time history analyses, Newmark's constant average acceleration method is used for time-stepping with increments of $\Delta t = 0.03$ s. The time step size is selected following a convergence study to ensure a large time step size that provides adequate accuracy. All of the measured acceleration response time histories are also resampled at $\Delta t = 0.03$ s for the model updating phase.

4. A Two-Step System Identification Approach

A two-step FE model updating approach is used for model and input identification of the soil–structure system. In the first step, the measured acceleration responses at the foundation level are used to calculate six components of the foundation-level motions. The foundation-level motions are then used as uniform base input excitations, and the measured responses of the structure are used as output measurements to update the FE model of the "fixed-base" structure. The objective is to estimate the model parameters characterizing the structural model, regardless of the soil subsystem. During this process, we discovered inconsistencies in the recordings obtained from two of the foundation sensors, as is discussed further in the next section. These sensors are installed inside a utility tunnel—referred to as the steam tunnel—next to the foundation slab. Therefore, the acceleration response time histories of these two sensors are assumed as unknown input motions and are estimated jointly with the structural model parameters. This step results in FE model updating with partially unknown inputs.

The second step is an output-only FE model updating procedure, wherein the identified structural model parameters are fixed at their mean estimates obtained from the first step, and the three translational components of FIM and other parameters characterizing the soil–structure model are estimated jointly. Since the Millikan Library foundation is not deep and does not have large dimensions (comparable with the lengths of incident waves), the rotational components of FIM are not considered in this study.

To decide about and select the estimation parameters, an identifiability analysis is performed, as is outlined in the next section for each identification step. Before proceeding with the real-life data, we verified the two-step FE model updating procedure using numerically simulated data to examine the effectiveness of the proposed estimation algorithm and the accuracy of its results. The verification studies are documented in [33] and are not included here for brevity.

5. Step 1 System Identification: FE Model Updating of the Structural System with Partially Unknown Inputs

5.1. Model Identifiability and Parameter Selection

The identifiability of model parameters is investigated using an information-theoretic approach based on [34]. In this approach, a set of candidate parameters are selected first, and then, the entropy gain of each parameter is calculated. The entropy gain is used as a quantitative metric to measure the information each model parameter receives from the measurement data. The entropy gain is compared between the estimation parameters to assess their relative identifiability. Moreover, the mutual entropy gain (or mutual gain) between parameter pairs is used to investigate the dependence between each pair of model parameters. The entropy gain and mutual gain are then used to determine the most identifiable parameter set. This is an alternative procedure to prior methods that, for example, adaptively group updating parameters based on the sensitivities of a model updating objective function with respect to the parameters (e.g., see [35,36]).

Fifteen (15) structural model parameters characterizing the material, inertial, and damping properties were initially selected as the estimation parameters for identifiability assessment at this step. These parameters were selected initially based on our judgment on their potential contribution in the structural response. For a given input motion, the entropy gain for each parameter is a function of the parameter values, which are unknown in advance. Without the knowledge of correct parameter values, their nominal values are used for prior identifiability assessment, following the recommendation in [34]. These parameters and their nominal values are listed in Table 2. The logic based on which the nominal values are estimated is further described below.

The material parameters consist of the effective elastic modulus of the floor system (beams and slabs) and the vertical/lateral system (walls and columns). The material parameters are grouped at the story level and along the height of the building. With the exception of the first three stories, the effective elastic modulus of structural components is grouped together at the upper stories. The first three stories are considered separately because previous studies have suggested minor structural damage in the first stories of the building after the 1971 San Fernando earthquake (e.g., [26,37,38]). The effective elastic modulus for floor slabs and beams is calculated by applying a 35% reduction factor on the nominal elastic modulus of lightweight aggregate concrete, which is derived based on ACI 318-14 [39] as $E_c = 0.043(\rho_{c-LW})^{1.5}\sqrt{f'_c}$, where $f'_c = 27.5$ MPa is the nominal compressive strength. The effective elastic modulus for columns and walls is calculated by applying a 70% reduction factor on the nominal elastic modulus calculated as $E_c = 4700\sqrt{f'_c}$, where $f'_c = 27.5$ MPa. For foundation, a 35% reduction factor is applied on the elastic modulus, which is calculated considering $f'_c = 20.7$ MPa, to account for flexural cracking.

To estimate the elastic modulus of diagonal brace elements representing the precast claddings, the results concluded from modal analysis performed during the construction period are used. According to [6,26], the installation of precast panels resulted in a 25% increase in the lateral stiffness of the building structure in the EW direction. The initial values for the elastic modulus of the diagonal brace elements in the model are tuned iteratively and manually to yield a similar increase in the stiffness of the first mode in the EW direction. Consequently, the estimated elastic modulus of diagonal braces is found to be 20 GPa. The initial value of the distributed floor mass (to account for live load and mass of nonstructural elements) is estimated to be about 250 kg/m^2. Estimating the accurate weight and mass contribution of mechanical equipment on the roof was not possible; therefore, the equivalent distributed roof mass is approximately estimated as 300 kg/m^2. Finally, the initial values of the Rayleigh damping parameters are estimated by assuming a 5% damping ratio for the first and second modes.

Table 2. Fifteen model parameters used for prior model identifiability assessment.

Parameter ID	Description	Value
1	Elastic modulus of brace elements representing precast claddings	20 GPa
2	Effective elastic modulus of beam and slab concrete at 1st floor	5.8 GPa
3	Effective elastic modulus of beam and slab concrete at 2nd floor	5.8 GPa
4	Effective elastic modulus of beam and slab concrete at 3rd floors	5.8 GPa
5	Effective elastic modulus of beam and slab concrete at 4th, 5th, and 6th floors	5.8 GPa
6	Effective elastic modulus of beam and slab concrete at 7th, 8th, 9th, and roof floors	5.8 GPa
7	Effective elastic modulus of column and wall concrete at basement	17.3 GPa
8	Effective elastic modulus of column and wall concrete at 1st story	17.3 GPa
9	Effective elastic modulus of column and wall concrete at 2nd story	17.3 GPa
10	Effective elastic modulus of column and wall concrete at 3rd, 4th, and 5th stories	17.3 GPa
11	Effective elastic modulus of column and wall concrete at 6th, 7th, 8th, 9th, and roof stories	17.3 GPa
12	Mass-proportional Rayleigh damping coefficient	0.4
13	Stiffness-proportional Rayleigh damping coefficient	5.3×10^{-3}
14	Distributed floor mass on 1st to 9th floors	$250\ \text{kg/m}^2$
15	Distributed floor mass on roof	$300\ \text{kg/m}^2$

The identifiability of model parameters is assessed using the measured foundation-level motion, including all translational and rotational components. As mentioned earlier, the recordings of two acceleration channels on the foundation are later treated as unknowns and estimated through the model updating process. Nevertheless, these measured recordings are used for the identifiability analysis as approximate inputs. This approximation is expected to have negligible effects on the overall identifiability results. As described in [34], evaluating the entropy gain and mutual gain also requires the prior covariance matrix of the unknown parameter vector (or the prior variance of parameters). A 10% coefficient of variation is assumed for all parameters to derive the a priori covariance matrix.

Figure 3 displays the relative entropy gain of the fifteen candidate estimation parameters. The presented entropy gains are relative, which means that the entropy gain values are scaled with respect to the largest value. This plot presents the relative information that the model responses (which represent the measurements) carry about each model parameter. Those parameters that receive small information from the measurement data (e.g., parameters #2 to #6) are likely to be unidentifiable. Furthermore, Figure 4 shows the relative mutual entropy gain between the parameter pairs. This figure is used to investigate the dependence between each two parameters. The diagonal components in this figure represent the entropy gain of each parameter. The dark off-diagonal colors in this figure indicate strong relative dependence between parameter pairs. For example, Figure 4 suggests that there is a mutual dependence between parameters #7 and #8, and some mutual dependence between parameters #7, #8, and #10. Moreover, there are some competing effects between these parameters and parameter #14. These observations are physically meaningful since parameters #7, #8, and #10 characterize the stiffness of the building, and parameter #14 contributes to the building mass. Figure 3 along with Figure 4 can be used to guide the selection of the estimation parameters. The parameters that gain relatively small amounts of information from the measurement data or have strong dependencies on other estimation parameters can be put aside (i.e., fixed) or merged with other parameters, if physically meaningful, to enhance the identifiability. Indeed, the process is not definitive and requires some engineering judgment in selecting the final estimation parameter set.

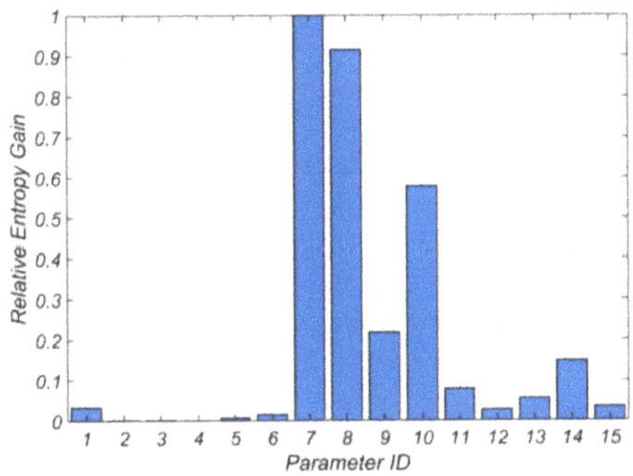

Figure 3. Relative entropy gain of the fifteen model parameters (see Table 2 for parameter IDs).

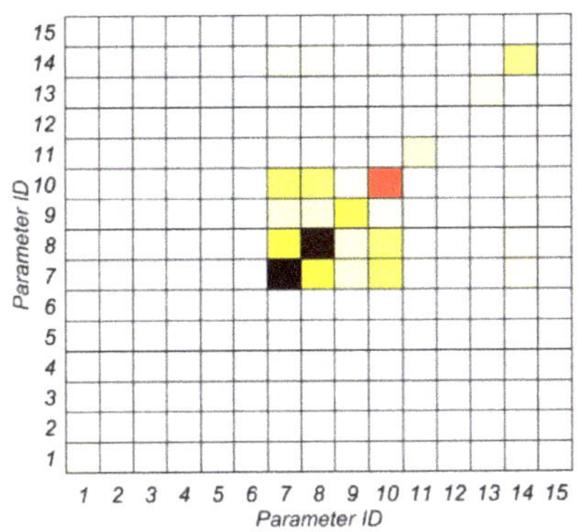

Figure 4. Relative mutual entropy gain between the parameter pairs (see Table 2 for parameter IDs).

The strong mutual gain between parameters #7 and #8 suggests merging these two parameters into a single unknown parameter. This is justifiable as these parameters characterize the effective elastic modulus of columns and walls at the basement level and the first story, which are expected to be close. The mutual gains between other stiffness-related parameters are not significant; nevertheless, their small entropy gains in Figure 3 suggest merging them together. Furthermore, since the initial estimate of mass-related parameters (#14 and #15) are close and their individual entropy gains are small, they are also merged together to improve their identifiability. Finally, while the entropy gains of parameters #1 and #12 are small, we still consider them as unknown parameters to be estimated; however, the resulting estimates are expected to be inaccurate and include large uncertainties. As a result of the outlined identifiability assessment process, we end up with six estimation parameters, as listed in Table 3.

Table 3. Final selected model parameters for Step 1 system identification.

Parameter ID	Description	Value
1	Elastic modulus of brace elements representing precast claddings (E_{Clad})	20 GPa
2	Effective elastic modulus of column and wall concrete at basement and 1st story ($E_{W\&C}1$)	17.3 GPa
3	Effective elastic modulus of column and wall concrete at 2nd story to roof ($E_{W\&C}2$)	17.3 GPa
4	Mass-proportional Rayleigh damping coefficient (α)	0.4
5	Stiffness-proportional Rayleigh damping coefficient (β)	5.3×10^{-3}
6	Distributed floor mass on 1st to roof floors (m)	250 kg/m^2

5.2. Joint System and Partial Input Identification Using Yorba Linda Earthquake Data

The seismic data recorded at the Millikan Library building during the 2002 Yorba Linda earthquake are utilized for the FE model updating. Here, we present two model updating cases. In Case 1, the recordings of the six accelerometers installed on the foundation are averaged to find the time histories of translational and rotational components of the foundation-level motion. These input excitations along with the measured structural responses obtained from all the channels are used for an "input–output" FE model updating to estimate the six model parameters listed in Table 3. In Case 2, we assume that the recordings of the two sensors located in the steam tunnel, i.e., Channels #4 and #5, are unknown. This resulted in FE model updating with partially unknown inputs to jointly estimate the six model parameters and the acceleration response time history at two sensor locations.

Figure 5 presents the relative root mean square error (RRMSE) between the measured and FE predicted acceleration responses for both Case 1 and Case 2. The figure also shows the RRMSE values obtained from the initial FE model (i.e., based on the initial parameter values). As can be seen in this figure, the RRMSE is consistently reduced from the initial to the updated FE model in both cases. However, Case 2 shows more reduction in RRMSE for all measurement channels compared to Case 1 except for Channels #4 and #5, in which RRMSEs are increased. This suggests that these two sensor measurements do not comply with the other measurements on the foundation level. Hence, the updated FE model in Case 2 is most likely a better representation of the real-world building. As shown in Appendix A, these two channels correspond to the sensors installed in the steam tunnel—Channel #4 measures the acceleration response in the NS-direction and Channel #5 in the Up–Down or Z-direction. The steam tunnel is a rectangular reinforced concrete utility access tunnel built next to the foundation level and is not part of the basement structure. After the 1971 San Fernando Earthquake, a 0.5 mm crack around the steam tunnel, at the point where the tunnel connects to the basement wall, was reported [37]. Based on these observations, it is concluded that the tunnel is most likely not fully connected to the structural system, and thus, the readings of the accelerometers located in there may not correctly represent the foundation level motions. Figure 6 compares the measured and estimated acceleration response time histories at Channels #4 and #5. The differences between the measured and estimated acceleration time histories are minor and more prominent for Channel #5, which is the Z-direction measurement.

The initial and final estimates of the six model parameters, along with their final estimated coefficients of variation (COVs), are listed in Table 4. The coefficient of variation quantifies the estimation uncertainty. In an unbiased estimation, the COV should merge to zero. The non-zero values for the COV mean that the estimation might be uncertain. The larger the COV, the more uncertainty there is in the estimated model parameter values. As can be seen in Table 4, the COV of the two parameters with the highest COV in case 1, i.e., E_{Clad} and m, is significantly reduced in Case 2. It is also seen that the estimate of these parameters is unrealistically low in Case 1, while their estimates in Case 2 seem to be more realistic. These results, along with the RRMSE results shown in Figure 5, indicate that Case 2 is likely to be more reliable than Case 1, and hence, the parameter estimates from this case will be used for the next step.

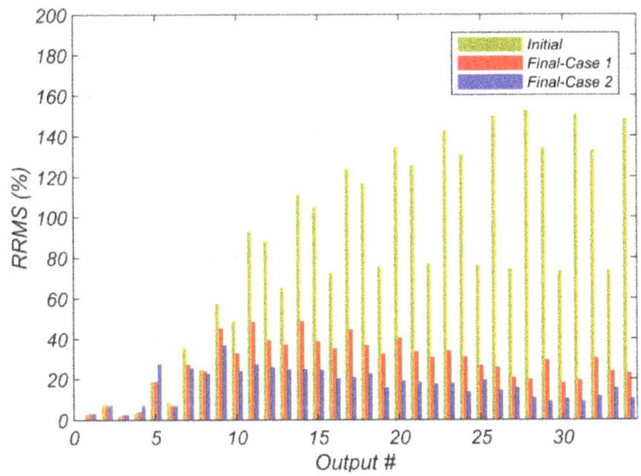

Figure 5. Relative root mean square error (RRMSE) between the measured and the FE predicted structural responses using the initial and final FE models.

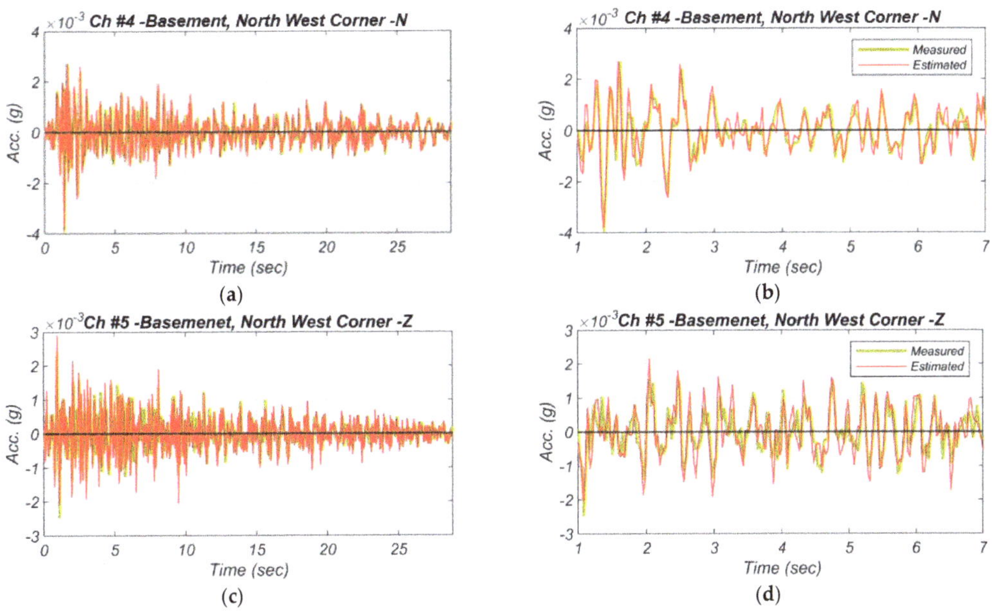

Figure 6. Comparison of the measured and estimated acceleration response time histories at Channels #4 and #5. The right-hand-side plots magnify the response time history between 1 and 7 s. Each figure's title shows the sensor location and recording direction, e.g., "Basement, North West Corner-Z" means acceleration response time history in the Z-direction, recorded by the sensor located on the north-west corner of the basement.

Table 4. Initial and final estimates of the model parameters, along with their final estimated coefficient of variation (COV). Parameters are defined in Table 3.

Parameter	E_{Clad}	$E_{W\&C}1$	$E_{W\&C}2$	α	β	m
Initial Estimate	20 GPa	17.3 GPa	17.3 GPa	0.40	5.3×10^{-3}	250 kg/m^2
Case 1—Final Estimate (COV)	0.25 GPa (9.87%)	16.2 GPa (0.41%)	29.7 GPa (0.68%)	0.24 (2.64%)	2.6×10^{-3} (1.4%)	2.8 kg/m^2 (8.97%)
Case 2—Final Estimate (COV)	6.17 GPa (2.87%)	19.6 GPa (0.66%)	26.1 GPa (0.84%)	0.29 (2.61%)	1.7×10^{-3} (2%)	161 kg/m^2 (2.44%)

Figure 7 shows how well the updated model predictions match the measurement records for both model updating cases, wherein selected measured acceleration response time histories are shown together with those estimated using the initial and final updated FE models. The figure shows that while the initial FE model response predictions have remarkable discrepancies with the measured responses, the updated model predictions have good agreement with the measurements. It also shows that the updated model predictions in Case 2 have better agreement with the measurements than Case 1, which further proves the reliability of Case 2 over Case 1. In the next paragraph, we describe the details of the estimation algorithm setup.

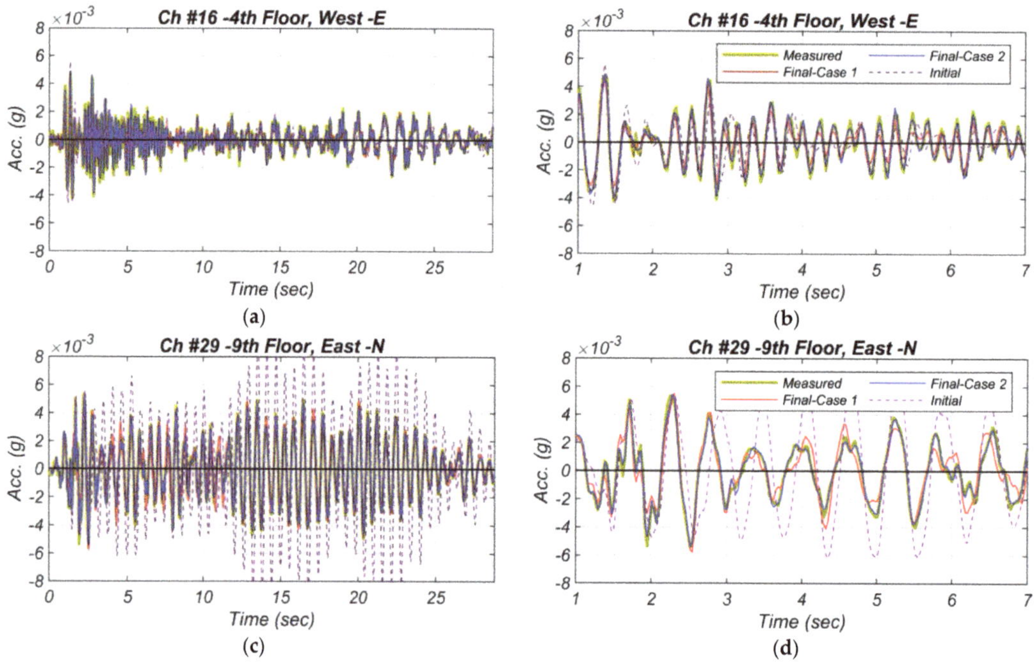

Figure 7. Comparison of the measured structural responses with the structural responses predicted using the initial and final-updated FE models for both Case 1 and Case 2. The right-hand-side plots magnify the response time history between 1 and 7 s. Each figure's title shows the sensor location and recording direction, e.g., "9th floor, East-N" means acceleration response time history in the NS direction, recorded by the sensor located on the east side of the 9th floor.

The initial values of model parameters ($\hat{\theta}_0$) are listed in Table 3, and their initial coefficient of variation is selected as 10%. The estimation constraints for model parameters are selected as $0.1\hat{\theta}_0 \leq \theta \leq 5\hat{\theta}_0$. The initial estimate of the discrete values of the unknown ground motion time history is selected to be zero, and their initial standard deviation is selected as $0.05\frac{\text{Rad}}{\text{s}^2}$. A variable estimation window configuration (length and overlap) is considered. During the first 6 s of the earthquake (which corresponds to the strong motion part), a small window length (60 time steps = 1.8 s) with 50% overlap is used. As we march through the time history, the estimation window length is gradually increased to 100 time steps (3 s) with 20% overlap. Selecting smaller lengths and larger overlaps for the estimation windows at the beginning of the earthquake ensures the incremental absorption of the information when the model has large uncertainties and the measurements have large information content (due to strong motion). The window length is subsequently increased and the overlap is reduced to improve the computational efficiency. The estimation window configuration can affect the computation demand and the estimation accuracy and should be selected based on experience and adjusted by observing the performance of the estimation algorithm.

The process noise covariance matrix **Q** (see Table 1) is selected as a diagonal matrix. The diagonal entry corresponding to the ith model parameter is selected as $\left(q\hat{\theta}_i\right)^2$, where $q = 0.001$ and $\hat{\theta}_i$ is the mean estimate of the ith model parameter. This means that the RMS of the process noise corresponding to the model parameters is taken as 0.1% of the mean estimate of the associated model parameter. Thus, this part of the matrix **Q** would be time-varying. The diagonal entries corresponding to the unknown ground motion time histories are constant, and are taken as $(0.001\text{g})^2$. This means that the root mean square (RMS) of the process noise corresponding to the input excitation is time-invariant and is equal to 0.1% g. This selection is performed based on our previous experience with the estimation algorithm to ensure stability and proper convergence rate (more discussions are provided in [11]). Finally, the simulation error covariance matrix **R** is also selected as a time-invariant diagonal matrix, whose diagonal entries are selected as $(0.001\text{g})^2$. In other words, the simulation error is modeled as a 0.1% g RMS Gaussian white noise process.

6. Step 2 System Identification: FE Model Updating of the Soil–Structure System with Unknown Inputs

At this stage, we use the updated model of the superstructure (Case 2) for soil–structure system identification. The objective is to estimate the FIMs and the parameter characterizing the soil–structure system, including the stiffness and viscosity of the soil springs and dashpots used for modeling the inertial soil–structure interaction effects. For this purpose, and similar to the previous step, we start with an identifiability analysis by evaluating the information contained in the model responses. Then, we use the real measurement data to estimate the soil–structure model parameters.

6.1. Model Identifiability and Parameter Selection

Distributed linear soil springs and dashpots are included underneath the foundation slab of the updated FE model, obtained from the previous step. Three linear springs and three linear viscous dashpots are modeled independently in the x-, y-, and z-direction (corresponding to EW, NS, and Up–Down directions, respectively) at each nodal point of the foundation slab. The stiffness of soil springs and viscosity of dashpots are computed using the subgrade modulus (i.e., soil stiffness per unit area) and viscosity modulus (i.e., soil viscosity per unit area), respectively. The spring stiffness and dashpot viscosity at each nodal point are calculated by multiplying the tributary area of the nodal point by the corresponding subgrade modulus and viscosity modulus, respectively.

The Millikan Library has a two-level foundation system consisting of a central pad and two north and south foundation strips, as shown in Figure 8. Six unknown subgrade moduli—namely k_x, k_y1, k_y2, k_z1, k_z2, and k_z3—are defined for different foundation regions, where, for example, the parameter k_z2 characterizes the vertical subgrade modulus for the interior area of the central pad. Two strips along the east and west edges of the central pad, which are underneath the two box shear walls at the east and west sides of the building, are assigned a different vertical subgrade modulus (k_z3) based on the recommendation provided in [3] to account for the rotational stiffness of the soil subsystem. Similarly, a different vertical subgrade modulus (k_z1) is assigned to the two north and south foundation strips. Likewise, six (unknown) parameters—namely c_x, c_y1, c_y2, c_z1, c_z2, and c_z3—are used to define the viscosity modulus. Different foundation regions and their corresponding subgrade and viscosity modulus are shown in Figure 9. The twelve unknown subgrade and viscosity moduli along with the Rayleigh damping parameters (α and β) and the effective elastic modulus of foundation slabs (E_{Found}) are selected as parameters for the identifiability analysis. Although estimated in the previous step, the Rayleigh damping parameters are included for the soil–structure system identification since the damping model behavior of the fixed-based structure is expected to be different from the flexible-base structure.

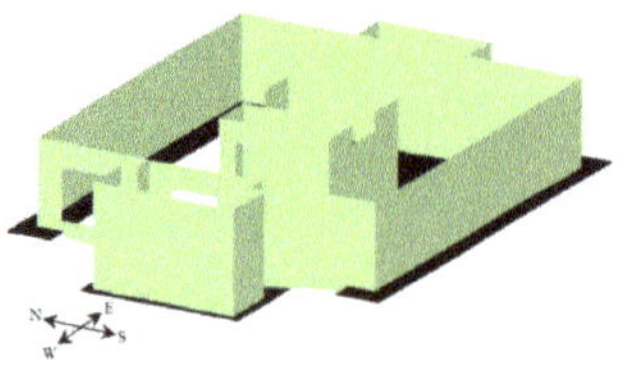

Figure 8. Structural configuration of the Millikan Library basement showing the structural walls and two-level foundation details, including the central pad (at EL.−6 m from the ground surface), and the two north and south foundation beams (at EL.−5 m from the ground surface).

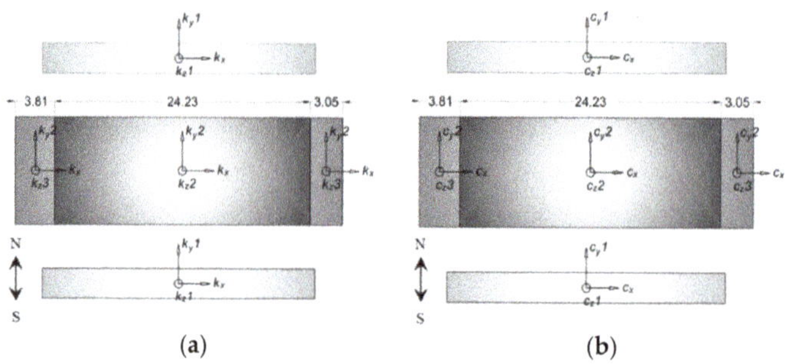

Figure 9. (**a**) Six unknown subgrade modulus parameters, and (**b**) six unknown viscosity modulus parameters defined for different foundation regions. The figures show the foundation plan of the Millikan library, including the central pads, and the two foundation beams (dimensions are in meters).

To assess the identifiability of these fifteen (15) unknown parameters, we pursue the same identifiability analysis approach presented before. Initial estimates of soil model parameters are assigned using the NIST standard guidelines [3]. The FIMs are unknown at this stage and will be estimated through the model updating process. In the absence of the correct FIMs, the three translational components of the foundation-level motion, estimated from the previous step, are used as the FIMs for identifiability assessment. The parameters used for identifiability analysis and their corresponding initial values are listed in Table 5. The initial values of the Rayleigh damping parameters are selected based on the estimation results obtained in the previous step. Moreover, a 20% coefficient of variation is assumed for all parameters to derive the a priori covariance matrix.

Figure 10 displays the relative entropy gain of the fifteen estimation parameter candidates. Figure 11 displays the mutual entropy gain between parameter pairs. In Figure 11, the diagonal terms (i.e., entropy gain of estimation parameters) are found to be much larger than the off-diagonal terms (i.e., mutual entropy gain between parameter pairs); hence, including both of them in the figure would underrepresent the off-diagonal terms since they would appear with a hard-to-see, light color. Therefore, the diagonal terms are excluded from this figure. These two figures can be used together for selecting the estimation parameters. Figure 10 shows that some of the parameters (e.g., parameters #2, #8, #11, and #12) receive little information from the measurements. These parameters are, therefore, unlikely to be identifiable. Based on this figure, parameters #2 and #3, which represent the subgrade modulus in the north direction for the foundation beams and central pad, respectively, are merged together. This is due to the small relative entropy gain of parameter #2. Similarly, parameters #8 and #9, and #10 to #12 are merged together to improve their identifiability. Moreover, Figure 11 shows mutual dependence between parameter #15 and parameters #4 to #6. This implies that the estimation of the effective elastic modulus of foundation slabs will likely affect the estimation accuracy of the vertical and rocking soil stiffnesses. Hence, parameter #15 is fixed at its nominal value given the relatively large thickness of foundation slab, which is expected to result in linear elastic behavior. Based on this identifiability study, the estimation parameters characterizing the soil–structure model are reduced to ten, which are k_x, k_y, k_z1, k_z2, k_z3, c_x, c_y, c_z, α, and β.

Table 5. Fifteen model parameters used for prior model identifiability assessment.

Parameter ID	1	2	3	4	5	6	7	8
Parameter	k_x	k_y1	k_y2	k_z1	k_z2	k_z3	c_x	c_y1
Value	65 MN/m^3	100 MN/m^3	150 MN/m^3	20 MN/m^3	22.5 MN/m^3	37.5 MN/m^3	700 kN.s/m^3	700 kN.s/m^3
Parameter ID	9	10	11	12	13	14	15	
Parameter	c_y2	c_z1	c_z2	c_z3	α	β	E_{Found}	
Value	700 kN.s/m^3	1000 kN.s/m^3	1000 kN.s/m^3	1000 kN.s/m^3	0.29	1.7×10^{-3}	7.5 GPa	

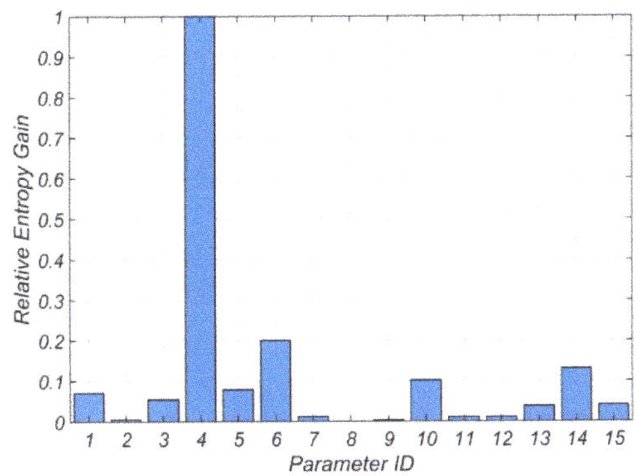

Figure 10. Relative entropy gain of the fifteen model parameters (see Table 5 for parameter IDs).

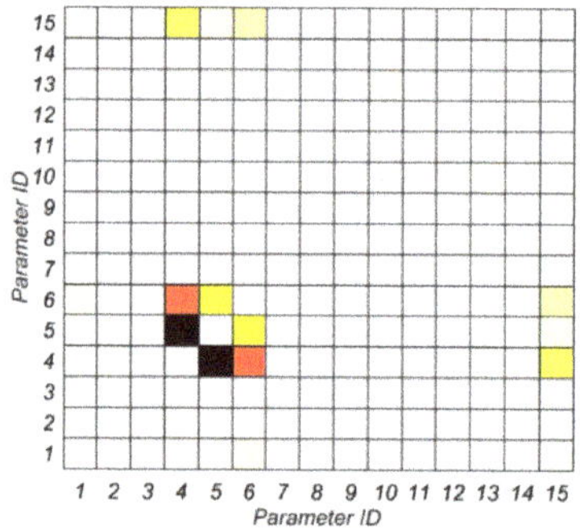

Figure 11. Relative mutual entropy gains between the parameter pairs (see Table 5 for parameter IDs).

6.2. *Joint System and Input Identification Using the Recorded Yorba Linda Earthquake Data*

The seismic data recorded at the Millikan Library building during the 2002 Yorba Linda earthquake are now utilized in an output-only FE model updating procedure to estimate the eight soil parameters, two Rayleigh damping parameters, and the three components of the FIM in EW, NS, and Up–Down directions (no rotational component is considered for the FIM). The model parameters characterizing the superstructure (except the Rayleigh damping parameters) are kept constant based on the estimated results obtained from the previous system identification step. Except the initial coefficient of variation of estimation parameters, which is selected as 20%, and the estimation constraints for model parameters, which are selected as $0.1\hat{\theta}_0 \leq \theta \leq 50\hat{\theta}_0$, the setup of the estimation algorithm and its parameters are similar to the previous step (see Section 5.2).

Figure 12 shows the time history of the posterior mean and standard deviation (SD) of the three components of the FIM. The initial and final estimates of the ten soil–structure model parameters along with the estimated coefficient of variation (COV) are listed in Table 6. The estimated parameter values with larger COV include larger estimation uncertainties. Figure 13 compares the measured acceleration response time histories at the selected measurement channels with those estimated using the final estimates of the model parameters and FIMs. This figure indicates that there is a remarkable agreement between the estimated and measured acceleration responses. Furthermore, Figure 14 presents the RRMSE of the updated FE model responses at different measurement channels. Comparing Figure 14 with Figure 5, the discrepancies between the FE predictions and measurements in the NS direction are generally less than those in the EW direction. This is predictable since the soil–structure interaction effects are more dominant in the NS direction of the Millikan Library [28], and hence, the flexible-base model uncertainty is expected to be less in the NS direction than in the EW direction.

Table 6. Initial and final estimates of the soil–structure model parameters along with their final estimated coefficient of variation (COV).

Parameter	k_x	k_y	k_z1	k_z2	k_z3
Initial Estimate	200 MN/m^3	200 MN/m^3	225 MN/m^3	225 MN/m^3	225 MN/m^3
Final Estimate (COV)	47.9 MN/m^3 (2.7%)	418.2 MN/m^3 (4.9%)	386.8 MN/m^3 (0.9%)	614.4 MN/m^3 (3.4%)	310.5 MN/m^3 (3.4%)
Parameter	c_x	c_y	c_z	α	β
Initial Estimate	700 kN.s/m^3	700 kN.s/m^3	1000 kN.s/m^3	0.29	1.7×10^{-3}
Final Estimate (COV)	313.7 kN.s/m^3 (13.1%)	1187 kN.s/m^3 (10.7%)	4 kN.s/m^3 (906.5%)	0.15 (3.1%)	2.3×10^{-3} (2%)

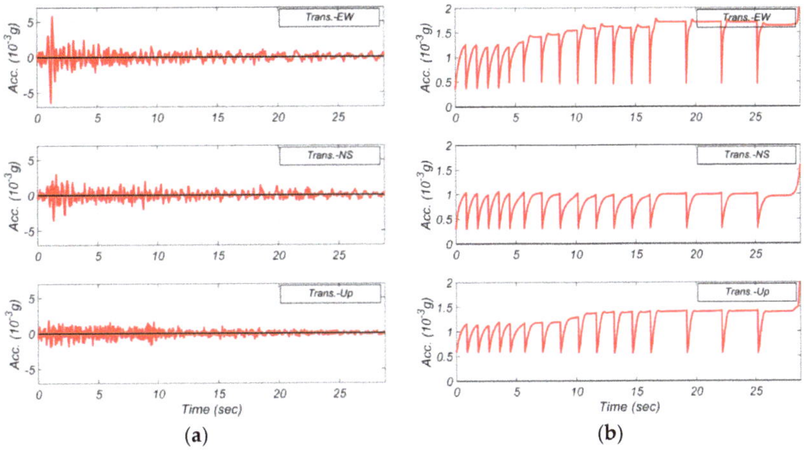

Figure 12. (**a**) Estimated foundation input motion (FIM) time histories and (**b**) their standard deviation (SD) values.

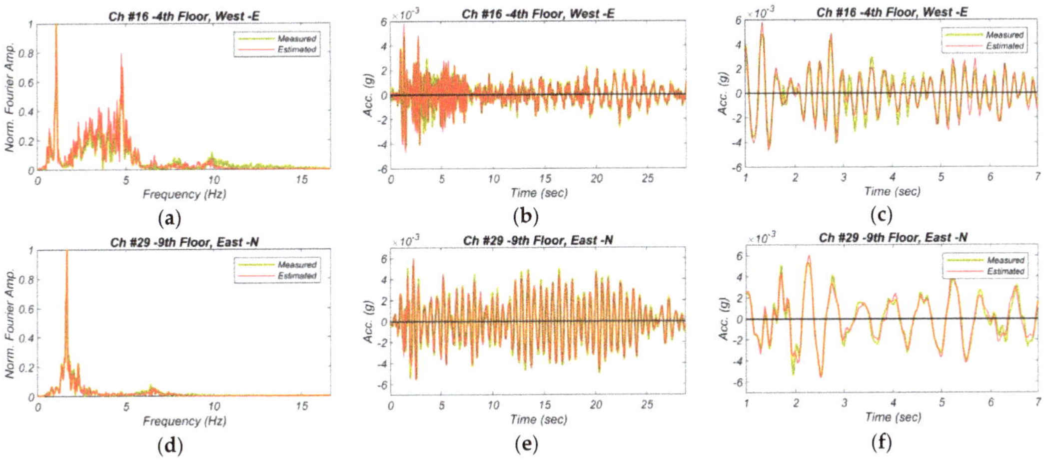

Figure 13. Comparison of the measured structural responses with the structural responses predicted using the final estimates of the soil-structure model parameters and FIMs. The left-hand-side plots compare the responses in frequency domain, the middle plots compare the time histories, and the right-hand-side plots magnify the response time history between 1 and 7 s.

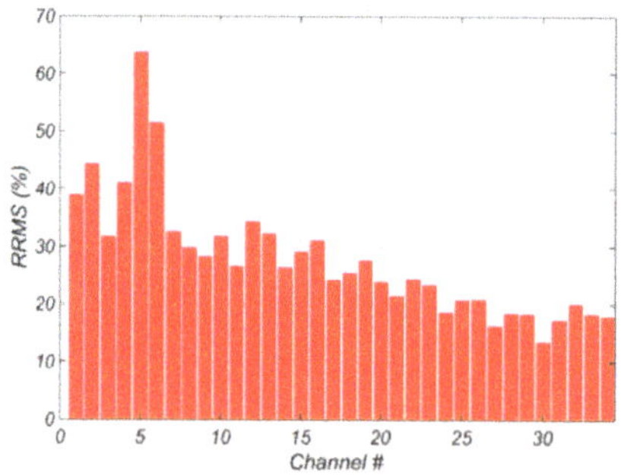

Figure 14. Relative root mean square error (RRMSE) of the FE predicted structural responses using the final estimates of the soil–structure model parameters.

7. Summary and Discussion

7.1. Effective Structural Stiffness Parameters

The estimated effective elastic moduli of columns and walls ($E_{W\&C}1$ = 19.6 GPa, $E_{W\&C}2$ = 26.1 GPa) are larger than the nominal value that is based on code recommendations, which is 17.3 GPa. Several factors can collectively account for this difference. (i) The effects of lateral stiffness of the backfill soil at the basement level are not modeled in the FE model. This is likely the reason behind the larger estimated value for $E_{W\&C}1$. It can also be a source of modeling error because the columns and walls at the first-story and basement level are parameterized with the same effective elastic modulus. (ii) Concrete aging, which will increase the stiffness of concrete material, is not accounted for in the nominal estimations. (iii) The lateral stiffness of nonstructural components and systems (i.e., partition walls and stairs) are not accounted for in the FE model. This can result in estimating larger values for the effective elastic moduli of the columns and walls, which characterize the lateral stiffness of the building. (iv) Finally, the nominal elastic modulus is calculated for the concrete material without accounting for the effects of steel reinforcement, which can partially explain the larger estimated effective elastic moduli.

7.2. Soil Subsystem Stiffness and Viscosity

Table 7 compares the estimated lumped stiffness and viscosity of the soil subsystem with the NIST recommendations [3] based on [40]. The lumped soil stiffness and viscosity should be evaluated at the fundamental frequency of soil–structure system using the flexible-base dimensionless frequency $a_0 = \tilde{\omega}b/V_s$, where $\tilde{\omega}$ is the flexible-base fundamental circular frequency, b = 11.5 m is the foundation half-width, and V_s is the shear wave velocity, which is averaged over an effective soil profile depth based on the recommendations provided in [3]. Using the shear wave velocity profile of the Millikan Library site presented in [41], the average shear wave velocity is calculated as 398.1 m/s for translation, 392.1 m/s for rocking in the NS direction, 407.7 m/s for rocking in the EW direction, and 422.3 m/s for torsion. The flexible-base frequency is a function of the soil stiffness coefficients as follows [3].

$$\frac{\tilde{T}_x}{T_x} = \sqrt{1 + \frac{k_x}{K_x} + \frac{k_x h^2}{K_{yy}}} \quad (9)$$

Equation (9) presents the flexible-base to fixed-base fundamental period ratio (i.e., period elongation) in the x-direction, where $\tilde{T}_x$ is the flexible-base fundamental period, T_x is the fixed-base fundamental period, k_x is the fixed-base structural stiffness, h is the effective fundamental modal height, K_x is the soil horizontal stiffness, and K_{yy} is the soil rocking stiffness. k_x is calculated as 411.4 MN/m given the effective fundamental modal mass (m_x = 5252.9 ton) and T_x = 0.71 s. Likewise, k_y is calculated as 1129.9 MN/m given m_y = 5541.2 ton and T_y = 0.44 s. The effective fundamental modal height is taken as 0.7 times the building height [1] ($h = 0.7 \times 43.9 = 30.73$ m). The process to calculate $\widetilde{T}$ is iterative [42]. T is intially utilized to evaluate the soil stiffnesses. Then, $\widetilde{T}$ is calculated using Equation (9), and the soil stiffnesses are reevaluated using $\widetilde{T}$. Then, $\widetilde{T}$ and the soil stiffnesses are iteratively updated until no considerable change in $\widetilde{T}$ is observed. This procedure is used for calculating the soil parameters corresponding to the two translational vibration modes in the NS and EW directions. For the torsional mode, the fixed-base torsional period ($T_{zz} = 0.33$ s) is used for evaluating the torsional soil parameters.

Table 7. Comparison of the normalized estimated lumped stiffness and viscosity of the soil–foundation subsystem with the recommended values in [3]. The translational components are normalized by GL and the rotational components by GL^3, where $G = 268$ MPa and $2L = 27.4$ m.

Foundation Vibration Mode	*Trans.-EW*	*Trans.-NS*	*Trans.-Up*	*Rock.-EW*	*Rock.-NS*	*Torsion*
Estimated Stiffness	5.8	50.65	59.08	2.15	1.38	2.14
NIST-based Stiffness	5.14	5.27	6.08	6.16	3.98	7.08
Estimated Viscosity	0.04	0.14	0.0005	0.00002	0.00001	0.007
NIST-based Viscosity	0.13	0.13	0.23	0.005	0.004	0.038

The estimated values in Table 7 are the integrated stiffness and viscosity of the soil subsystem with respect to the center of the foundation pad using the mean estimates of subgrade and viscosity moduli obtained in Section 6.2. To maintain compatibility with previous studies [28], we normalize the sway stiffness and viscosity by GL, where $G = 268$ MPa and $2L = 27.4$ m are the soil shear modulus and a reference foundation length, respectively. For rocking and torsional components, the normalizing factor is GL^3.

Table 7 indicates that the estimated stiffness and viscosity coefficients have a non-negligible difference with the NIST-based recommended values. One potential reason for the observed difference is the assumption of rigid foundation used in deriving the approximate formulas of the impedance function reported in [3], which was shown to be violated in the Millikan Library, especially in the EW direction [26]. In contrast, the foundation flexibility is explicitly modeled in this study. Furthermore, the estimated impedance function in this study is obtained using the building's seismic response that contains effective high-frequency components, as shown in the recording of Channel #16 presented in Figure 13, which is typical for the response of mid-level and lower-level floors. Since the impedance function is frequency-dependent, the presence of multiple frequencies in the response will likely make it difficult to exclusively tune the estimated impedance function to the flexible-base fundamental frequency. Therefore, some deviation of the estimated impedance function using multiple-frequency responses from the NIST recommendations, in which the impedance function is evaluated at a single frequency, is expected.

Comparing the stiffness coefficients, an especially significant discrepancy can be observed between the estimated and NIST-based recommended values of the translational stiffness in the NS direction and Up–Down direction. A similarly significant discrepancy between the theoretical and experimental estimates of soil translational stiffness in the NS direction was reported in [41], and it was attributed to either an error in the measured foundation translational motion or the crude simplification of the foundation model. To our knowledge, no studies have reported the soil stiffness in the Up–Down direction; therefore, we could not compare our estimated values to other reference values.

Comparing the viscosity coefficients, almost all estimated viscosity coefficients deviate considerably from the corresponding NIST-based values, which is possibly a manifestation of the high uncertainty with which these parameters are estimated, as shown in Table 6. Estimating damping experimentally has always been a key problem in system identification of building structures [43]. This is likely the case with the Millikan Library since there were challenges with accurately determining the phase of the response (with respect to the applied force) in forced vibration tests [41]. Moreover, the estimated viscosities are almost always less than their NIST-based counterparts. This can be explained by examining the soil shear wave velocity profile of the Millikan Library site reported in [41]. Looking at this profile, the shear wave velocity increases almost linearly with depth until it reaches 944.8 m/s at 118.57 m depth. This rapid increase in the shear wave velocity with depth most likely causes multiple wave reflections at the different soil layer interfaces, hence the reduction in soil radiation damping. We can also observe that the reduction in radiation damping due to the nonuniform soil profile is more significant in rocking than in horizontal translation, which is reported in [3]. Thus, we can state that although using an average shear wave velocity can fairly simulate the nonuniform soil stiffness, it cannot simulate the nonuniform soil radiation damping at the same level of accuracy.

The lumped stiffnesses of the soil subsystem for the Millikan Library have also been estimated using other system identification methods in the literature. Table 8 presents the estimated stiffness coefficients in six different previous studies compared to those identified in this study.

Table 8. Comparison of the normalized lumped stiffness of soil–foundation subsystem between different studies. The translational components are normalized by GL and the rotational components by GL^3, where G = 268 MPa and $2L$ = 27.4 m.

Study	Foundation Vibration Mode				
	Trans.-EW	*Trans.-NS*	*Rock.-EW*	*Rock.-NS*	*Torsional*
[6]	31.1	8.7	207.2	145.9	9.9–10.5
[44]	6.9	4.6	10.0	5.4	—
[41]	7.2	6.7	10.2	5.9	—
[26]	5.5	5.2	7.4	5.6	11.1
[45]	2.7	2.7	1.8	1.8	—
[28]	—	6.2	2.5	3.9	0.98
Current Study	5.14	50.65	2.15	1.38	2.14

As can be seen, the translational stiffness in the EW direction is close to the value identified by Luco et al. [26]; the NS translational stiffness is not comparable to any other study; the EW and NS rocking stiffnesses are close to the static values used by Chen et al. [45] and adopted from Balendra et al. [46]; and the torsional stiffness is close to the value estimated by Ghahari et al. [28]. The observed discrepancy between the identified model parameters and the corresponding parameters identified in previous studies highlights the challenging nature of the problem at hand. It also draws attention to the importance of widening the application of structural sensing to real-life structures to develop accurate numerical models for structural health monitoring, condition assessment, and damage prognosis, especially in high-seismicity areas [47].

7.3. Period Elongation

Table 9 compares the period of the first five structural modes between the updated flexible-base and fixed-base models. The table, moreover, reports the period elongations. The estimated modal periods of the flexible-base structure have good agreement with the results reported in [28].

Table 9. Comparison of the flexible-based and fixed-base modal periods of the updated FE models.

Structural Vibration Mode	*1st Trans.-EW*	*1st Trans.-NS*	*1st Torsional*	*2nd Trans.-EW*	*2nd Trans.-NS*
Flexible-base Period (s)	0.95	0.6	0.42	0.22	0.16
Fixed-base Period (s)	0.88	0.5	0.37	0.2	0.16
Period Elongation	1.08	1.2	1.14	1.1	1.0

7.4. Rayleigh Damping

Figure 15 compares the Rayleigh damping ratio as a function of frequency between the initial, updated fixed-base, and updated flexible-base FE models. Table 6 shows that the estimated mass-proportional Rayleigh damping coefficient for the flexible-base structure is relatively small, which results in a nearly stiffness-proportional linear Rayleigh damping curve for the updated flexible-base model. This is a well-known effect of SSI on effective modal damping [48].

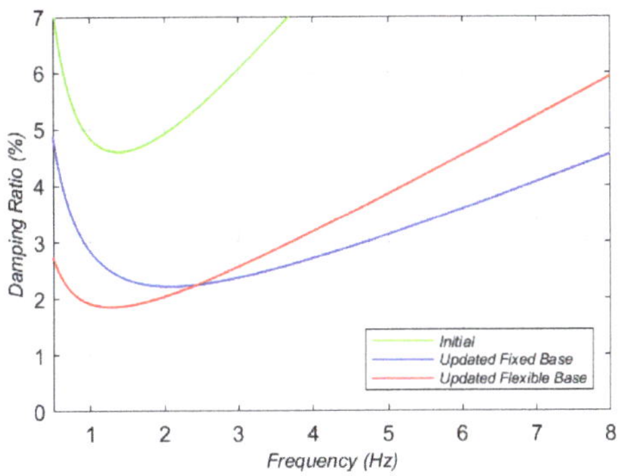

Figure 15. The Rayleigh damping ratio as a function of frequency in the initial, updated fixed-base, and updated flexible-base FE models.

7.5. Possible Sources of Estimation Uncertainty

The estimated model parameters include uncertainties, which are quantified relatively using the estimated coefficient of variations (COVs). Aside from parameter identifiability issues, which are discussed earlier in the paper, modeling errors are most likely the most important source of estimation uncertainties. Modeling errors result from the inherent imperfections, approximations, and idealizations in the mathematical FE model. As a result, the selected class of FE models does not contain the real-world structure, and the estimated parameters characterize the closest possible model, in the model class, to the real structure [49]. Geometric approximations, material nonlinearities (even during low-amplitude earthquake excitations), effects of nonstructural components and systems, and modeling of damping energy dissipation mechanisms are some examples of imperfections in the numerical FE model. Model parameterization (i.e., selecting the estimation parameters in the model) is another source of modeling error. Modeling errors lead to biased estimation results. This means that by changing the earthquake event, the estimation result may vary, preferably slightly. Addressing the effects of modeling error requires repeating this study using more refined FE models and/or different earthquake data, which is beyond the scope of the present paper and can be the subject of future studies.

8. Conclusions

In this study, we developed a finite element (FE) model updating framework using a sequential Bayesian estimation approach to identify the soil and structural model parameters as well as the input motions from recorded seismic response of the Millikan Library building during the 2002 Yorba Linda earthquake. The newly developed Bayesian FE model updating procedure was used in a two-step system identification approach. In the first step, a fixed-base structural model was identified using partially unknown inputs. The structural model parameters included the effective stiffness of columns and shear walls (grouped along the height of the structure), the Rayleigh damping parameters, and the distributed floor mass. In the second step, the parameters characterizing the soil–structure model and the time histories of the Foundation Input Motions (FIMs) were estimated. The soil–structure model parameters included the soil subgrade and viscosity moduli at different foundation regions and the Rayleigh damping parameters. For each step, an identifiability analysis based on an information theoretic approach was performed to study the dependencies between model parameters and select the most identifiable parameter sets to be estimated. In Section 7 of the paper, the estimated model parameters were compared with the state-of-practice recommendations and other previous studies on the Millikan Library, and the differences were highlighted.

Minor differences were observed between the measured and FE-predicted response time histories obtained from the updated FE models. Considering the various sources of modeling error in this identification problem, the agreement between the updated FE model responses and the measurements was remarkable. Future studies are suggested to extend this process with more refined FE models (e.g., including material nonlinearities) and different earthquake data to further quantify the effects of modeling error. Applying this process to nonlinear SSI models using the recorded structural responses to strong ground motions may introduce additional complications due to the damage in nonstructural components, the nonlinear kinematic interaction between the structural and nonstructural systems, and the nonlinearities in the soil response. As a result, the identification of the nonlinear structural model parameters and the Raleigh damping model parameters is expected to include uncertainties and estimation biases. Furthermore, the identified soil–foundation subsystem model parameters, i.e., the impedance function, are expected to deviate more considerably from the state-of-practice recommendations. Despite these difficulties, undertaking such a research effort is crucial for developing a better understanding of the nonlinear interaction between the non-structure, structure, and soil subsystems. By utilizing the presented model inversion framework for different civil structures and earthquake records, key structural model parameters can be estimated from real-world seismic data. These are valuable information that can guide—along with theoretical results—the selection of engineering analysis and design parameters to improve state-of-the-art seismic analysis and design procedures.

Author Contributions: Conceptualization, H.E., F.G. and E.T.; methodology, H.E. and F.G.; software, H.E. and F.G.; val-idation, H.E. and A.T.; formal analysis, H.E., A.T., F.G. and E.T.; investigation, H.E., A.T. and F.G.; data curation, H.E. and A.T.; writing—original draft preparation, H.E. and A.T.; writing—review and editing, F.G. and E.T.; visualization, H.E. and A.T.; project administration, H.E. and D.A.; funding acquisition, H.E., D.A. and E.T. All authors have read and agreed to the published version of the manuscript.

Funding: This study was supported by the California Geological Survey through contracts no. 1016-987 and 1019-15, and the U.S. Geological Survey through grant no. G20AP00115-01. The supports are gratefully acknowledged. Any opinions, findings, conclusions, or recommendations expressed in this study are those of the authors and do not necessarily reflect the views of the sponsoring agencies.

Data Availability Statement: All data, models, or codes that support the findings of this study are available from the corresponding author upon reasonable request.

Conflicts of Interest: The authors declare no conflict of interest.

Appendix A. Instrumentation Plan for the Millikan Library

Table A1. Sensors' locations and measurement directions.

Channel #	Sensor Location	Measurement Direction
1	Basement, North East Corner	N
2	Basement, North East Corner	Up
3	Basement, North East Corner	E
4	Basement, North West Corner	N
5	Basement, North West Corner	Up
6	Basement, South East Corner	Up
7	1st Floor, East	N
8	1st Floor, West	E
9	1st Floor, West	N
10	2nd Floor, West	E
11	2nd Floor, West	N
12	3rd Floor, East	N
13	3rd Floor, West	E
14	3rd Floor, West	N
15	4th Floor, East	N
16	4th Floor, West	E
17	4th Floor, West	N
18	5th Floor, East	N
19	5th Floor, West	E
20	5th Floor, West	N
21	6th Floor, East	N
22	6th Floor, West	E
23	6th Floor, West	N
24	7th Floor, East	N
25	7th Floor, West	E
26	7th Floor, West	N
27	8th Floor, West	E
28	8th Floor, West	N
29	9th Floor, East	N
30	9th Floor, West	E
31	9th Floor, West	N
32	Roof, East	N
33	Roof, West	E

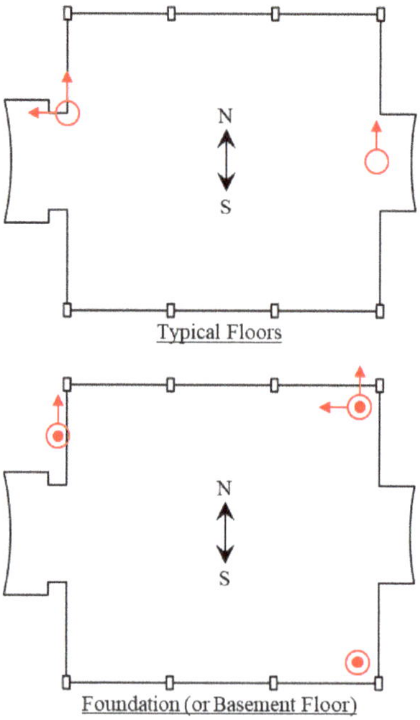

Figure A1. Floor plans showing the different sensors' locations and measurement directions.

References

1. *ASCE/SEI 7-10*; Minimum Design Loads for Buildings and Other Structures. American Society of Civil Engineers: Reston, VA, USA, 2010.
2. PEER. *Guidelines for Performance-Based Seismic Design of Tall Buildings*; Pacific Earthquake Engineering Research Center as part of the Tall Buildings Initiative, University of California: Berkeley, CA, USA, 2010.
3. Stewart, J.; Crouse, C.; Hutchinson, T.; Lizundia, B.; Naeim, F.; Ostadan, F. *Soil-Structure Interaction for Building Structures*; Grant/Contract Reports (NISTGCR); National Institute of Standards and Technology (NIST): Gaithersburg, MD, USA, 2012.
4. Roesset, J.M. A review of soil-structure interaction. In *Soil-Structure Interaction: The Status of Current Analysis Methods and Research, Rep. No. NUREG/CR-1780 and UCRL*; Nuclear Regulatory Commission: Washington, DC, USA, 1980.
5. Tileylioglu, S.; Stewart, J.P.; Nigbor, R.L. Dynamic Stiffness and Damping of a Shallow Foundation from Forced Vibration of a Field Test Structure. *J. Geotech. Geoenviron. Eng.* **2011**, *137*, 344–353. [CrossRef]
6. Jennings, P.C.; Kuroiwa, J.H. Vibration and soil-structure interaction tests of a nine-story reinforced concrete building. *Bull. Seism. Soc. Am.* **1968**, *58*, 891–916. [CrossRef]
7. Wolf, J.P.; Deeks, A.J. *Foundation Vibration Analysis: A strength of Materials Approach*; Elsevier: Amsterdam, The Netherlands, 2004.
8. Stewart, J.P.; Seed, R.B.; Fenves, G.L. *Empirical Evaluation of Inertial Soil-Structure Interaction Effects*; Pacific Earthquake Engineering Research Center: Berkeley, CA, USA, 1998.
9. Pais, A.L.; Kausel, E. On rigid foundations subjected to seismic waves. *Earthq. Eng. Struct. Dyn.* **1989**, *18*, 475–489. [CrossRef]
10. Luco, J.E.; Westmann, R.A. Dynamic Response of a Rigid Footing Bonded to an Elastic Half Space. *J. Appl. Mech.* **1972**, *39*, 527–534. [CrossRef]
11. Ebrahimian, H.; Astroza, R.; Conte, J.P.; Papadimitriou, C. Bayesian optimal estimation for output-only nonlinear system and damage identification of civil structures. *Struct. Control. Health Monit.* **2018**, *25*, e2128. [CrossRef]
12. Ebrahimian, H.; Astroza, R.; Conte, J.P. Extended Kalman filter for material parameter estimation in nonlinear structural finite element models using direct differentiation method. *Earthq. Eng. Struct. Dyn.* **2015**, *44*, 1495–1522. [CrossRef]
13. Astroza, R.; Ebrahimian, H.; Conte, J.P. Material Parameter Identification in Distributed Plasticity FE Models of Frame-Type Structures Using Nonlinear Stochastic Filtering. *J. Eng. Mech.* **2015**, *141*. [CrossRef]
14. Ebrahimian, H.; Astroza, R.; Conte, J.P.; de Callafon, R.A. Nonlinear finite element model updating for damage identification of civil structures using batch Bayesian estimation. *Mech. Syst. Signal Process.* **2016**, *84*, 194–222. [CrossRef]
15. Beck, J.L.; Katafygiotis, L.S. Updating models and their uncertainties. I: Bayesian statistical framework. *J. Eng. Mech. ASCE* **1998**, *124*, 455–462. [CrossRef]
16. Beck, J.L. Bayesian system identification based on probability logic. *Struct. Control. Health Monit.* **2010**, *17*, 825–847. [CrossRef]
17. Simon, D. *Optimal State Estimation: Kalman, H infinity, and Nonlinear Approaches*; John Wiley & Sons: Hoboken, NJ, USA, 2006.

18. Astroza, R.; Ebrahimian, H.; Conte, J.P. *Batch and Recursive Bayesian Estimation Methods for Nonlinear Structural System Identification*; Springer: Berlin/Heidelberg, Germany, 2017; pp. 341–364. [CrossRef]
19. Conte, J.P.; Vijalapura, P.K.; Meghella, M. Consistent Finite-Element Response Sensitivity Analysis. *J. Eng. Mech.* **2003**, *129*, 1380–1393. [CrossRef]
20. McKenna, F. OpenSees: A Framework for Earthquake Engineering Simulation. *Comput. Sci. Eng.* **2011**, *13*, 58–66. [CrossRef]
21. MathWorks. *MATLAB 2022a*; MathWorks: Natick, MA, USA, 2022.
22. Simon, D.; Simon, D.L. Constrained Kalman filtering via density function truncation for turbofan engine health estimation. *Int. J. Syst. Sci.* **2010**, *41*, 159–171. [CrossRef]
23. Ebrahimian, H.; Kohler, M.; Massari, A.; Asimaki, D. Parametric estimation of dispersive viscoelastic layered media with application to structural health monitoring. *Soil Dyn. Earthq. Eng.* **2018**, *105*, 204–223. [CrossRef]
24. Kuroiwa, J.H. *Vibration Test of a Multistory Building*; Caltech: Pasadena, CA, USA, 1967.
25. Crouse, C.B.; Jennings, P.C. Soil-structure interaction during the San Fernando earthquake. *Bull. Seism. Soc. Am.* **1975**, *65*, 13–36. [CrossRef]
26. Luco, J.E.; Trifunac, M.D.; Wong, H.L. On the apparent change in dynamic behavior of a nine-story reinforced concrete building. *Bull. Seism. Soc. Am.* **1987**, *77*, 1961–1983. [CrossRef]
27. Bradford, S.C.; Clinton, J.F.; Favela, J.; Heaton, T.H. *Results of Millikan Library Forced Vibration Testing*; Caltech: Pasadena, CA, USA, 2004.
28. Ghahari, S.F.; Abazarsa, F.; Avci, O.; Çelebi, M.; Taciroglu, E. Blind identification of the Millikan Library from earthquake data considering soil-structure interaction. *Struct. Control Health Monit.* **2015**, *23*, 684–706. [CrossRef]
29. Taciroglu, E.; Ghahari, S.; Abazarsa, F. Efficient model updating of a multi-story frame and its foundation stiffness from earthquake records using a timoshenko beam model. *Soil Dyn. Earthq. Eng.* **2017**, *92*, 25–35. [CrossRef]
30. Shirzad-Ghaleroudkhani, N.; Mahsuli, M.; Ghahari, S.F.; Taciroglu, E. Bayesian identification of soil-foundation stiffness of building structures. *Struct. Control Health Monit.* **2017**, *25*, e2090. [CrossRef]
31. CESMID. Center for Engineering Strong Motion Data. Available online: https://www.strongmotioncenter.org/ (accessed on 10 October 2022).
32. Computers and Structures. SAP2000. Available online: https://www.csiamerica.com/products/sap2000 (accessed on 10 October 2022).
33. Ebrahimian, H.; Asimaki, D.; Kusanovic, D.; Ghahari, S.F. Untangling the Dynamics of Soil-Structure Interaction Using Nonlinear Finite Element Model Updating. In Proceedings of the SMIP17 Conference on Utilization of Strong-Motion Data, Berkeley, CA, USA, 19 October 2017.
34. Ebrahimian, H.; Astroza, R.; Conte, J.P.; Bitmead, R.R. Information-Theoretic Approach for Identifiability Assessment of Nonlinear Structural Finite-Element Models. *J. Eng. Mech.* **2019**, *145*. [CrossRef]
35. Brun, R.; Reichert, P.; Künsch, H.R. Practical identifiability analysis of large environmental simulation models. *Water Resour. Res.* **2001**, *37*, 1015–1030. [CrossRef]
36. Yu, E.; Taciroglu, E.; Wallace, J.W. Parameter identification of framed structures using an improved finite element model-updating method—Part I: Formulation and verification. *Earthq. Eng. Struct. Dyn.* **2006**, *36*, 619–639. [CrossRef]
37. Foutch, D.A.; Jennings, P.C. A study of the apparent change in the foundation response of a nine-story reinforced concrete building. *Bull. Seism. Soc. Am.* **1978**, *68*, 219–229. [CrossRef]
38. Todorovska, M.I. Soil-Structure System Identification of Millikan Library North-South Response during Four Earthquakes (1970-2002): What Caused the Observed Wandering of the System Frequencies? *Bull. Seism. Soc. Am.* **2009**, *99*, 626–635. [CrossRef]
39. *ACI 318-14*; Building Code Requirements for Structural Concrete and Commentary. American Concrete Institute: Farmington Hills, MI, USA, 2014.
40. Pais, A.; Kausel, E. Approximate formulas for dynamic stiffnesses of rigid foundations. *Soil Dyn. Earthq. Eng.* **1988**, *7*, 213–227. [CrossRef]
41. Wong, H.; Trifunac, M.; Luco, J. A comparison of soil-structure interaction calculations with results of full-scale forced vibration tests. *Soil Dyn. Earthq. Eng.* **1988**, *7*, 22–31. [CrossRef]
42. Veletsos, A.S.; Meek, J.W. Dynamic behaviour of building-foundation systems. *Earthq. Eng. Struct. Dyn.* **1974**, *3*, 121–138. [CrossRef]
43. He, X.; Moaveni, B.; Conte, J.P.; Elgamal, A.; Masri, S.F. System Identification of Alfred Zampa Memorial Bridge Using Dynamic Field Test Data. *J. Struct. Eng.* **2009**, *135*, 54–66. [CrossRef]
44. Luco, J.E.; Trifunac, M.D.; Wong, H.L. Isolation of soil-structure interaction effects by full-scale forced vibration tests. *Earthq. Eng. Struct. Dyn.* **1988**, *16*, 1–21. [CrossRef]
45. Cheng, M.H.; Heaton, T.H. Simulating Building Motions Using Ratios of the Building's Natural Frequencies and a Timoshenko Beam Model. *Earthq. Spectra* **2015**, *31*, 403–420. [CrossRef]
46. Balendra, T.; Tat, C.W.; Lee, S.-L. Modal damping for torsionally coupled buildings on elastic foundation. *Earthq. Eng. Struct. Dyn.* **1982**, *10*, 735–756. [CrossRef]
47. Freddi, F.; Galasso, C.; Cremen, G.; Dall'Asta, A.; Di Sarno, L.; Giaralis, A.; Gutiérrez-Urzúa, F.; Málaga-Chuquitaype, C.; Mitoulis, S.A.; Petrone, C.; et al. Innovations in earthquake risk reduction for resilience: Recent advances and challenges. *Int. J. Disaster Risk Reduct.* **2021**, *60*, 102267. [CrossRef]

48. Cruz, C.; Miranda, E. Evaluation of soil-structure interaction effects on the damping ratios of buildings subjected to earthquakes. *Soil Dyn. Earthq. Eng.* **2017**, *100*, 183–195. [CrossRef]
49. Ghahari, S.F.; Sargsyan, K.; Çelebi, M.; Taciroglu, E. Quantifying modeling uncertainty in simplified beam models for building response prediction. *Struct. Control Health Monit.* **2022**, *29*, e3078. [CrossRef]

buildings

MDPI

Article

Automated Detection for Concrete Surface Cracks Based on Deeplabv3+ BDF

Yonggang Shen [1,2], Zhenwei Yu [1,3], Chunsheng Li [4], Chao Zhao [4] and Zhilin Sun [1,*]

1 College of Civil Engineering and Architecture, Zhejiang University, Hangzhou 310058, China
2 Center for Balance Architecture, Zhejiang University, Hangzhou 310028, China
3 The Architectural Design & Research Institute of Zhejiang University Co., Ltd., Hangzhou 310027, China
4 Zhejiang Communications Construction Group Co., Ltd., Hangzhou 310051, China
* Correspondence: oceansun@zju.edu.cn; Tel.: +86-10-13606705667

Abstract: Concrete cracks have always been the focus of research because of the serious damage they cause to structures. With the updating of hardware and algorithms, the detection of concrete structure surface cracks based on computer vision has received extensive attention. This paper proposes an improved algorithm based on the open-source model Deeplabv3+ and names it Deeplabv3+ BDF according to the optimization strategy used. Deeplabv3+ BDF first replaces the original backbone Xception with MobileNetv2 and further replaces all standard convolutions with depthwise separable convolutions (DSC) to achieve a light weight. The feature map of a shallow convolution layer is additionally fused to improve the detail segmentation effect. A new strategy is proposed, which is different from the two-stage training. The model training is carried out in the order of transfer learning, coarse-annotation training and fine-annotation training. The comparative test results show that Deeplabv3+ BDF showed good performance in the validation set and achieved the highest mIoU and detection efficiency, reaching real-time and accurate detection.

Keywords: damage detection; non-destructive evaluation; deep learning; concrete structure; crack segmentation

1. Introduction

In recent years, many concrete infrastructures have suffered from structural degradation due to long-term, high-load operation, or are close to the end of their natural service life, resulting in safety problems. Therefore, it is necessary to regularly inspect the health status of infrastructure, and the identification and evaluation of structural surface cracks are the tasks that managers and researchers are focusing on. However, traditional manual crack detection methods are inefficient and subjective. How to develop and promote more effective and reliable detection methods is the current research direction.

In view of the limitations of artificial crack detection, Yeum et al. [1] have carried out a lot of research on intelligent crack identification methods in the last decade. Initially, image processing techniques (IPTs) were used to carry out detection tasks, but this method requires additional pre-processing and post-processing technologies, thus, reducing the degree of intelligence. Deep learning algorithms [2], which can automatically extract the sensitive features of the target in the training process, were subsequently proposed and widely studied. Among them, the most representative algorithms are of two types: objection detection [3] and semantic segmentation [4]. Objection detection gives the category and position of the target in an image in the form of a rectangular box, and some models with excellent performance have been proposed, such as the YOLO series and SSD [5] one-stage models and Fast R-CNN [6] and Faster R-CNN [7] two-stage models. Improved models have also been put forward according to specific task requirements. Park et al. [8] proposed a structural crack detection and quantification method in which YOLOv3-tiny is used to locate concrete cracks in real time. Zhao et al. [9] proposed a crack feature pyramid network (Crack-FPN), which has superior feature extraction capability and reduced computational cost. Some research or reviews on the application of objection detection algorithms to crack detection have also been carried out [10–16]. However, due to the simplicity of the result form, the target detection algorithm is only applicable to simple target existence determination.

Crack images and datasets are highly class imbalanced, and cracks usually have complex textures. According to these characteristics, the research has further developed from objection detection to semantic segmentation algorithm, that is, all pixels belonging to the same type of target are represented by a monochrome mask, and the picture is simplified into a combination of multiple different color masks. Some research has also been carried out on the application of the semantic segmentation algorithm in crack detection [17–22]. Xiang et al. [23] proposed a crack segmentation method based on super-resolution reconstruction, which achieved a more than 10% performance improvement compared with previous models but could not meet the real-time requirements. Ren et al. [24] proposed a new end-to-end crack segmentation method based on a fully convolutional network which uses dilated convolution, spatial pyramid pooling, skip connection and an optimization loss function to obtain higher efficiency and accuracy. However, researchers have ignored or avoided some aspects of research, such as: (1) The computation amount required by the semantic segmentation algorithm is very large, and it even takes a few seconds to detect an image in the early stage. If the semantic segmentation is intended to be used in an actual scene, the data type is usually video with a frame rate of 60. When considering frame extraction or reducing the frame rate, real-time detection requires that the model processes images at a speed of 0.033 to 0.04 s per image, that is, 25 to 30 frames per second (FPS). At present, the detection efficiency of many models is difficult to achieve in actual projects. (2) Due to the limitations of manual labeling, the division between the cracks and the background boundary in the label is relatively vague, which makes the segmentation results given by the trained model show a large number of false positives and false negatives on the boundary [25]. (3) Mei et al.'s study [26] and many other studies deployed a transfer learning [27] strategy in a model, using initial weights trained on a large dataset containing many categories. These datasets have many objects of different classes from cracks (for example, ImageNet [28] has more than 5000 classes), and the extracted features are not highly related to cracks. Even if the models are continuously trained with carefully prepared datasets after transfer learning, it usually takes a lot of time to complete the production of segmentation labels. It seems to be a method to automatically label targets with computers, but the marking model still requires an initial dataset to complete training before it can be put into use.

In this paper, a new pixel-level semantic segmentation model for crack detection based on Deeplabv3+ [29] is proposed to solve the above problems and is named Deeplabv3+ BDF. This model can overcome the interference of background and crack-like features, extract the crack boundary quickly and accurately and, thereby, prepare for the intelligent detection of fine indicators such as crack width across complex background, so that the management and maintenance department can concentrate resources to study cracks and ignore the background or other objects. In addition, this paper also attempts to use a new training strategy to reduce the common labeling cost problem of semantic segmentation models, which provides a potential solution for researchers with a large amount of data but not enough resources to fine-label all data. The main contributions of this paper are as follows:

(1) A lightweight network MobileNetv2 is used as the backbone, and all standard convolutions are replaced by DSC to reduce the number of parameters and realize real-time detection;

(2) On the basis of the characteristics of semantic hierarchy and cracks, during the up-sampling process, the shallow feature map after one down-sampling is fused to improve the segmentation accuracy at the boundary between the foreground and background;

(3) Focusing on the problem of the labeling cost of semantic segmentation model being too high, a three-step training strategy according to the sequence of transfer learning, coarse-annotation (CA) training and fine-annotation (FA) training is designed and proposed, which can enhance the learning and extraction of crack features. This training strategy can train a better segmentation model with a large number of CA images on the premise of only a few FA images, saving a lot of human and material resources.

2. Models and Methodology

2.1. Deeplabv3+

The Deeplab series was developed on the basis of FCN [30]. Its main feature is to expand the receptive field by using atrous spatial pyramid pooling (ASPP) to obtain more image feature information. Deeplabv3+ achieves 87.8 mIoU on the PASCAL VOC-2012 dataset, and its image segmentation effect is superior to other Deeplab series models. Compared with Deeplabv3, the main feature of Deeplabv3+ is that it adds a decoder module with transposed convolution as the main unit, which can gradually restore high-dimensional feature vectors to the feature map of the same size as the input image. Figure 1 shows the network diagram of Deeplabv3+.

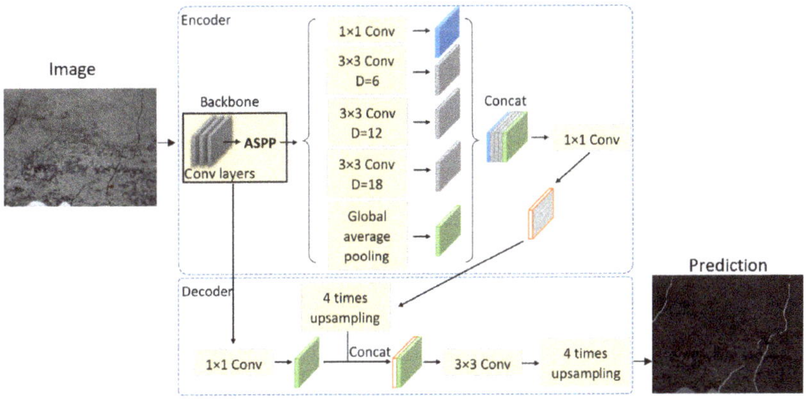

Figure 1. The structure of Deeplabv3+.

The encoder consists of backbone network Xception [31] and ASPP. Xception extracts two feature maps of high semantic information and low semantic information at the same time. The former usually represents an abstract concept and is the information expressed by the image closest to human understanding, while the latter is the color, texture and shape. The high semantic information feature map conducts multi-scale, dilated convolution sampling in the ASPP module, generates and fuses multiple feature maps of different scales and, finally, uses 1×1 convolution for dimension reduction. Low semantic information is transferred into the decoder part for 1×1 convolution and is fused with the high semantic information feature map after bilinear up-sampling four times to enhance the network learning effect and improve the segmentation accuracy. Then, the feature is extracted through 3×3 convolution, and the final semantic segmentation map is obtained by up-sampling four times.

2.2. Deeplabv3+ BDF and Optimization Strategies

According to some defects in the current research described in Section 1, including defects relating to detection efficiency, boundary ambiguity and initial weight mismatch, we propose a method to improve the model accordingly and rename the model according to the optimization strategy used, namely Deeplabv3+ BDF.

2.2.1. Backbone

MobileNetv2 is used to replace Xception as the backbone. MobileNetv2 is the same as MobileNetv1, which is a lightweight CNN, and it uses DSC. By adjusting the number of channels in each convolution layer, MobileNetv2 does not affect the performance but can reduce the amount of computation. Taking it as a backbone can effectively improve the detection speed and decrease the occupation, making the model oriented to real-time detection. For more details, please refer to [32].

2.2.2. DSC

All standard convolutions other than the backbone are replaced by DSC [33], including standard convolutions in the decoder to accelerate the detection. DSC can be divided into depthwise convolution (DWC) and pointwise convolution (PWC) [34]. The comparison of these convolution operations is shown in Figure 2. Take the convolution operation in Figure 2 as an example. There are four filters in the standard convolution, and each filter has three convolution kernels, which correspond to three channels of the image. After convolution, the feature maps with the same number of filters is obtained, and the parameter quantity is $4 \times 3 \times 3 \times 3 = 108$. There is only one convolution kernel in each filter of DWC, which is responsible for one channel, respectively. The number of channels before and after convolution remains unchanged, and the parameter quantity is $3 \times 3 \times 3 = 27$. The convolution kernel size of PWC is 1×1, and its function is to generate a new feature map by weighted combination of the output feature maps of the upper layer. It is a special case of standard convolution when convolution kernel size is 1×1 and the parameter quantity is $1 \times 1 \times 3 \times 4 = 12$. After DWC and PWC, a four-channel output can also be obtained, which is the same as in standard convolution. Moreover, compared with the standard convolution, the parameter quantity of DSC is $27 + 12 = 39$, which is only 36.1% of the standard convolution, and the calculation cost is significantly reduced.

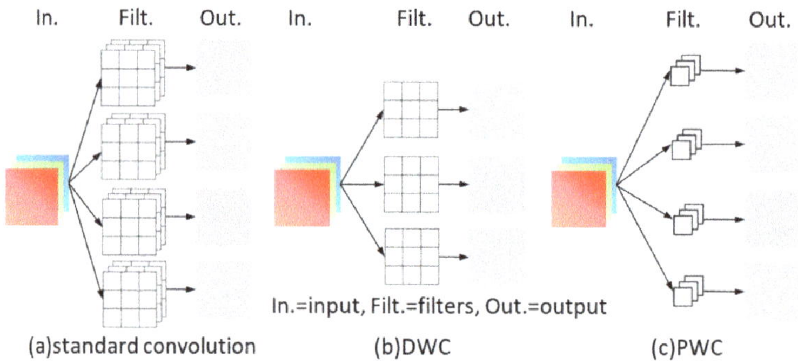

Figure 2. Comparison of convolution operations.

2.2.3. Feature Fusion

The feature map after one down-sampling is fused additionally, as shown in Figure 3. Convolution features are hierarchical and choosing different layers may achieve completely different results. Shallow features of CNN focus on detail features, such as edges and corners, which are usually associated with individual segmentation results. Middle features are a part of the object, and deep features focus on deeper semantic information. Intuitively, deep feature maps can represent a complete object, which is usually related to the accuracy of classification results. Only when the receptive field size of the feature map is larger than the object can the correct detection be carried out. Correspondingly, shallow features can only cover small objects, while deep features can cope with larger objects.

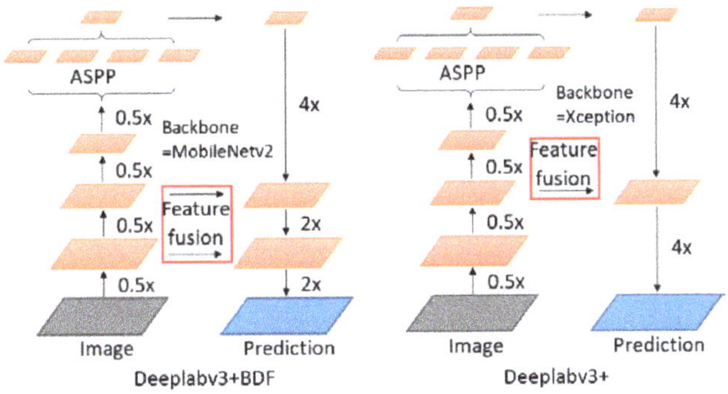

Figure 3. Comparison of network structure changes.

The boundary of the crack belongs to edge or corner features, so the feature map after one down-sampling is added to the up-sampling process to enhance the ability of the model to deal with the demarcation. For the feature map after three down-samplings, because the crack itself is a tiny object, even if it runs through the whole image, the number of pixels belonging to the crack is very small. The cracks contained in the receptive field corresponding to the feature map after two down-samplings can meet the training requirements, because a part of a long crack can still be regarded as a crack with complete features. However, it is also unreasonable to only fuse the feature map after one down-sampling, because it needs a larger field of vision to judge whether it is a crack or a crack-like object. Only focusing on the edges or corners of the object may mistake some black, slender objects for cracks, thus, we retain the strategy of fusing the feature map after two down-samplings in Deeplabv3+.

3. Establishment of Dataset

The biggest difference between the actual project and the laboratory scene is the environment around the crack. Generally, images obtained in the field experience interference due to handwriting, template lines and other crack-like objects, while the crack images obtained in the laboratory have a monotonous background and no sundries; so, the trained model is difficult to extend to practical applications. Therefore, we take 82 images from multiple scenes. The illumination conditions, exposure intensity and acquisition equipment of these images are different, so the dataset has enough diversity. Because the dataset is collected by mobile phones or high-definition cameras, the image capture distance varies in a large range (0.2 m to 5 m), and the image scene also has enough complexity; the trained model has good recognition effect on common crack images. However, the model has the potential to improve the recognition effect of fine cracks, especially for cracks where the width is less than 1 pixel, which are very easy to be missed in detection. This is also the difficulty of the semantic segmentation model used for crack detection at present. Examples of the dataset are shown in Figure 4. Four images at 3840 × 2880 pixels are from a composite plate failure experiment, eight images at 1920 × 1080 pixels are from another composite plate experiment, four images at 4608 × 3456 pixels are from a bridge, seven images at 4608 × 3456 pixels are from some cracked walls or structures in Zhejiang University, five images at 1920 × 1080 pixels are from a concrete beam bending experiment and five images at 1920 × 1080 pixels are from a concrete column bending experiment. These images are manually labeled at pixel level using the Labelme program. Another 49 images at 1920 × 1080 pixels are obtained from a concrete beam bending test, and CAs are made to enable the model to be pretrained. Although transfer learning is an effective strategy, its initial weights are usually trained by multi-class objects, and there are a lot of irrelevant or weak correlation features. After transfer learning, more pretraining for cracks can weaken these irrelevant or weak correlation features and strengthen the recognition and extraction of crack features.

Figure 4. Examples of the dataset.

The comparison of CA and FA is shown in Figure 5. CAs use more long lines and obvious angles to represent the irregular contour of the crack. The form shown in the figure shows that the mosaic area is wider, the jagged boundary of the crack is ignored and the final label can be approximately regarded as a polygon, which greatly reduces the time for CA. FAs make the marking points fit the crack as much as possible so that the width of the mosaic and the outline of the label basically match the crack. Therefore, obvious bending does not appear in the figure, and the overall appearance of FA is smoother. It takes about 6 min for an image to be coarsely labeled and 15 min to be finely labeled. This method of training based on two types of labels does not require fine labeling of all images and can reduce labeling costs.

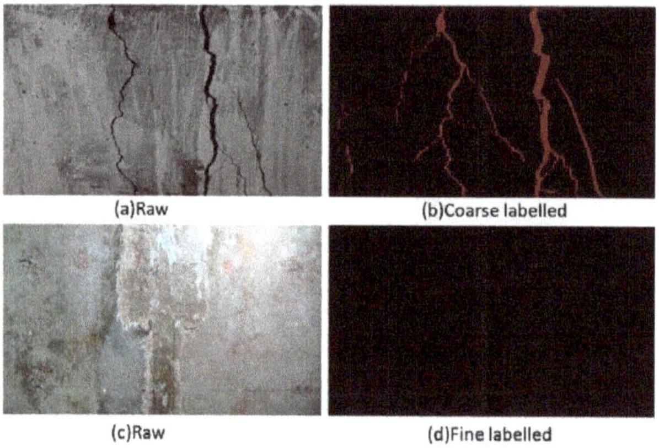

Figure 5. Comparison of CA and FA.

Firstly, all images are divided into sub-images at 576 $\times$ 576 pixels; 588 coarse-labeled images and 944 fine-labeled images are obtained. Then, images without cracks are removed, and 550 coarse-labeled images and 676 fine-labeled images are obtained. Due to the small amount of data, in order to make the test results more reliable, we refer to a data configuration strategy similar to K-fold verification in [35] so as to avoid over-fitting. The dataset is divided into five subsets for cross-training and validation. The division of sub-datasets is shown in Table 1, and the dataset not used in each training process is used as the code of this training process; so, each sub-dataset includes 540/541 images (including only FA images) or 980/981 images. CA images are not used for testing.

Table 1. Division of sub-datasets.

Sub-Dataset No. or Training Process	CAs	FAs
Sd1	110	135
Sd2	110	135
Sd3	110	135
Sd4	110	135
Sd5	110	136
Total	550	676

Due to the small amount of data, the model proposed in this paper should be limited to detecting concrete cracks, and the prediction effect of surface cracks in extreme environments (such as earthquake) or other materials needs further research.

4. Model Training and Results

The contents of this chapter include the details and experimental results of training Deeplabv3+ BDF. The optimization of the model is implemented using Python and the open-source framework Deeplabv3+. The computing workstation is configured with four 1080 Ti GPUs. In the following paper, transfer learning is referred to as process T, training on CA images is referred to as process C and training on FA images is referred to as process F.

4.1. Training Strategy and Experimental Results

Training is divided into four types: F, C + F, T + F and T + C + F, i.e., three-step training. Process T does not need to be specific, and using the Cityscapes initial weights already available in the Deeplabv3+ model package can be considered an alternative to this process. Deeplabv3+ BDF is trained on CA images firstly and then on FA images as the model converges. Experiments show that process C converges after 15,000 epochs, and the weight of any subsequent epoch can be used as the initial weight of process F. In this paper, process C is iterated with 20,000 epochs. The image input size is set to 577×577 resolution, the loss function is binary cross-entropy loss, the initial learning rate is 0.0001, the learning rate attenuation coefficient is 0.1, the number of attenuation steps is 2000, the batch size is 32, the dropout rate is set to 0.5 and the total number of epochs is 50,000. That is, when process C with 20,000 epochs exists, process F continues to iterate 30,000 epochs, and the loss is recorded every 10 epochs.

In order to present the figures clearly, a simple moving average (SMA) of every 500 steps is used to describe the loss curve, as shown in Figure 6. The SMA is calculated according to Equation (1):

$$\text{SMA} = \begin{cases} (L_i + L_{i-1} + L_{i-2} + \cdots + L_1)/i, & if\ i \in [1,500) \\ (L_i + L_{i-1} + L_{i-2} + \cdots + L_{i-499})/500, & if\ i \in [500,\ 50000] \end{cases} \quad (1)$$

where i is the number of iterations, and L_i is the loss value of the ith iteration. Since the number of iterations is less than 500, SMA calculation needs to follow another variant form in the first 499 iterations, while SMA is normally calculated after 500 iterations. After two pretraining sessions of process T and process C, the three-step training model still shows the potential to be optimized in process F and further decreases in the loss value and finally converges to 0.25, which is the lowest of the four training strategies. Even for process C + F without process T this phenomenon also appears, that is, after the process C training weight excellence, it still has room to be improved, and its loss converges to 0.39.

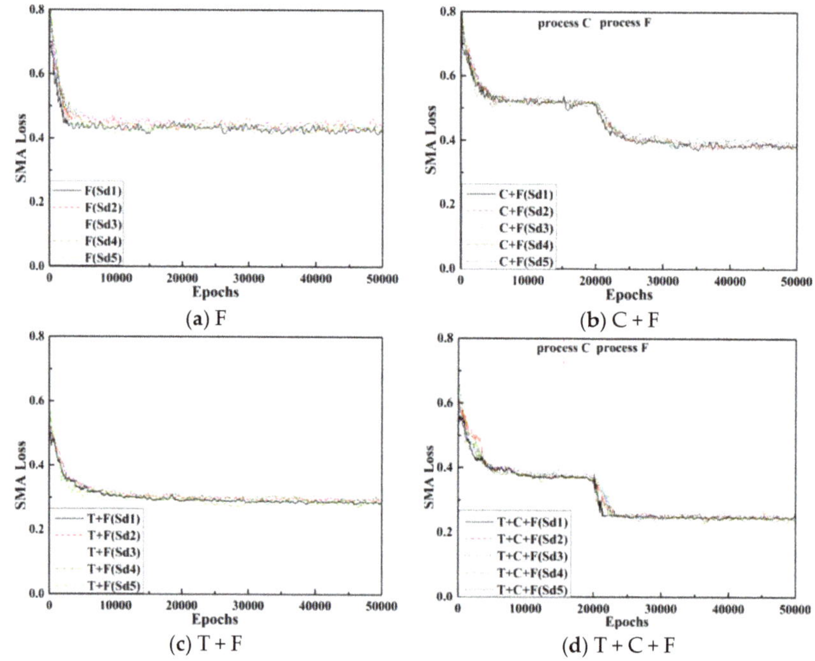

Figure 6. SMA loss curve.

Process T + F and process F, which are two conventionally used training strategies with convergence values of about 0.30 and 0.42, respectively, without process C can also converge but are higher than T + C + F. In the two strategies without process T, the SMA curve fluctuates in different degrees. Note that this is the average curve, and its fluctuation is somewhat mitigated, but it still has disadvantages compared with the corresponding curves of the two strategies with process T. Therefore, process T is necessary.

The effectiveness of our proposed training strategy can also obtain the same conclusion from the validation indices. The precision (P), recall (R), F1-score and the most commonly used index in semantic segmentation task, mIoU, are used for evaluation, and they are calculated according to Equations (2)–(5):

$$P = TP/(TP + FP), \tag{2}$$

$$R = TP/(TP + FN), \tag{3}$$

$$\text{F1-score} = 2 \times P \times R/(P + R), \tag{4}$$

$$\text{mIoU} = (1/k) \times TP/(FN + FP + TP), \tag{5}$$

where TP, FP and FN represent true positive, false positive and false negative, respectively, and K is the number of target categories in all images. Figure 7 shows that, when the training strategy is T + C + F and the dataset is Sd2, the change of mIoU curve is opposite to the loss curve, but the change trend is the same, that is, when process F is carried out after the convergence of process C, the curve has a certain mutation. The mIoU curves of the four training types on the Sd1 subset are shown in Figure 8. From the comparison between F and C + F, and the comparison between T + F and T + C + F, it can be seen that, although the curves of C + F and T + C + F are lower at the initial stage, the inversion is achieved after 20,000 iterations, which indicates that the models with CA and FA training have significant optimization in terms of mIoU.

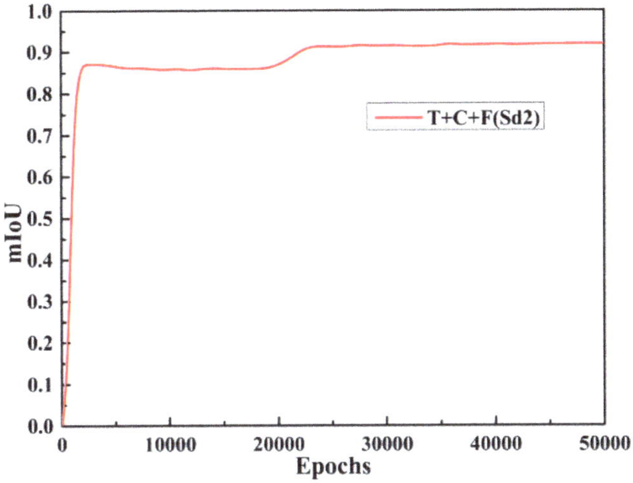

Figure 7. Example of mIoU curve.

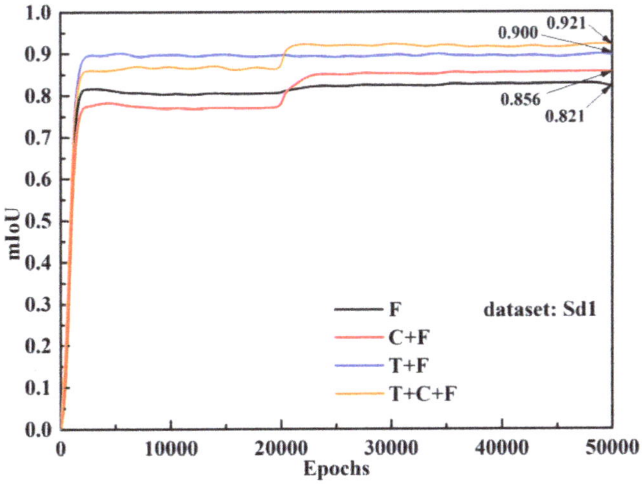

Figure 8. mIoU curves of four training types.

The performance of the four training strategies on the validation set is shown in Table 2, which shows the indices of different training strategies on different sub-datasets. The results of process T + C + F is the highest in the four indices. From the comparison of T + F and F, T + C + F and C + F, it is revealed that transfer learning is still a very effective strategy when dealing with small datasets. It can provide better initial performance, optimization rate and convergence for the model. The effectiveness of establishing a training strategy for a specific task can be proved by the comparison of T + F and T + C + F and the comparison of F and C + F. Deeplabv3+ BDF that is pretrained with CA data for a specific task and then trained normally is better at indicators than the model that is trained directly on FA data. The conventional training strategy, i.e., process T + F, is 0.019 lower than the secondary pretraining method proposed in this paper in terms of mIoU.

Table 2. Indicators obtained by different training strategies.

Training Strategy	Training Process	P	R	F1	mIoU
F	Sd1	0.837	0.841	0.839	0.821
	Sd2	0.843	0.838	0.840	0.824
	Sd3	0.831	0.828	0.829	0.816
	Sd4	0.834	0.829	0.831	0.818
	Sd5	0.839	0.835	0.837	0.820
	Average	0.837	0.834	0.835	0.820
C + F	Sd1	0.871	0.874	0.872	0.856
	Sd2	0.863	0.859	0.861	0.841
	Sd3	0.854	0.865	0.859	0.838
	Sd4	0.861	0.868	0.864	0.845
	Sd5	0.859	0.851	0.855	0.843
	Average	0.862	0.863	0.862	0.845
T + F	Sd1	0.883	0.880	0.881	0.900
	Sd2	0.875	0.874	0.875	0.893
	Sd3	0.881	0.877	0.879	0.902
	Sd4	0.879	0.877	0.878	0.899
	Sd5	0.874	0.879	0.876	0.895
	Average	0.881	0.877	0.879	0.898
T + C + F	Sd1	0.892	0.903	0.897	0.921
	Sd2	0.889	0.897	0.893	0.917
	Sd3	0.883	0.887	0.885	0.912
	Sd4	0.887	0.882	0.884	0.914
	Sd5	0.892	0.896	0.894	0.919
	Average	0.889	0.893	0.891	0.917

Some experiments are carried out on the proportion of data required by process C and process F, and the final mIoU is taken as the evaluation index. A total of 550 CA images are divided into five sub-datasets. Trial training is conducted according to the number of sub-datasets from 1 to 5, and the mIoU of each experiment is recorded, which can be seen in Figure 9. The results show that the difference of mIoU is within ± 0.03 after using three or more sub-datasets, i.e., 330 CA images, which can be regarded as the fully developed optimization potential. However, this result is only for the task of this paper. In an actual project, the number ratio of CA and FA images should be determined according to the complexity of the task, the characteristics of the object and other factors. Section 3 describes the time spent in annotation, and an FA image is 2.5 times a CA image. If 330 or more CA images are labeled according to the FA image standard, it takes more time to achieve the same result, and this problem can be solved using the three-step training strategy proposed in this paper.

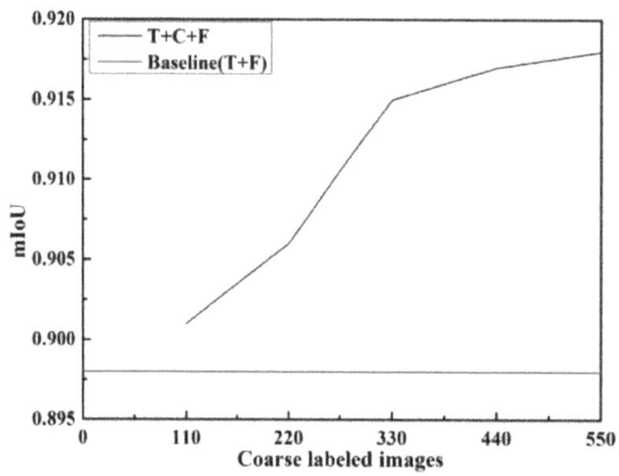

Figure 9. The change of mIoU with the number of CA images.

4.2. Comparison Results

The excellent performance of Deeplabv3+ BDF can also be shown in the comparison experiment, and we compare Deeplabv3+ BDF with a variety of representative semantic segmentation models, including Deeplabv3+, U-Net [36], PSP-Net [37], DeepCrack [38] and DeepCrack-Aug [38], among which DeepCrack is a model specially designed for crack detection. Table 3 shows the results of the comparative experiment. Due to the use of a lighter backbone and the replacement of all standard convolutions with DSCs, the number of parameters in Deeplabv3+ BDF is greatly reduced, the physical occupation and running video memory occupation are reduced and the detection speed is greatly improved. In addition, even Deeplabv3+ BDF without process C intensive training is superior to other models in various evaluation indicators, while Deeplabv3+ BDF with process C training further expands its advantages. The mIoU of Deeplabv3+ BDF (T + C + F) takes the lead over Deeplabv3+ with better performance at 0.102, and the detection speed reaches 26.132FPS, 2.9 times faster than U-Net. The comparison experiment proves that our optimization measures are effective. Deeplabv3+ BDF has both accuracy and speed and can meet real-time detection requirements (generally 20–25 FPS). If Deepalabv3+ BDF (T + C + F), Deeplabv3+ and DeepCrack-Aug are used simultaneously to detect 100 images at 576 × 576 pixels, which takes 3.83 s, 26.56 s and 11.43 s, respectively, our model can complete the same task with high accuracy and speed, and this advantage is more obvious when the number of images is larger.

Table 3. Comparison results using the dataset of this paper.

Model	Deeplabv3+	U-Net	PSP-Net	DeepCrack	DeepCrack-Aug	Deeplabv3+ BDF (T + F)	Deeplabv3+ BDF (T + C + F)
P	0.821	0.813	0.801	0.467	0.512	0.880	0.888
R	0.783	0.779	0.724	0.532	0.502	0.877	0.897
F1	0.802	0.796	0.761	0.497	0.507	0.878	0.892
mIoU	0.815	0.782	0.746	0.612	0.538	0.898	0.917
FPS	3.765	8.934	7.824	2.784	8.752	25.783	26.132

The results of crack semantic segmentation detection can be used to measure the width and length of cracks, and its application in the field of structural damage detection is not only as demonstrated here, but is also for corrosion detection and calculation and statistics of corrosion area, which will be carried out in our future work.

4.3. Typical Inference Results

The proposed three-step training method is proved to be effective. In this section, we use representative inference results to show that Deeplabv3+ BDF with additional feature map fusion has better performance. Another training session is conducted using the same strategy and dataset Sd1, but the training object is changed to Deeplabv3+ BDF, which only fuses the feature map twice, and it is named Deeplabv3+ BDF-single for differentiation. These two models are inferred from the same test set, three representative images are selected and the mIoU values after detection are attached. The pixels in the segmentation result can be divided into four categories, namely, TP in red, FN in blue, FP in green and TN representing the background, as shown in Figure 10.

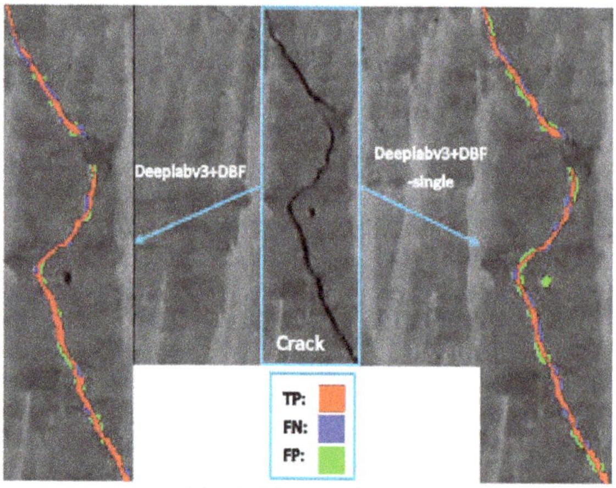

(**a**) mIoU = 0.908/0.831

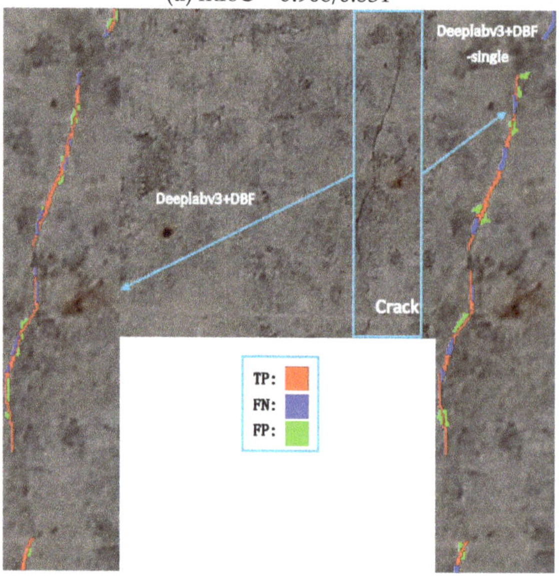

(**b**) mIoU = 0.823/0.672

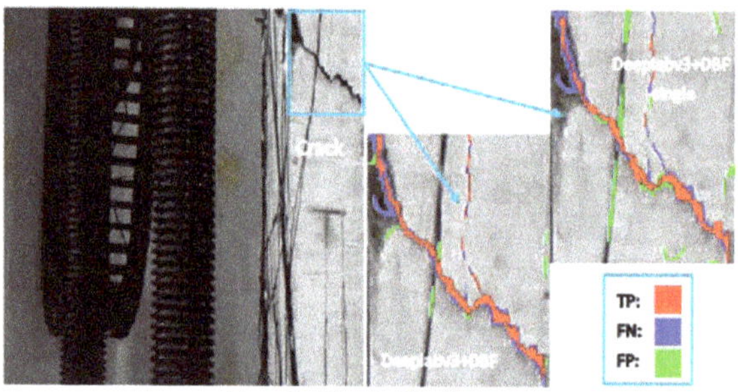

(**c**) mIoU = 0.876/0.752

Figure 10. Typical segmentation results.

Deeplabv3+ BDF gives more accurate segmentation results of all fractures and obtains significant advantages over the comparison model in mIoU. Figure 10a is a crack image containing a small cavity with simple background. Deeplabv3+ BDF-single incorrectly identifies the cavity as a crack, but Deeplabv3+ BDF does not. The segmentation results show that the mIoU of Deeplabv3+ BDF-single is only 0.831 even in simple background, and more FP pixels appear on the dark concrete background at the crack edge. Figure 10b contains a fuzzy fine crack. Fine cracks are the key and difficult task of crack detection. However, due to the accuracy of human eyes, the edges of cracks cannot be accurately marked, which makes it difficult to completely segment the fine cracks. Deeplabv3+ BDF has less truncation on the whole fine crack and less error expansion at the crack edge, while Deeplabv3+ BDF-single detects a complete fine crack as nearly 10 cracks, and the mIoU also reflects the performance difference from the data level. Figure 10c contains a spalling concrete surface, and the crack that passes through it can be confusing. Deeplabv3+ BDF overcomes the problem of crack area expansion caused by spalling to a certain extent, but there are a lot of FN results in the spalling area, which still has the potential for improvement. Deeplabv3+ BDF-single has a similar problem in dealing with the problem of spalling concrete. The whole crack is segmented more thinly, and the detection of the next thin cracks still fails, and the instrument circuit is classified as crack, which is unacceptable. It can be seen from Figure 10 that most of the FP and FN values of the crack segmentation task appear at crack boundary, which is the reason why we choose to fuse the shallow feature map, because it corresponds to the edges and corners of crack. Although the judgment of crack boundary is subjective due to the different degrees of image blur, at an overall level, however, the additional feature map fusion strategy reduces the proportion of FP and FN and has advantages in filtering various complex backgrounds or processing cracks with different widths.

Figure 11 shows the segmentation results of our model, Deeplabv3+, U-Net and DeepCrack-Aug. Figure 11a shows a crack with an average width of about 3.7 pixels. Our model has almost the same result as the groundtruth. The worst one is DeepCrack-Aug. As mentioned earlier, the dataset used by DeepCrack is relatively simple, and it is easy to make mistakes in some common images with general complexity. Figure 11b shows a crack with an average width of only 1.6 pixels. In this image, all models have made many errors, but, from a comprehensive perspective, our model still achieves the best results. The other three models have different degrees of truncation and error expansion for this fine crack. It can also be inferred from Figure 11b that our model can identify cracks only 1 pixel wide and separate them from the background. It is proved that our model is superior to the recent work in both data and segmentation visualization.

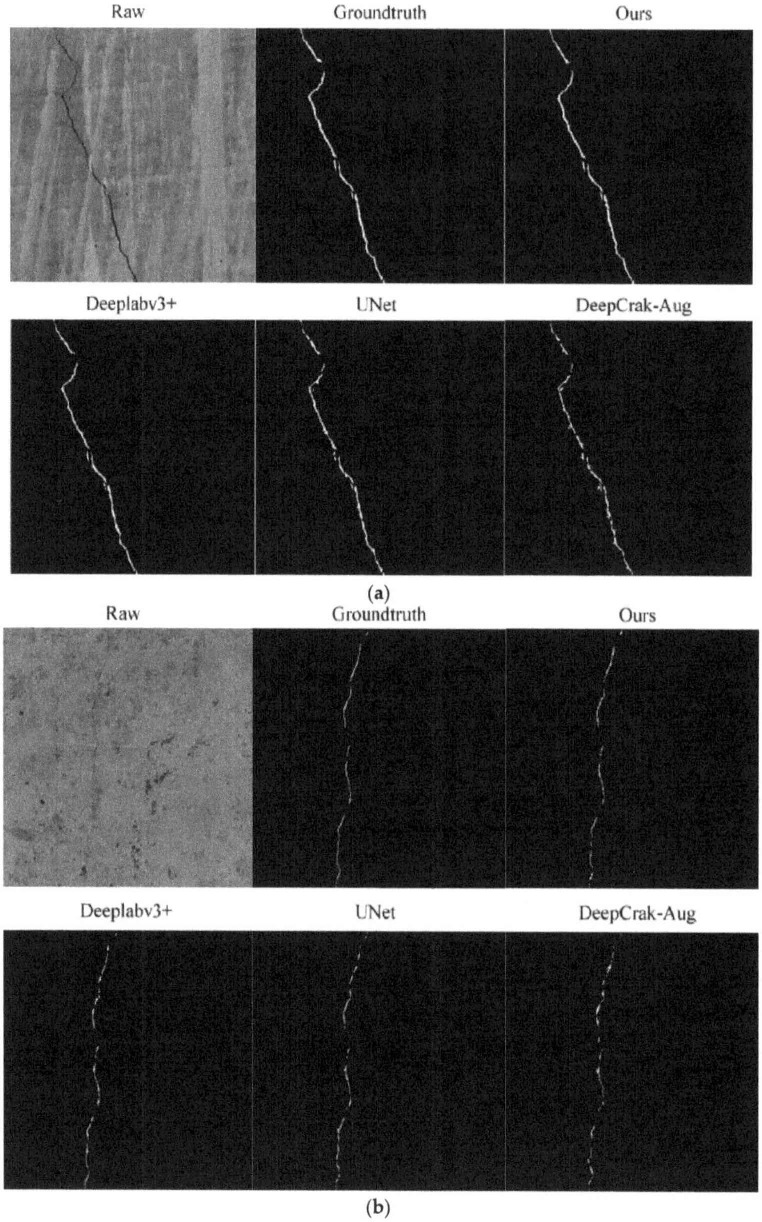

Figure 11. Segmentation results comparison of multiple models. Subfigure (**a**,**b**) are two randomly selected crack images.

5. Conclusions

In this study, we noticed that the current research has some defects or ignores some problems, so we proposed a semantic segmentation model improving on Deeplabv3+ and named the model Deeplabv3+ BDF according to optimization strategies. The identification of cracks is the most critical task in structural health detection, so we think it is necessary and beneficial to propose a deep learning network dedicated to crack detection according to the unique characteristics of cracks. We adjusted the training strategy of Deeplabv3+ BDF and optimized the network structure and established a dataset including CA images and FA images. After the training, we evaluated the performance of the proposed Deeplabv3+ BDF by comparing it with other models. The results show that Deeplabv3+ BDF solves the three problems mentioned in the paper well, especially realizing real-time detection. The conclusions are as follows:

(1) The network structure of Deeplabv3+ BDF is made lightweight by using MobileNetv2 as the backbone network, so that its FPS of 576 × 576 pixels image is 26.132, which meets the real-time requirements and is 2.9 times faster than in recent works;

(2) Because of the additional fusion of shallow feature map, Deeplabv3+ BDF can reduce the number of FN and FP values in detection results and improve the processing ability of the boundary between foreground and background under the same conditions;

(3) After the second pretraining, that is, the proposed three-step training strategy, the potential of Deeplabv3+ BDF is further developed. Compared with the conventional training strategy, the mIoU of Deeplabv3+ BDF is increased to 0.917, which is at least 0.102 ahead of other models.

Author Contributions: Conceptualization, Y.S., Z.Y. and Z.S.; methodology, Y.S. and Z.Y.; validation, Y.S. and Z.Y.; investigation, Z.Y., C.L. and C.Z.; resources, Y.S. and Z.S.; data curation, Z.Y.; writing—original draft preparation, Y.S. and Z.Y.; visualization, Z.Y.; supervision, Y.S. and Z.S.; project administration, C.L. and C.Z. All authors have read and agreed to the published version of the manuscript.

Funding: This research was funded by ZJU-ZCCC Institute of Collaborative Innovation (No. ZDJG2021009).

Institutional Review Board Statement: Not applicable.

Informed Consent Statement: Not applicable.

Data Availability Statement: Not applicable.

Acknowledgments: We acknowledge Tutor Huang Yifang at the Training Platform of Construction Engineering at the Polytechnic Institute of Zhejiang University.

Conflicts of Interest: The authors declare no conflict of interest.

References

1. Yeum, C.M.; Dyke, S.J. Vision-based automated crack detection for bridge inspection. *Comput. Aided Civ. Infrastruct. Eng.* **2015**, *30*, 759–770. [CrossRef]
2. Lecun, Y.; Bengio, Y.; Hinton, G. Deep learning. *Nature* **2015**, *521*, 436–444. [CrossRef]
3. Redmon, J.; Divvala, S.; Girshick, R.; Farhadi, A. You only look once: Unified, real-time object detection. In Proceedings of the IEEE Conference on Computer Vision and Pattern Recognition, Las Vegas, NV, USA, 27–30 June 2016. [CrossRef]
4. Tabernik, D.; Šela, S.; Skvarč, J.; Skočaj, D. Segmentation-based deep-learning approach for surface-defect detection. *J. Intell. Manuf.* **2020**, *31*, 759–776. [CrossRef]
5. Liu, W.; Anguelov, D.; Erhan, D.; Szegeedy, C.; Reed, S.; Fu, C.Y.; Berg, A.C. SSD: Single shot multibox detector. In Proceedings of the European Conference on Computer Vision, Amsterdam, The Netherlands, 11–14 October 2016. [CrossRef]
6. Girshick, R.B. Fast R-CNN. In Proceedings of the IEEE International Conference on Computer Vision, Santiago, Chile, 13–16 December 2015. [CrossRef]
7. Ren, S.Q.; He, K.M.; Girshick, R.B.; Sun, J. Faster R-CNN: Towards real-time object detection with region proposal networks. In Proceedings of the Conference and Workshop on Neural Information Processing Systems, Montreal, QC, Canada, 7–12 December 2015. [CrossRef]
8. Park, S.E.; Eem, S.H.; Jeon, H. Concrete crack detection and quantification using deep learning and structured light. *Constr. Build. Mater.* **2015**, *252*, 119096. [CrossRef]
9. Zhao, W.J.; Liu, Y.Y.; Zhang, J.W.; Shao, Y.; Shu, J.P. Automatic pixel-level crack detection and evaluation of concrete structures using deep learning. *Struct. Control. Health Monit.* **2022**, *29*, e2981. [CrossRef]
10. Yu, Z.W.; Shen, Y.G.; Shen, C.K. A real-time detection approach for bridge cracks based on YOLOv4-FPM. *Autom. Constr.* **2021**, *122*, 103514. [CrossRef]
11. Cha, Y.J.; Choi, W.; Büyüköztürk, O. Deep learning-based crack damage detection using convolutional neural networks. *Comput. Aided Civ. Infrastruct. Eng.* **2017**, *32*, 361–378. [CrossRef]

12. Wu, C.S.; Zhang, J.Q.; Qi, L.L.; Zhuo, D.B. Defect identification of concrete piles based on numerical simulation and convolutional neural network. *Buildings* **2022**, *12*, 664. [CrossRef]
13. Nhat-Duc, H.; Nguyen, Q.L.; Tran, V.D. Automatic recognition of asphalt pavement cracks using metaheuristic optimized edge detection algorithms and convolution neural network. *Autom. Constr.* **2018**, *94*, 203–213. [CrossRef]
14. Ali, L.; Alnajjar, F.; Khan, W.; Serhani, M.A.; Jassmi, H.A. Bibliometric analysis and review of deep learning-based crack detection literature published between 2010 and 2022. *Buildings* **2022**, *12*, 432. [CrossRef]
15. Dung, C.V.; Sekiya, H.; Hirano, S.; Okatani, T.; Miki, C. A vision-based method for crack detection in gusset plate welded joints of steel bridges using deep convolutional neural networks. *Autom. Constr.* **2019**, *102*, 217–229. [CrossRef]
16. Zhang, L.; Yang, F.; Zhang, Y.D.; Zhu, Y.J. Road crack detection using deep convolutional neural network. In Proceedings of the IEEE International Conference on Image Processing, Phoenix, AZ, USA, 25–28 September 2016. [CrossRef]
17. Li, S.Y.; Zhao, X.F.; Zhou, G.Y. Automatic pixel-level multiple damage detection of concrete structure using fully convolutional network. *Comput. Aided Civ. Infrastruct. Eng.* **2019**, *34*, 616–634. [CrossRef]
18. Ji, A.K.; Xue, X.L.; Wang, Y.N.; Luo, X.W.; Xue, W. An integrated approach to automatic pixel-level crack detection and quantification of asphalt pavement. *Autom. Constr.* **2020**, *114*, 103176. [CrossRef]
19. Liu, Z.C.; Li, X.D.; Li, J.H.; Teng, S. A new approach to automatically calibrate and detect building cracks. *Buildings* **2020**, *12*, 1081. [CrossRef]
20. Su, H.F.; Wang, X.; Han, T.; Wang, Z.Y.; Zhao, Z.X.; Zhang, P.F. Research on a U-Net bridge crack identification and feature-calculation methods based on a CBAM attention mechanism. *Buildings* **2020**, *12*, 1561. [CrossRef]
21. König, J.; Jenkins, M.D.; Mannion, M.; Barrie, P.; Morison, G. Optimized deep encoder-decoder methods for crack segmentation. *Digit. Signal Prog.* **2021**, *108*, 102907. [CrossRef]
22. Adán, A.; Quintana, B.; Prieto, S.A.; Bosché, F. An autonomous robotic platform for automatic extraction of detailed semantic models of buildings. *Autom. Constr.* **2020**, *109*, 102963. [CrossRef]
23. Xiang, C.; Wang, W.; Deng, L.; Shi, P.; Kong, X. Crack detection algorithm for concrete structures based on super-resolution reconstruction and segmentation network. *Autom. Constr.* **2022**, *140*, 104346. [CrossRef]
24. Ren, Y.P.; Huang, J.S.; Hong, Z.Y.; Lu, W.; Yin, J.; Zou, L.J.; Shen, X.H. Image-based concrete crack detection in tunnels using deep fully convolutional networks. *Constr. Build. Mater.* **2020**, *234*, 117367. [CrossRef]
25. Choi, W.; Cha, Y.J. SDDNet: Real-time crack segmentation. *IEEE Trans. Ind. Electron.* **2019**, *67*, 8016–8025. [CrossRef]
26. Mei, Q.P.; Gül, M. A cost effective solution for pavement crack inspection using cameras and deep neural networks. *Constr. Build. Mater.* **2020**, *256*, 119397. [CrossRef]
27. Pan, S.J.; Yang, Q. A survey on transfer learning. *IEEE Trans. Knowl. Data Eng.* **2010**, *22*, 1345–1359. [CrossRef]
28. Deng, J.; Dong, W.; Socher, R.; Li, L.J.; Li, K.; Li, F.-F. ImageNet: A large-scale hierarchical image data-base. In Proceedings of the IEEE Conference on Computer Vision and Pattern Recognition, Miami, FL, USA, 20–25 June 2009. [CrossRef]
29. Chen, L.C.; Zhu, Y.K.; Papandreou, G.; Schroff, F.; Adam, H. Encoder-decoder with atrous separable convolution for semantic image segmentation. In Proceedings of the European Conference on Computer Vision, Munich, Germany, 8–14 September 2018. [CrossRef]
30. Long, J.; Shelhamer, E.; Darrell, T. Fully convolutional networks for semantic segmentation. In Proceedings of the IEEE Conference on Computer Vision and Pattern Recognition, Boston, MA, USA, 7–12 June 2015. [CrossRef]
31. Chollet, F. Xception: Deep learning with depthwise separable convolutions. In Proceedings of the IEEE Conference on Computer Vision and Pattern Recognition, Honolulu, HI, USA, 21–26 July 2017. [CrossRef]
32. Sandler, M.; Howard, A.; Zhu, M.L.; Zhmoginov, A.; Chen, L.C. MobileNetV2: Inverted residuals and linear bottlenecks. In Proceedings of the IEEE Conference on Computer Vision and Pattern Recognition, Salt Lake City, UT, USA, 18–23 June 2018. [CrossRef]
33. Kaiser, L.; Gomez, A.N.; Chollet, F. Depthwise separable convolutions for neural machine translation. In Proceedings of the International Conference of Learning Representation, Vancouver, BC, Canada, 30 April–3 May 2017.
34. Ma, L.F.; Li, Y.; Li, J.; Tan, W.K.; Yu, Y.T.; Chapman, M.A. Multi-scale point-wise convolutional neural networks for 3D object segmentation from LiDAR point clouds in large-scale environments. *IEEE Trans. Intell. Transp. Syst.* **2021**, *22*, 821–836. [CrossRef]
35. Dorafshan, S.; Azari, H. Evaluation of bridge decks with overlays using impact echo, a deep learning approach. *Autom. Constr.* **2020**, *113*, 103133. [CrossRef]
36. Ronneberger, O.; Fischer, P.; Brox, T. U-Net: Convolutional networks for biomedical image segmentation. In Proceedings of the Medical Image Computing and Computer-Assisted Intervention, Munich, Germany, 5–9 October 2015. [CrossRef]
37. Zhao, H.S.; Shi, J.P.; Qi, X.J.; Wang, X.G.; Jia, J.Y. Pyramid Scene Parsing Network. In Proceedings of the IEEE Conference on Computer Vision and Pattern Recognition, Honolulu, HI, USA, 21–26 July 2017. [CrossRef]
38. Liu, Y.H.; Yao, J.; Lu, X.H.; Xie, R.P.; Li, L. DeepCrack: A deep hierarchical feature learning architecture for crack segmentation. *Neurocomputing* **2019**, *338*, 139–153. [CrossRef]

buildings

MDPI

Article

A New Drive-by Method for Bridge Damage Inspection Based on Characteristic Wavelet Coefficient

Tingpeng Zhang [1], Jin Zhu [1,*], Ziluo Xiong [1,2], Kaifeng Zheng [1] and Mengxue Wu [3]

1 Department of Bridge Engineering, Southwest Jiaotong University, Chengdu 610031, China
2 Department of Civil and Environmental Engineering, Colorado State University, Fort Collins, CO 80523, USA
3 School of Civil Engineering and Architecture, Southwest Petroleum University, Chengdu 610500, China
* Correspondence: zhujin@swjtu.edu.cn

Abstract: The drive-by method has become a popular indirect approach for bridge damage inspection (BDI) because of its simplicity in deployment by evaluating the bridge health status solely via the vehicle dynamic response. Derived from the vehicle dynamic response, the recent proposed contact-point response involves no vibration signal with the vehicle frequency, bearing great potential for drive-by BDI. However, an appropriate methodology for the application of contact-point response in drive-by BDI remains lacking. The present study proposes a novel drive-by method, in which a new damage factor index, i.e., the characteristic wavelet coefficient (CWC), is established for bridge damage identification in an efficient and accurate manner. The CWC is obtained by analyzing the contact-point response via the continuous wavelet transform (CWT) and complete ensemble empirical mode decomposition with adaptive noise (CEEMDAN) techniques. CEEMDAN is introduced to overcome the issue of modal aliasing and pseudo-frequency. First, the general framework of the proposed drive-by BDI method is introduced. Then, a demonstration case study is carried out to examine the effectiveness of the proposed method. Subsequently, a parametric study is carried out to explore the effects of several parameters on the performance of BDI including the scale factor, vehicle speed, environmental noise, and boundary effect. The results indicate that the proposed drive-by BDI method can better eliminate the mode mixing and pseudo-frequency problems during the extraction of the CWC, compared with the traditional ensemble empirical mode decomposition method. The extracted CWC curve is smooth, convenient for damage inspection, and has strong anti-noise performance. After adding white noise with a signal-to-noise ratio of 20, a bridge girder with a damage severity of 20% can be identified successfully. In addition, the selection of the scale factor is critical for bridge damage inspection based on the extracted CWC. The effective scale factor of the CWC extracted using the proposed method has a wide range, which improves the inspection efficiency. Finally, a low vehicle speed is beneficial to alleviate the adverse effect of the boundary effect on the damage inspection of bridge girder ends.

Keywords: vehicle–bridge interaction; drive-by method; bridge damage inspection; characteristic wavelet coefficient; complete ensemble empirical mode decomposition with adaptive noise

1. Introduction

Due to the increase in aging infrastructures worldwide, structural system inspection and damage detection have become increasingly important for structural health status assessment. For bridges in particular, it is also crucial to ensure their structural integrity and safety, because bridges usually serve as the critical links in transportation networks. In general, bridge health inspection methods are mainly divided into two categories, i.e., direct inspection methods and indirect inspection methods. Direct inspection methods aim to detect damage by installing sensors on bridges and analyzing vibration data collected by sensors [1–4]. Although direct inspection methods have been widely used, they often require a large-scale deployment of sensors on the bridge, leading to demanding costs and cumbersome maintenance. Therefore, a structural health motoring (SHM) system is typically installed on long-span bridges with a relatively higher budget than short- or medium-span bridges [5].

In order to overcome the limitations of direct inspection methods, indirect inspection methods have been proposed for the damage assessment of bridge structures, such as drive-by-based methods, global navigation satellite system (GNSS)-based methods [6], and unmanned aerial vehicle (UAV)-based methods [7]. Among them, the drive-by method was first proposed by Yang et al. [8], and it aims to assess the bridge health status indirectly via the vehicle dynamic response when traveling on the bridge, based on the vehicle–bridge interaction (VBI). The feasibility of the drive-by method was later verified through a real bridge test [9]. Since then, the drive-by method has been widely used by many scholars for detecting damage of bridge girders. In general, the drive-by method constructs the bridge damage index through either the modal related parameters or the non-modal related parameters. As the name implies, the modal-parameter-based drive-by method needs to first identify the frequency, vibration mode, or damping [10–13] of the bridge, and then further construct the bridge damage index. For example, OBrien et al. [14] presented a novel algorithm for bridge damage detection based on the mode shapes, in which the bridge response was measured from an instrumented vehicle with laser vibrometers and accelerometers. The algorithm was proved to be effective and accurate when the vehicle speed was less than 8 m/s. Similarly, Oshima et al. [15] evaluated two types of damage of a bridge, i.e., immobilization of a support, and a decrease in the stiffness of the bridge girder at the mid-span, via mode shapes estimated from the responses of passing vehicles. In a recent study conducted by Robert and Abdollah [16], the contact-point responses were utilized to extract the bridge frequencies for bridge damage detection, which avoids the effects of the vehicle frequency. It is noteworthy that, although much progress has been made in the last two decades, key challenges remain with the modal-parameter-based drive-by method for bridge damage inspection, including how to extract the modal parameters efficiently and accurately, and how to eliminate the effect of road roughness and noise.

Different from the modal-parameter-based method, the non-modal-parameter-based method does not rely on modal parameters and typically utilizes machine learning and signal processing techniques, which have been proven effective in detecting and locating damage. The advantage of machine learning algorithms is that they can make full use of the massive data collected by the drive-by system. Mei et al. [17] proposed a novel damage detection technique by combining the mel-frequency cepstrum (MFC) and principal component analysis (PCA) for bridge damage detection using drive-by data measured from vehicles. With the proposed method, the influence of temperature and road roughness under operating conditions can be adequately considered. Corbally and Malekjafarian [18] proposed a new data-driven approach for the drive-by monitoring of bridge condition using an artificial neural network (ANN), which is subsequently trained to predict the bridge behavior using acceleration measurements from multiple passes of a traversing vehicle. Sarwar and Cantero [19] proposed a vehicle-assisted bridge damage assessment approach based on a deep autoencoder (DAE) architecture, considering multiple convolutional layers and an LSTM layer. With the proposed approach, the relationship between vehicle responses and bridge dynamics is established successfully. Locke et al. [20] proposed a drive-by health monitoring technique to detect bridge damage considering environmental (temperature) and operational effects (road roughness and random traffic flow). One major challenge of machine-learning-based indirect inspection methods is that massive training data are required, and sometimes the data need to be manually classified or marked. In addition, machine-learning-based indirect inspection methods may not be applicable under complex operational conditions, e.g., road roughness, temperature, and random traffic flow, in practical applications.

Compared with machine-learning-based indirect inspection methods, signal-processing-based methods, e.g., wavelet transform (WT) and empirical mode decomposition (EMD), are more convenient and relatively easy to implement, and they have drawn increasing attention recently. For example, Abdulkareem et al. [21–23] proposed a series of wavelet analysis methods to extract the modal shapes of plate structures and to construct structural damage factors. Zhu and Law [24] employed the continuous wavelet transform (CWT) to identify cracks in bridge beam structures under a moving load. Hester and González [25] successfully detected the damage in a 40 m simply supported bridge beam model by analyzing its acceleration response under a moving load via the WT. Nguyen and Hai [26] proposed an approach for detecting a multi-cracked beam-like structure subjected to a moving vehicle based on the on-vehicle vibration signal and wavelet analysis. By using the continuous wavelet method, Khorram et al. [27] compared the effect of the signal obtained by the bridge mid-span sensor and the sensor installed at the contact point between the mobile vehicle and the bridge for damage inspection. The analysis results showed that the inspection method of the bridge contact sensor was more effective. Tan et al. [28] employed Shannon entropy to select the optimal scale factor in the process of the CWT, which improves the efficiency of bridge damage inspection. The aforementioned WT-based methods are mainly implemented on vehicle dynamic responses, while their performance on the vehicle–bridge contact-point response remains unclear. Yang and Chang [29] employed the EMD technique to decompose the vehicle response into a set of intrinsic mode functions (IMFs), with which the bridge frequencies, especially the higher modes, were successfully extracted. Obrien et al. [30] also utilized EMD to decompose the acceleration of a car body's mass center (vehicle acceleration) to construct the damage index for accurate bridge damage detection. In recent years, ensemble empirical mode decomposition (EEMD) and extreme-point symmetric mode decomposition (ESMD) have been successively used by Zhu and Malekjafarian [31] and Yang et al. [32] to improve the efficiency and accuracy of the frequency inspection of bridges.

It is reported that the above signal decomposition methods such as EEMD still have modal aliasing and pseudo-modal problems [33]. In addition, the contact-point response is proved to be more efficient and accurate than the vehicle dynamic response for bridge damage inspection, yet there is currently a lack of related research in this field [34]. The present study proposes a novel drive-by-based method, in which a new damage factor index, i.e., the characteristic wavelet coefficient (CWC), is established for bridge damage identification in an efficient and accurate manner. The CWC is obtained by analyzing the contact-point response via the CWT and complete ensemble empirical mode decomposition with adaptive noise (CEEMDAN) techniques. CEEMDAN is introduced to overcome the issues of modal aliasing and pseudo-frequency. The remainder of this paper is organized as follows. First, the general framework of the proposed drive-by bridge damage inspection methodology is introduced. Then, a demonstration case study is carried out to examine the effectiveness of the proposed drive-by damage inspection methodology. Subsequently, a parametric study is carried out to explore the effects of several parameters on the performance of the proposed damage inspection methodology including the scale factor, vehicle speed, environmental noise, and boundary effect.

2. Bridge Damage Inspection Methodology

The proposed drive-by bridge damage inspection methodology consists of three parts, i.e., VBI analysis (part I), preliminary damage inspection (part II), and thorough damage inspection (part III), as illustrated in Figure 1. In part I, the VBI analysis is performed, in which the dynamic responses of the vehicle–bridge system are obtained. Subsequently, the dynamic responses of the vehicle–bridge system are analyzed using the CWT, in which the wavelet coefficient (WC) is extracted and used for the preliminary damage inspection of the bridge, as detailed in part II. It is noteworthy that the WC contains a variety of high-order frequency contents of bridge vibration, which has adverse effects on the damage inspection accuracy. To overcome this issue, the WC in part II is further processed by CEEMDAN, with which the CWC is extracted and used for thorough damage inspection, i.e., identifying the damage location and severity of the bridge accurately, as introduced in detail in part III. Each one of the three parts of the proposed drive-by bridge damage inspection methodology will be elaborated in the following subsections.

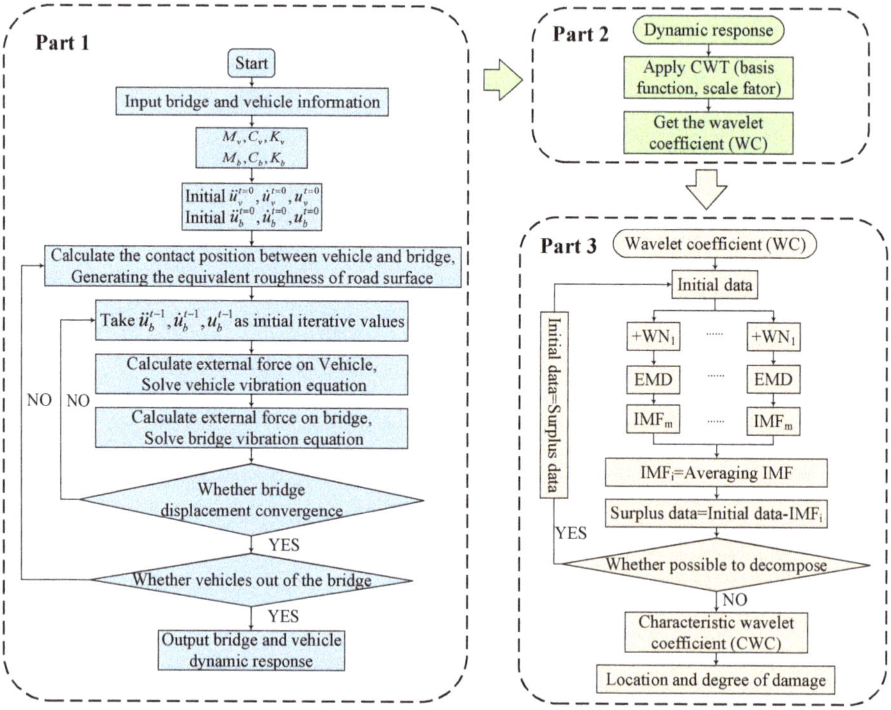

Figure 1. Flowchart of the proposed drive-by bridge damage inspection methodology.

2.1. Vehicle–Bridge Interaction (VBI) Analysis

2.1.1. Analytical Solution

For illustration purposes, the coupled vehicle–bridge system is simplified as a lumped sprung mass moving on a simply supported beam, as shown in Figure 2. By neglecting the damping effects of the vehicle–bridge system, the corresponding equations of motion can be written as

$$\begin{aligned} m_v \ddot{u}_v + K_v(u_v - u_b|_{x=vt}) &= 0 \\ m_b \ddot{u}_b + EIu_b'''' &= f_c(t)\delta(x - vt) \end{aligned} \quad (1)$$

where u_v and u_b are the vertical displacement of the vehicle body (lumped sprung mass) and the bridge (simply supported beam), respectively; the two dots represent quadratic differentials with respect to time t; m_v and m_b are the mass of the vehicle and bridge, respectively; K_v is the stiffness matrix of the vehicle; EI is the flexural rigidity of the bridge; $x = vt$ is the position of the moving vehicle on the bridge; v is the moving speed of the vehicle; $\delta(t)$ is the Dirac function; and $f_c(t)$ is the contact force between the vehicle and bridge, which is given by

$$f_c(t) = k_v(u_v - u_b|_{x=vt}) + m_v g \quad (2)$$

where g is the gravitational acceleration.

Figure 2. Lumped sprung mass moving on a simply supported beam.

Based on the modal superposition method, the bridge displacement response $u_b(x,t)$ is expressed in terms of modal shapes sin $(n\pi x/L)$ and generalized coordinates $q_{b,n}(t)$:

$$u_b(x,t) = \sum_n \sin\frac{n\pi x}{L} q_{b,n}(t) \tag{3}$$

where L is the length of the bridge.

One can obtain the theoretical solution of the acceleration time histories of the bridge, vehicle, and contact point between the bridge and vehicle as [8,35]

$$\ddot{u}_b(x,t) = \sum_n \sin\frac{n\pi x}{L}\frac{\Delta_{st,n}}{1-S_n^2}\left[-\left(\frac{n\pi v}{L}\right)^2 \sin\left(\frac{n\pi vt}{L}\right) + S_n\omega_{b,n}^2 \sin\omega_{b,n}t\right] \tag{4}$$

$$\ddot{u}_v(t) = \sum_n\left\{\overline{\overline{A}}_{v,n}\cos\omega_v t + \overline{\overline{A}}_{d,n}\cos\frac{2n\pi vt}{L} + \overline{\overline{A}}_{br,n}\cos\left(\omega_{b,n}+\frac{n\pi v}{L}\right)t + \overline{\overline{A}}_{bl,n}\cos\left(\omega_{b,n}-\frac{n\pi v}{L}\right)t\right\} \tag{5}$$

$$\ddot{u}_c(t)|_{x=vt} = \sum_n \sin\frac{n\pi vt}{L}\frac{\Delta_{st,n}}{1-S_n^2}\left[-\left(\frac{n\pi v}{L}\right)^2 \sin\left(\frac{n\pi vt}{L}\right) + S_n\omega_{b,n}^2 \sin\omega_{b,n}t\right] \tag{6}$$

in which the amplitude coefficients $\overline{\overline{A}}_{v,n}$, $\overline{\overline{A}}_{d,n}$, $\overline{\overline{A}}_{br,n}$, and $\overline{\overline{A}}_{bl,n}$ are

$$\begin{cases} \overline{\overline{A}}_{v,n} = -\omega_v^2\frac{2\Delta_{st,n}\mu_n^2 S_n^2}{1-S_n^2}\left\{\frac{1}{1-4\mu_n^2 S_n^2} - \frac{1}{\left[1-\mu_n^2(1-S_n)^2\right]\left[1-\mu_n^2(1+S_n)^2\right]}\right\} \\ \overline{\overline{A}}_{d,n} = \omega_{d,n}^2\frac{\Delta_{st,n}}{2\left(1-S_n^2\right)\left(1-4\mu_n^2 S_n^2\right)} \\ \overline{\overline{A}}_{br,n} = -\omega_{br,n}^2\frac{\Delta_{st,n}S_n}{2\left(1-S_n^2\right)\left[1-\mu_n^2(1+S_n)^2\right]} \\ \overline{\overline{A}}_{bl,n} = \omega_{bl,n}^2\frac{\Delta_{st,n}S_n}{2\left(1-S_n^2\right)\left[1-\mu_n^2(1-S_n)^2\right]} \end{cases} \tag{7}$$

where $\Delta_{st,n} = 2m_v gL^3/EI\pi^4$; $\mu_n = \omega_b/\omega_v$; $S_n = n\pi v/L\omega_b$; $\omega_{bl,n}$ and $\omega_{br,n}$ are the left and right frequencies of the bridge, respectively; ω_v is the frequency of the vehicle; and $\omega_{d,n}$ is the driving frequency.

2.1.2. Numerical Simulation

Generally, the bridge can be discretized as several beam elements, and the vehicle is idealized as a spring-mass model. To consider the vehicle–bridge interaction, it is common to formulate the governing equation of the coupled vehicle–bridge system as [8]

$$\begin{bmatrix} m_v & 0 \\ 0 & [m_b] \end{bmatrix}\begin{Bmatrix} \ddot{q}_v \\ \{\ddot{u}_b\} \end{Bmatrix} + \begin{bmatrix} k_v & -k_v\{N\}^{\mathrm{T}} \\ -k_v\{N\} & [k_b] + v^2 m_v\{N\}\frac{\partial^2\{N\}^{\mathrm{T}}}{\partial x^2} + k_v\{N\}\{N\}^{\mathrm{T}} \end{bmatrix}\begin{Bmatrix} u_v \\ \{u_b\} \end{Bmatrix} = \begin{Bmatrix} 0 \\ -m_v g\{N\}_c \end{Bmatrix} \tag{8}$$

where m_v and k_v denote the mass and stiffness of the vehicle; $[m_b]$ and $[k_b]$ are the mass and stiffness matrixes of the bridge element supporting the vehicle; $\{u_b\}$ represents the displacement vector of the bridge element; $\{N\}$ contains the cubic Hermitian interpolation functions associated with the transverse displacement of the element (i.e., vertical displacement for the bridge); and $\{N\}_c$ represents the realization of $\{N\}$ at the vehicle–bridge contact point, as given below:

$$\{N(x_c)\} = \left\{1-3\left(\frac{x_c}{l}\right)^2+2\left(\frac{x_c}{l}\right)^3 \; x_c\left(1-\frac{x_c}{l}\right)^2 \; 3\left(\frac{x_c}{l}\right)^2+2\left(\frac{x_c}{l}\right)^3 \; \frac{x_c^2}{l}\left(\frac{x_c}{l}-1\right) \; 1-\frac{x_c}{l} \; \frac{x_c}{l}\right\}^T \tag{9}$$

in which l is the length of the bridge element. The numerical solution of the coupling between the vehicle and the bridge system can be realized by the separation iteration between the vehicle and the bridge system. In the solving process of the vehicle–bridge interaction program, the geometric and mechanical coupling relationship between the vehicle and bridge systems is satisfied through the Newmark-β method and separation iteration, so as to realize the independent solution of the bridge and vehicle motion equations (Figure 1, part I). For more information about the numerical solution system in this article, see [36–38].

2.2. Preliminary Damage Inspection Based on CWT

The WT is a mathematical approach that enables the extraction of the prominent characteristics of the original dataset, such as discontinuities, trends, and breakdown points, which has been widely used for signal processing applications. The WT is divided into the CWT and discrete WT (DWT). In the present study, the CWT is employed to process the acceleration response of the vehicle–bridge system, which is given by

$$WT_f(a,b) = \frac{1}{\sqrt{a}}\int_0^{\infty} f(t)\psi^*\left(\frac{t-b}{a}\right)\mathrm{d}t \tag{10}$$

$$\psi_{a,b}(t) = \frac{1}{\sqrt{a}}\psi\left(\frac{t-b}{a}\right) \tag{11}$$

where $f(t)$ is the signal (the acceleration response in the present study); $\psi_{a,b}(t)$ is the wavelet basis function; $\psi^*(t)$ represents the conjugate function of the basic wavelet function; t is the time; a is the scale factor; and b is the time shift factor. When the parameters t, a, and b are continuous variables, the above transformation is the CWT, and $WT_f(a,b)$ is the wavelet coefficient under scale factor a and shift factor b. The WC of the signal at a certain scale can be obtained by applying the CWT to the acceleration response of the vehicle–bridge system obtained in the previous section, which is subsequently used for the damage inspection of the bridge girder.

2.3. Thorough Damage Inspection Based on CWC

As mentioned previously, the WC contains a variety of high-order frequency contents of bridge vibration, which has adverse effects on the damage inspection accuracy. To improve the identification accuracy, the CEEMDAN technique is applied to the WC to obtain the CWC for thorough damage inspection.

Huang et al. [39] first developed the EMD method to examine the nonlinear and non-stationary signals in an adaptive time–frequency–amplitude space. With the EMD method, the signal can be decomposed into several intrinsic mode functions (IMFs) and a residue. Despite its wide applications, EMD experiences some problems, such as the presence of oscillations of very disparate amplitudes in a mode, or mode mixing. To overcome these problems, ensemble EMD (EEMD) has been proposed, which performs the EMD over an ensemble of the signal plus Gaussian white noise [40]. Based on EEMD, the CEEMDAN method has recently been proposed, in which a particular noise is added at each stage of the decomposition and a unique residue is computed to obtain each mode [41]. This helps to further reduce the modal aliasing and pseudo-modal problems as well as enhancing the anti-noise ability. The flowchart of CEEMDAN is illustrated in Figure 1. Given a set of measured data $X(t)$, the process of the CEEMDAN algorithm decomposing each order's IMF is described as follows:

(1) The Gaussian white noise is added to the signal $X(t)$ to obtain a new signal $X(t) + \sum\limits_{n=1}^{n} WN_n$, and the new signal is decomposed through EMD to obtain the first-

order IMF and remainder (Equation (12)). The above process is repeated N times, and N first-order IMFs are obtained because the white noise added is different.

$$E(X(t) + \sum_{n=1}^{n} WN_n) = C_1^j(t) + r^j(t) \tag{12}$$

where $E(\cdot)$ refers to the EMD process; $\sum_{n=1}^{n} WN_n$ refers to the white noise; n is the number of times the white noise is added; $C_1^j(t)$ is the first-order IMF; $r^j(t)$ is the remaining margin; and j is the number of decompositions.

(2) The average of the resulting N first-order IMFs yields the first IMF $\overline{C}_1(t)$ decomposed by CEEMDAN:

$$\overline{C}_1(t) = \frac{1}{N}\sum_{j=1}^{N} C_1^j(t) \tag{13}$$

(3) The residual of the original signal $X(t)$ is calculated after removing the first modal component:

$$X_r(t) = X(t) - \overline{C}_1(t) \tag{14}$$

(4) The signal obtained in the previous step is treated as a new signal, which is processed by the first three steps to obtain the second-order IMF. Subsequently, the procedure is applied repeatedly to extract the third- and fourth-order IMF components (if they exist), until the residual signal is a monotonic function. Finally, assuming the number of extracted IMFs is K, the original signal $X(t)$ can be expressed as the summation of all IMF components and the residual signal, as given by

$$X_r(t) = X(t) - \overline{C}_1(t) \tag{15}$$

3. Numerical Case

A demonstration case study is carried out to examine the effectiveness of the proposed drive-by damage inspection methodology.

3.1. VBI of Simply Supported Bridge Beam

The demonstration case involves a single-degree-of-freedom (DOF) sprung mass moving along a simply supported bridge beam model, as shown in Figure 3. The total length of the bridge beam is 15 m, which is divided into 30 elements with an equal length of 0.5 m. The damage of the bridge girder is simulated by means of a reduction in the elastic modulus. The time step during the VBI analysis is set to 0.005 s. In addition, the speed of the vehicle is set to 4 m/s, and the key parameters of the vehicle–bridge system are shown in Table 1. The VBI is realized using a self-compiled program in the MATLAB software. For validation purposes, the simulated dynamic responses of the vehicle–bridge system are compared with the analytical solutions, as shown in Figure 4. It is clearly shown in Figure 4 that the vertical acceleration of the bridge mid-span, the vertical acceleration of the vehicle body, and the vertical acceleration of the contact point obtained from the numerical simulations are in good agreement with those of the analytical solutions.

Figure 3. Single-DOF sprung mass moving along a simply supported bridge beam model.

Table 1. Essential parameters of the vehicle–bridge system.

Parameter	Definition	Value
m_v	Vehicle mass	1000 kg
C_v	Vehicle rigidity	500 kN/m
m_b	Mass per meter of bridge	4000 kg
L	Length of bridge	15 m
I_b	Inertia moment of bridge section	0.12 m^4
w_v	Vehicle frequency	3.56 Hz
w_d	Driving frequency	0.266 Hz
w_{b1}	First-order frequency of bridge	6.623 Hz
w_{b2}	Second-order frequency of bridge	26.492 Hz

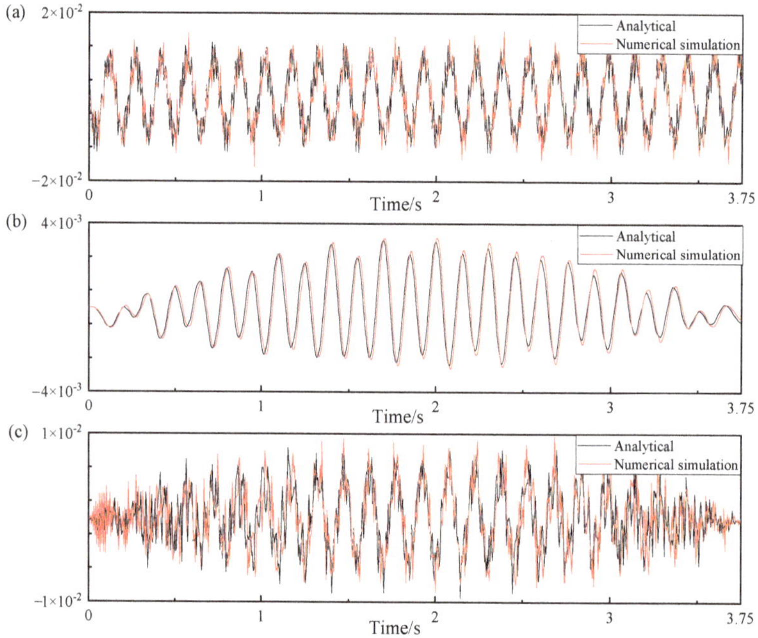

Figure 4. Comparison between the simulated dynamic response of the vehicle-bridge system and the analytical solutions: (**a**) bridge mid-span acceleration; (**b**) vehicle acceleration; (**c**) contact-point acceleration.

In total, 14 cases are considered, as listed in Table 2. In cases 1~5, the damage is assumed to be located at seventh element with various degrees of severity, i.e., 0%, 5%, 10%, 20%, and 40%. The vehicle mass and vehicle speed remain unchanged in cases 1~5, i.e., mv = 1000 kg and V = 4 m/s. To examine the influence of the damage location on the feasibility of the proposed damage inspection methodology, the damage location is shifted from the seventh element to the second element of the bridge in cases 6~8, with a damage severity of 0%, 20%, and 40%. The vehicle mass and vehicle speed in cases 6~8 are the same as those in cases 1~5. To further investigate the influence of the vehicle speed, the vehicle speed is reduced from 4 m/s to 1 m/s in cases 9~11. The damage in cases 9~11 is located at the second element of the bridge with a damage severity of 0%, 20%, and 40%. It is noted that cases 1~11 only contain one damage location, while cases 12~13 assume two damage locations. To be specific, in case 12, elements No. 4 and No. 14 are damaged with a severity of 10% and 30%, while in case 13, elements No. 4 and No. 14 are damaged with a severity of 30% and 10%.

Table 2. Case setting for VBI analysis.

Number	Vehicle Mass (kg)	Vehicle Speed (m/s)	Damage Location (No. of Element)	Degree of Damage
1	1000	4	7	0%
2	1000	4	7	5%
3	1000	4	7	10%
4	1000	4	7	20%
5	1000	4	7	40%
6	1000	4	2	0%
7	1000	4	2	20%
8	1000	4	2	40%
9	1000	1	2	0%
10	1000	1	2	20%
11	1000	1	2	40%
12	1000	4	4, 14	10%, 30%
13	1000	4	4, 14	30%, 10%

3.2. WC Extraction Based on CWT

Previous studies have shown that abnormal changes in the WC under a certain scale factor can be used to identify the bridge damage effectively. Figure 5a shows the time–frequency diagram after applying the CWT to the vehicle acceleration under case 5. The ordinate is the scale factor (Scale) of the CWT, and the abscissa is the time shift factor (Shift) of the CWT, which can be understood as the relative position of the bridge. The color shade indicates the magnitude of the WC. It is observed that the color of the WC varies significantly at the position of 6.5/30 = 0.21 (the center of the damaged No.7 element), and the scale factor ranges from approximately 30 to 60. To clearly observe the variation in the WC, the WC under the scale factor of 35 in case 5 is displayed in Figure 5b. For comparison purposes, the WC in the undamaged case (case 1) is also plotted in Figure 5b. As shown in Figure 5b, there is an obvious peak of the WC at the center of the damaged No.7 element, which suggests that the significant variation in the WC under an effective scale factor can help to identify the damage location of the bridge. It is also worth mentioning that the wavelet basis function could affect the damage identification results. The commonly used wavelet basis functions include the Haar wavelet, Morlet wavelet, Mexican hat wavelet, and Daubechies wavelet. A preliminary analysis was conducted to examine the effectiveness of the type of wavelet basis function on the damage identification results, where the Mexico hat wavelet outperformed the other selected wavelet basis functions and was selected for the subsequent analysis.

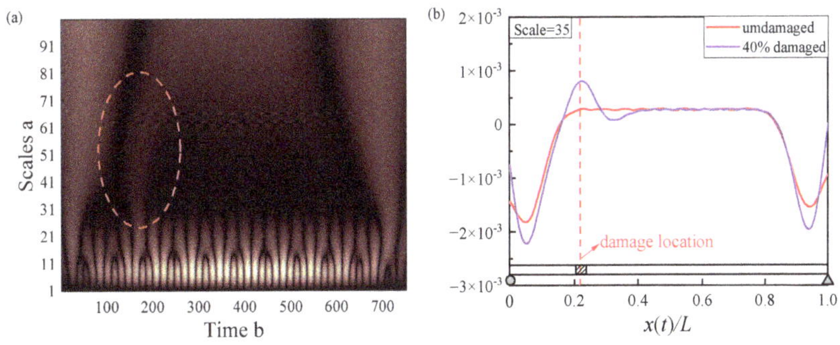

Figure 5. CWT of vehicle acceleration under case 5: (**a**) the time-frequency diagram; (**b**) wavelet coefficient graph with a scale factor of 35.

In addition to the WC of vehicle acceleration, the WC of the bridge mid-span acceleration and the contact-point acceleration in cases 2~5 is also extracted, as shown in Figure 6. It is observed from Figure 6a,b that although the trend of the WC shows a certain degree of increase at the damaged location, such an increase is not obvious, especially when the damage severity is less than 40%. This is because the overall trend of the WC is largely contaminated by its fluctuations, i.e., the high-frequency contents.

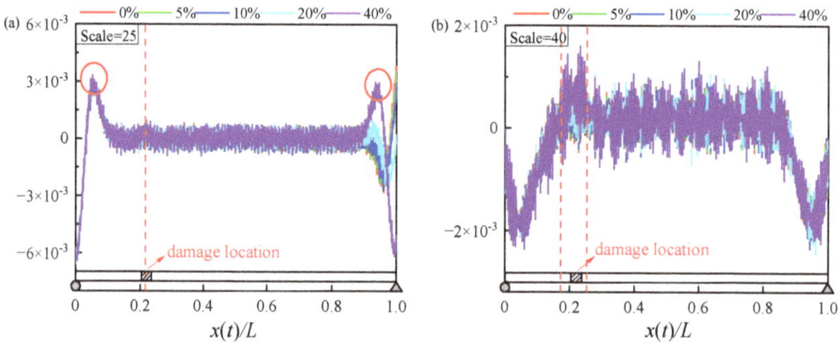

Figure 6. Wavelet coefficient diagram: (**a**) bridge mid-span acceleration; (**b**) contact-point acceleration.

3.3. *Extracting CWC Based on WC Using EEMD or CEEMDAN*

The previous discussion indicates that the extracted WC after applying the CWT to the acceleration response of the bridge mid-span and contact point has a poor damage inspection performance, due to the contamination of the high-frequency content of the WC. In order to improve the identification accuracy, it is natural to filter out the high-frequency contents of the WC to make the trend of the WC visible. In view of this, the extracted WC is further analyzed using two signal decomposition techniques, i.e., EEMD and CEEMADAN, in which the extracted WC is decomposed into a series of IMF components with different frequencies. Figures 7 and 8 show the IMFs and the associated FFT spectra after processing the WC (case 5) with EEMD and CEEMDAN, respectively. It is observed that the IMFs help to separate the first few bridge natural frequencies or the vehicle driving frequency. In addition, obvious mode mixing can be observed in IMF3 and IMF4 extracted using the EEMD method, as shown in Figure 7. Furthermore, our analysis showed that the IMFs with a dominant frequency higher than the first vertical natural frequency (6.623 Hz) of the bridge have adverse effects on the bridge damage identification results, while the opposite result is found for the IMFs with a frequency lower than the first vertical natural frequency of the bridge. Therefore, only the IMFs with a dominant frequency lower than the first vertical natural frequency of the bridge are selected to construct the new damage index, which is termed as the CWC in the present study.

The CWC decomposed by EEMD and CEEMDAN is shown in Figure 9. It is observed in Figure 9 that the CWC extracted by both EEMD and CEEMDAN shows an abrupt increase at the damage location of the bridge girder, and such an abrupt increase is more obvious under severer damage conditions. To be specific, the peak of the CWC extracted by EEMD is 2.13×10^{-4}, 2.87×10^{-4}, and 4.72×10^{-4} under damage of 0%, 20%, and 40%, while the peak of the CWC extracted by CEEMDAN is 1.41×10^{-4}, 3.35×10^{-4}, and 6.08×10^{-4} under damage of 0%, 20%, and 40%. This indicates that the CWC can be used to detect the damage location and to quantify the damage severity of the bridge girder. In addition, it is found that the CWC extracted using the CEEMDAN method is smoother than that extracted using the EEMD method. Therefore, the CEEMAN method is adopted for the subsequent analysis.

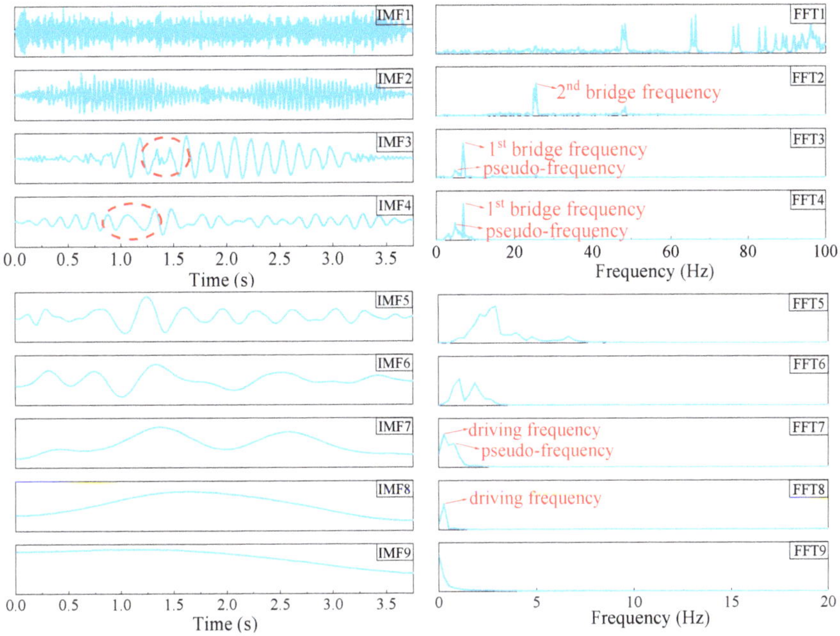

Figure 7. The IMF and FFT of the WC extracted by EEMD.

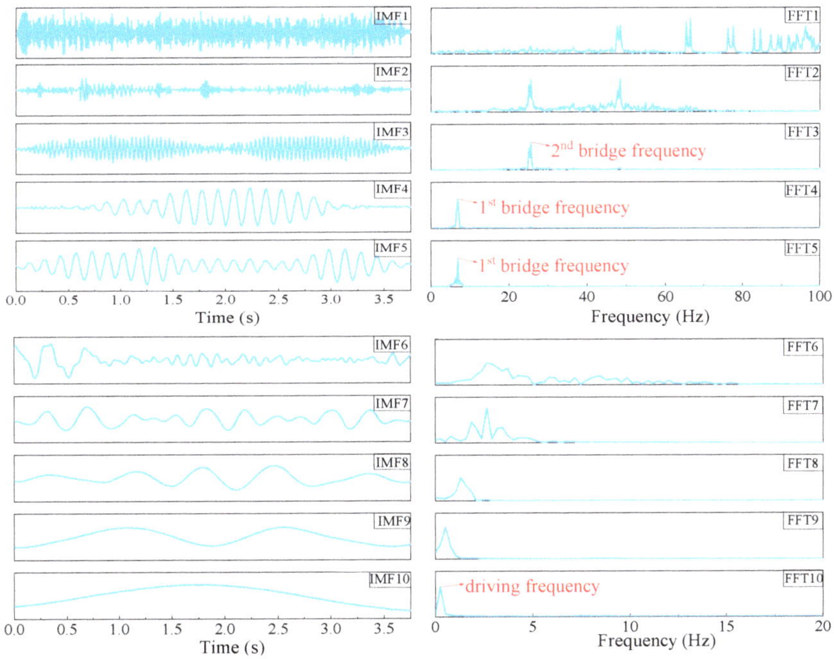

Figure 8. The IMF and FFT of the WC extracted by CEEMDAN.

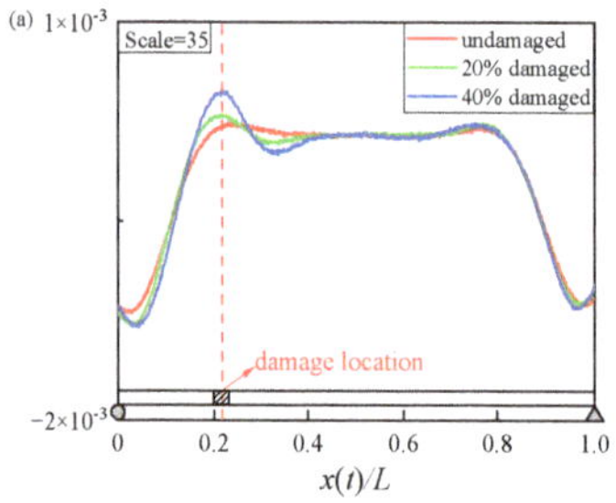

Figure 9. CWC extracted using different methods: (**a**) EEMD; (**b**) CEEMDAN.

3.4. CWC for Multi-Damage Case

In this section, the performance of the proposed drive-by damage inspection methodology in a multi-damage case is further evaluated. Two damage scenarios of the bridge girder with two damage locations are investigated. In the first damage scenario (case 12), elements No. 4 and No. 14 are damaged with a severity of 10% and 30%, while in the second damage scenario (case 13), elements No. 4 and No. 14 are damaged with a severity of 30% and 10%. Figure 10 shows the CWC by analyzing the bridge mid-span acceleration, vehicle acceleration, and contact-point acceleration. As shown in Figure 10a, the CWC extracted from the bridge mid-span acceleration has a poor damage inspection performance. In contrast, as for the CWC extracted from the vehicle acceleration and contact-point acceleration, two noticeable peaks can be observed at the two damaged elements, as shown in Figure 10b,c. In addition, for damage scenario one, the peak at element No. 4 with a 10% damage severity is less than that at element No. 14 with a 30% damage severity, while the opposite result is found for damage scenario two. The above observations clearly suggest that the CWC extracted from the vehicle acceleration and contact-point acceleration is also effective for damage identification in multi-damage scenarios.

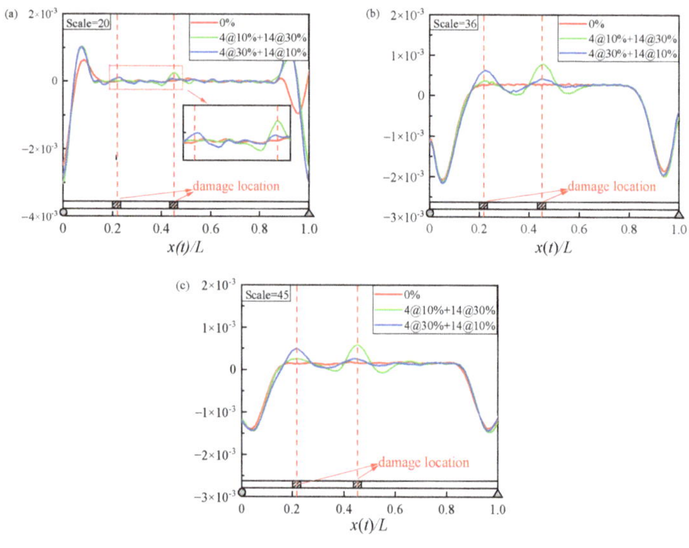

Figure 10. Inspection of multi-damage case: (**a**) CWC extracted from bridge mid-span acceleration; (**b**) CWC extracted from vehicle acceleration; (**c**) CWC extracted from contact-point acceleration. Where @ indicates the degree of damage.

4. Parametric Analysis

In this section, a parametric study is carried out to explore the effects of several parameters on the performance of the proposed damage inspection methodology including the scale factor, vehicle speed, environmental noise, and boundary effect.

4.1. *Effect of Scale Factor*

As discussed earlier, the scale of the CWT has a significant effect on the performance of the proposed drive-by damage inspection methodology. In practice, it is usually necessary to find a better scale factor through trial and error. Figure 11 shows the CWCs at scale factors of 15, 35, and 55, which are extracted from the contact-point accelerations in case 1, case 3, and case 5, respectively. It is observed that the CWC under a scale factor of 35 outperforms that under scale factors of 15 and 55.

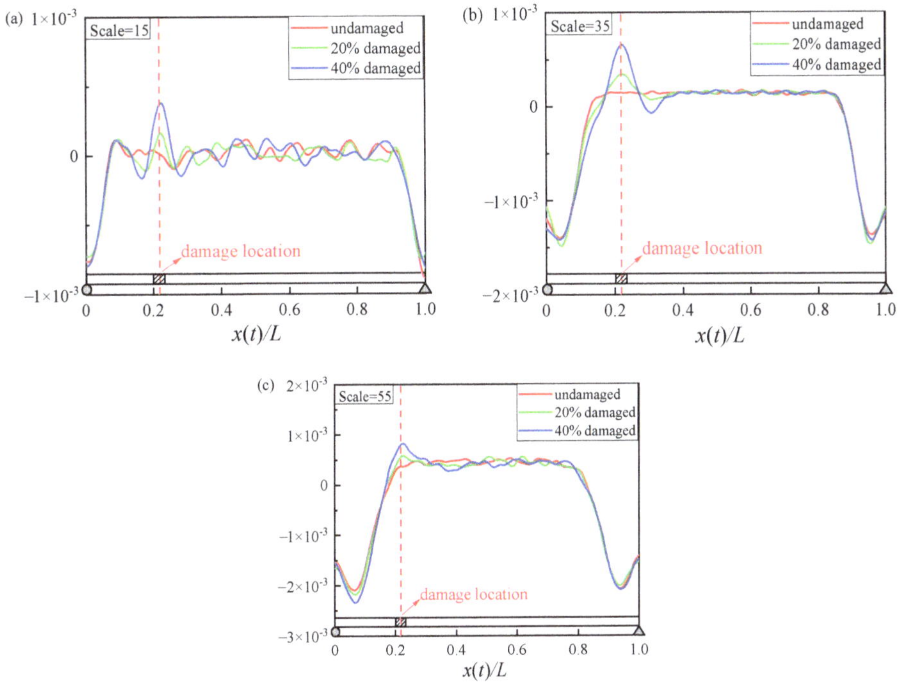

Figure 11. CWC under different scale factors: (**a**) scale = 15; (**b**) scale = 35; (**c**) scale = 55.

4.2. *Effect of Environmental Noise*

Vehicle signals are inevitably polluted by environmental noise during the acquisition process, resulting in signal distortion or some key information being missing. Therefore, in this section, the effect of noise on the performance of the proposed damage identification framework is investigated. During the analysis, Gaussian white noise with three different signal-to-noise ratios (SNR = 15, 20, and 50) is added to the vehicle acceleration response in cases 1~5. Figure 12 shows the CWC under SNR = 15, 20, and 50. It is observed that appreciable fluctuations are induced all along the bridge girder due to the Gaussian white noise. As such, the CWC can only identify the damage locations under a severity of 40%, below which it becomes quite difficult to successfully carry out damage identification.

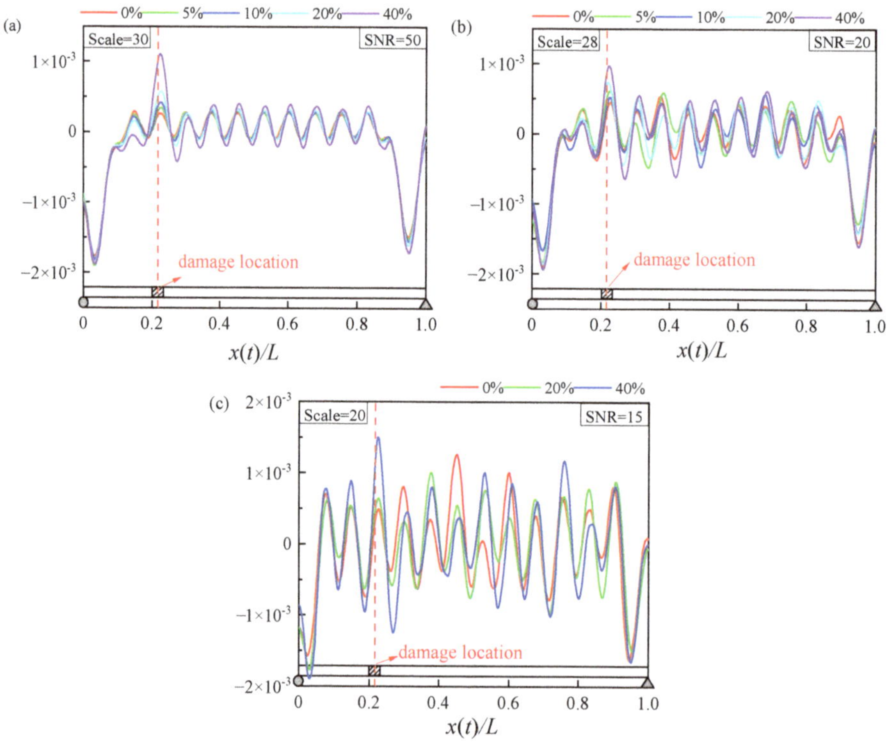

Figure 12. CWC extracted from signals with noise with different SNRs: (**a**) SNR = 50; (**b**) SNR = 20; (**c**) SNR = 15.

4.3. Vehicle Speed

The previous studies indicate that the performance of the conventional damage inspection method under a low vehicle speed is better than that under a high vehicle speed. In this section, the performance of the proposed damage inspection methodology in cases 1, 4, and 5 under two vehicle speeds, i.e., 8 m/s and 12 m/s, is investigated. It is observed in Figure 13 that the CWC is able to identify the damage location under both vehicle speeds, indicating that the proposed damage inspection methodology has a good performance under a relatively high vehicle speed.

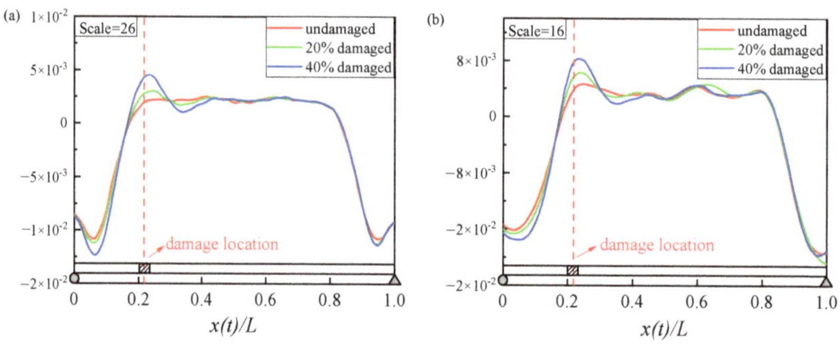

Figure 13. CWC at different speeds: (**a**) 8 m/s; (**b**) 12 m/s.

4.4. Boundary Effect

It is noted that in the process of the CWT, the WC is derived from the window function and wavelet convolution. When the window is located at the end of the signal, the signal is inevitably distorted because the signal is forced to zero at its end. This is referred to as the boundary effect of the CWT. In the time–frequency diagram, the quality of the signal becomes poor with the decrease in the frequency (scale a), as shown in Figure 14a. The boundary effect could inundate the damage information of the beam element near the bridge girder end. For example, when the vehicle speed is 4 m/s, it is impossible to extract the damage information of element 2, as shown in Figure 14a. The influence of the boundary effect is illustrated in the influence cone, in which the signal inside the influence cone is not affected by the boundary effect, while the signal outside the influence cone is significantly affected by the boundary effect. It is also noted that as the vehicle speed reduces from 4 m/s to 1 m/s, the influence cone becomes flatter and the influence of the boundary effect becomes less significant. As shown in Figure 14b, when the vehicle speed is 1 m/s, the second element can be successfully identified. Therefore, a low-speed test vehicle is expected to alleviate the adverse boundary effect. However, it remains impossible to identify the damage at the girder ends (i.e., 1st and 30th elements), even with a low-speed test vehicle, and more advanced algorithms may be considered in the future to resolve this limitation.

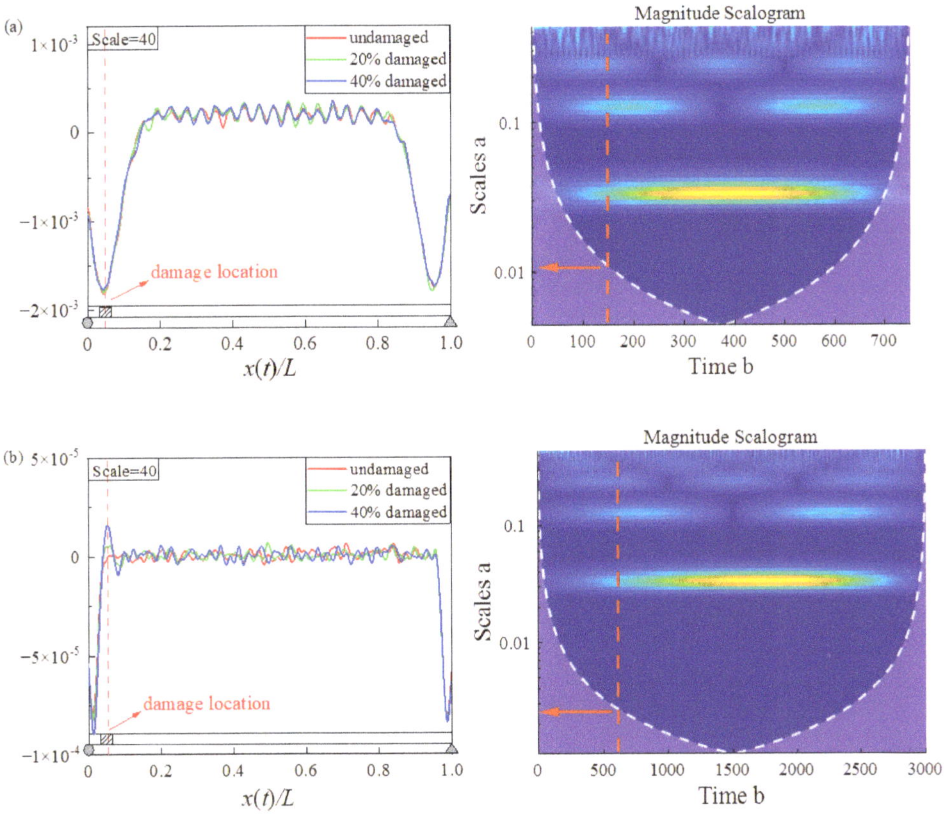

Figure 14. CWC and influence cone at different speeds: (**a**) 4 m/s; (**b**) 1 m/s.

5. Concluding Remarks

The present study proposes a novel drive-by-based method, in which a new damage factor index, i.e., the CWC, is established for bridge damage identification in an efficient and accurate manner. The following conclusions were obtained:

(1) Compared with the EEMD method, the CEEMAN algorithm can better eliminate the mode mixing and pseudo-frequency problems during the extraction of the CWC. The introduction of this method also makes the CWC curve smooth, convenient for damage inspection, with strong anti-noise performance. After adding white noise with a signal-to-noise ratio of 20, a bridge girder with a damage severity of 20% can be identified.
(2) The selection of the scale factor is critical for bridge damage inspection based on the extracted CWC. The effective scale factor of the CWC extracted using the proposed method has a wide range, which improves the inspection efficiency.
(3) A low vehicle speed is beneficial to alleviate the adverse effect of the boundary effect on the damage inspection of bridge girder ends.

It should be noted that, as with most existing studies on drive-by methods for bridge health inspection, the proposed method is numerically verified, demonstrating the potential for actual practice. However, there remain great challenges in achieving a satisfactory accuracy of bridge health inspection in real-world environments. In particular, the collected vibrational signals for damage identification are usually contaminated due to the interference of random traffic flow, environmental noises, etc. It is therefore recommended to carry out such a practical test when there is less adverse traffic excitation. Meanwhile, a specialized test vehicle with fewer internal noises (e.g., from motor vibration) is also beneficial to the performance of damage identification. To further improve the drive-by method, one may present further investigations on eliminating the common boundary effect through innovative algorithms and/or novel experimental designs.

Author Contributions: T.Z.: conceptualization, methodology, software, investigation, formal analysis, writing—original draft. J.Z.: conceptualization, funding acquisition, resources, supervision, writing—review and editing. Z.X.: conceptualization, methodology, formal analysis, writing—review and editing. K.Z.: resources, supervision. M.W.: resources, writing—review and editing. All authors have read and agreed to the published version of the manuscript.

Funding: This research were funded by the National Natural Science Foundation of China (Grant number 51908472), China Postdoctoral Science Foundation (Grant numbers 2019TQ0271 and 2019M663554), and Department of Science and Technology of Sichuan Province (2020YJ0080).

Institutional Review Board Statement: Not applicable.

Informed Consent Statement: Not applicable.

Data Availability Statement: All data, models, or codes that support the findings of this study are available upon reasonable request.

Conflicts of Interest: The authors declare no conflict of interest.

Abbreviations

WT	wavelet transform
CWT	continuous wavelet transform
WC	wavelet coefficient
CWC	characteristic wavelet coefficient
EMD	empirical mode decomposition
EEMD	ensemble empirical mode decomposition
CEEMDAN	complete ensemble empirical mode decomposition with adaptive noise

References

1. Ren, W.X.; Zhao, T.; Harik, I.E. Experimental and analytical modal analysis of steel arch bridge. *J. Struct. Eng.* **2004**, *130*, 1022–1031. [CrossRef]
2. Kim, C.W.; Chang, K.C.; Kitauchi, S.; Mcgetrick, P.J. A field experiment on a steel Gerber-truss bridge for damage detection utilizing vehicle-induced vibrations. *Struct. Health Monit.* **2016**, *15*, 421–429. [CrossRef]
3. Fan, W.; Qiao, P. Vibration-based Damage Identification Methods: A Review and Comparative Study. *Struct. Health Monit.* **2011**, *9*, 83–111. [CrossRef]
4. Amezquita-Sanchez, J.P.; Adeli, H. Signal Processing Techniques for Vibration-Based Health Monitoring of Smart Structures. *Arch. Comput. Methods Eng. State Art Rev.* **2016**, *23*, 1–15. [CrossRef]
5. Yang, Y.B.; Wang, Z.L.; Shi, K.; Xu, H.; Wu, Y.T. State-of-the-Art of Vehicle-Based Methods for Detecting Various Properties of Highway Bridges and Railway Tracks. *Int. J. Struct. Stab. Dyn.* **2020**, *20*, 1850025. [CrossRef]
6. Manzini, N.; Orcesi, A.; Thom, C.; Brossault, M.A.; Botton, S.; Ortiz, M.; Dumoulin, J. Machine Learning Models Applied to a GNSS Sensor Network for Automated Bridge Anomaly Detection. *J. Struct. Eng.* **2022**, *148*, 04022171. [CrossRef]
7. Reagan, D.; Sabato, A.; Niezrecki, C. Feasibility of using digital image correlation for unmanned aerial vehicle structural health monitoring of bridges. *Struct. Health Monit.* **2018**, *17*, 1056–1072. [CrossRef]
8. Yang, Y.B.; Lin, C.W.; Yau, J.D. Extracting bridge frequencies from the dynamic response of a passing vehicle. *J. Sound Vib.* **2004**, *272*, 471–493. [CrossRef]
9. Lin, C.W.; Yang, Y.B. Use of a passing vehicle to scan the fundamental bridge frequencies: An experimental verification. *Eng. Struct.* **2005**, *27*, 1865–1878. [CrossRef]
10. Yang, Y.B.; Chen, W.F. Extraction of Bridge Frequencies from a Moving Test Vehicle by Stochastic Subspace Identification. *J. Bridge Eng.* **2016**, *21*, 04015053. [CrossRef]
11. Malekjafarian, A.; OBrien, E.J. On the use of a passing vehicle for the estimation of bridge mode shapes. *J. Sound Vib.* **2017**, *397*, 77–91. [CrossRef]
12. González, A.; Obrien, E.J.; Mcgetrick, P.J. Identification of damping in a bridge using a moving instrumented vehicle. *J. Sound Vib.* **2012**, *331*, 4115–4131. [CrossRef]
13. Keenahan, J.; Obrien, E.J.; Mcgetrick, P.J.; Gonzalez, A. The use of a dynamic truck-trailer drive-by system to monitor bridge damping. *Struct. Health Monit.* **2013**, *13*, 143–157. [CrossRef]
14. OBrien, E.J.; Malekjafarian, A. A mode shape-based damage detection approach using laser measurement from a vehicle crossing a simply supported bridge. *Struct. Control. Health Monit.* **2016**, *23*, 1273–1286. [CrossRef]
15. Oshima, Y.; Yamamoto, K.; Sugiura, K. Damage assessment of a bridge based on mode shapes estimated by responses of passing vehicles. *Smart. Struct. Syst.* **2014**, *13*, 731–753. [CrossRef]
16. Robert, C.; Abdollah, M. Examining changes in bridge frequency due to damage using the contact-point response of a passing vehicle. *J. Struct. Integr. Maint.* **2021**, *6*, 148–158.
17. Mei, Q.; Gul, M.; Boay, M. Indirect health monitoring of bridges using Mel-frequency cepstral coefficients and principal component analysis. *Mech. Syst. Signal Proc.* **2018**, *119*, 523–546. [CrossRef]
18. Corbally, R.; Malekjafarian, A. A data-driven approach for drive-by damage detection in bridges considering the influence of temperature change. *Eng. Struct.* **2022**, *253*, 113783. [CrossRef]
19. Sarwar, M.Z.; Cantero, D. Deep autoencoder architecture for bridge damage assessment using responses from several vehicles. *Eng. Struct.* **2021**, *246*, 113064. [CrossRef]
20. Locke, W.; Sybrandt, J.; Redmond, L.; Safro, I.; Atamturktur, S. Using drive-by health monitoring to detect bridge damage considering environmental and operational effects. *J. Sound Vib.* **2020**, *468*, 115088. [CrossRef]
21. Abdulkareem, M.; Bakhary, N.; Vafaei, M.; Noor, N.M.; Padil, K.H. Non-probabilistic wavelet method to consider uncertainties in structural damage detection. *J. Sound Vib.* **2018**, *433*, 77–98. [CrossRef]
22. Abdulkareem, M.; Bakhary, N.; Vafaei, M.; Noor, N.M.; Mohamed, R.N. Application of two-dimensional wavelet transform to detect damage in steel plate structures. *Measurement* **2019**, *146*, 912–923. [CrossRef]
23. Abdulkareem, M.; Bakhary, N.; Ganiyu, A.; Nathaniel, O.; Jassam, T.M.; Al-Mansob, R.A. Consideration of uncertainty in damage detection using interval analysis wavelet without baseline data. *J. Struct. Integr. Maint.* **2021**, *6*, 99–109. [CrossRef]
24. Zhu, X.Q.; Law, S.S. Wavelet-based crack identification of bridge beam from operational deflection time history. *Int. J. Solids Struct.* **2006**, *43*, 2299–2317. [CrossRef]
25. Hester, D.; González, A. A wavelet-based damage detection algorithm based on bridge acceleration response to a vehicle. *Mech. Syst. Signal Proc.* **2012**, *28*, 145–166. [CrossRef]
26. Nguyen, K.V.; Hai, T.T. Multi-cracks detection of a beam-like structure based on the on-vehicle vibration signal and wavelet analysis. *J. Sound Vib.* **2010**, *329*, 4455–4465. [CrossRef]
27. Khorram, A.; Bakhtiari-Nejad, F.; Rezaeian, M. Comparison studies between two wavelet based crack detection methods of a beam subjected to a moving load. *Int. J. Eng. Sci.* **2012**, *51*, 204–215. [CrossRef]
28. Tan, C.; Elhattab, A.; Uddin, N. Wavelet-Entropy Approach for Detection of Bridge Damages Using Direct and Indirect Bridge Records. *J. Infrastruct. Syst.* **2020**, *26*, 04020037. [CrossRef]
29. Yang, Y.B.; Chang, K.C. Extraction of bridge frequencies from the dynamic response of a passing vehicle enhanced by the EMD technique. *J. Sound Vibr.* **2009**, *322*, 718–739. [CrossRef]

30. Obrien, E.J.; Malekjafarian, A.; González, A. Application of empirical mode decomposition to drive-by bridge damage detection. *Eur. J. Mech. A-Solids* **2017**, *61*, 151–163. [CrossRef]
31. Zhu, L.; Malekjafarian, A. On the Use of Ensemble Empirical Mode Decomposition for the Identification of Bridge Frequency from the Responses Measured in a Passing Vehicle. *Infrastructures* **2019**, *4*, 32. [CrossRef]
32. Yang, Y.B.; Xiong, F.; Wang, Z.L.; Xu, H. Extraction of Bridge Frequencies Inclusive of the Higher Modes by the ESMD Using the Contact-Point Response. *Int. J. Struct. Stab. Dyn.* **2020**, *20*, 718–723. [CrossRef]
33. Li, J.; Zhu, X.; Guo, J. Bridge modal identification based on successive variational mode decomposition using a moving test vehicle. *Adv. Struct. Eng.* **2022**, *25*, 2284–2300. [CrossRef]
34. Yang, Y.B.; Zhang, B.; Qian, Y.; Wu, Y. Further Revelation on Damage Detection by IAS Computed from the Contact-Point Response of a Moving Vehicle. *Int. J. Struct. Stab. Dyn.* **2018**, *18*, 1850137. [CrossRef]
35. Yang, Y.B.; Zhang, B.; Qian, Y.; Wu, Y. Contact-Point Response for Modal Identification of Bridges by a Moving Test Vehicle. *Int. J. Struct. Stab. Dyn.* **2018**, *18*, 1850073. [CrossRef]
36. Zhu, J.; Wu, M.X.; Zhang, W. Evaluation of Ride Comfort and Driving Safety for Moving Vehicles on Slender Coastal Bridges. *J. Vib. Acoust.-Trans. ASME* **2018**, *140*, 051012. [CrossRef]
37. Zhu, J.; Zhang, W. Numerical Simulation of Wind and Wave Fields for Coastal Slender Bridges. *J. Bridge Eng.* **2016**, *22*, 04016125. [CrossRef]
38. Xiong, Z.; Zhu, J.; Zheng, K.; Zhang, W.; Li, Y.; Wu, M. Framework of wind-traffic-bridge coupled analysis considering realistic traffic behavior and vehicle inertia force. *J. Wind Eng. Ind. Aerodyn.* **2020**, *205*, 104322. [CrossRef]
39. Huang, N.E.; Shen, Z.; Long, S.R.; Wu, M.C.; Shih, H.H.; Zheng, Q.; Yen, N.-C.; Tung, C.C.; Liu, H.H. The Empirical Mode Decomposition and the Hilbert Spectrum for Nonlinear and Non-Stationary Time Series Analysis. *Proc. R. Soc. A-Math. Phys. Eng. Sci.* **1998**, *454*, 903–995. [CrossRef]
40. Wu, Z.; Huang, N.E. Ensemble Empirical Mode Decomposition: A Noise-Assisted Data Analysis Method. *Adv. Adapt. Data Anal.* **2011**, *1*, 1–41. [CrossRef]
41. Torres, M.E.; Colominas, M.A.; Schlotthauer, G.; Flandrin, P. A complete ensemble empirical mode decomposition with adaptive noise. In Proceedings of the 2011 IEEE International Conference on Acoustics, Speech and Signal Processing (ICASSP), Prague, Czech Republic, 22–27 May 2011; pp. 4144–4147.

buildings

MDPI

Article

A Two-Step FE Model Updating Approach for System and Damage Identification of Prestressed Bridge Girders

Niloofar Malekghaini [1], Farid Ghahari [2], Hamed Ebrahimian [1,*], Matthew Bowers [3], Eric Ahlberg [2] and Ertugrul Taciroglu [2]

[1] Department of Civil and Environmental Engineering, University of Nevada, Reno, NV 89557, USA
[2] Department of Civil and Environmental Engineering, University of California, Los Angeles, CA 90095, USA
[3] SC Solutions Inc., Sunnyvale, CA 94085, USA
* Correspondence: hebrahimian@unr.edu

Abstract: This study presents a two-step FE model updating approach for health monitoring and damage identification of prestressed concrete girder bridges. To reduce the effects of modeling error in the model updating process, in the first step, modal-based model updating is used to estimate linear model parameters mainly related to the stiffness of boundary conditions and material properties. In the second step, a time-domain model updating is carried out using acceleration data to refine parameters accounting for the nonlinear response behavior of the bridge. In this step, boundary conditions are fixed at their final estimates using modal-based model updating. To prevent the convergence of updating algorithm to local solutions, the initial estimates for nonlinear material properties are selected based on the first-step model updating results. To validate the applicability of the two-step FE model updating approach, a series of forced-vibration experiments are designed and carried out on a pair of full-scale decommissioned and deteriorated prestressed bridge I-girders. In the first step, parameters related to boundary conditions, including stiffness of supports and coupling beams, as well as material properties, including initial stiffness of concrete material, are estimated. In the second step, concrete compressive strength and damping properties are updated. The final estimates of the concrete compressive strength are used to infer the extent of damage in the girders. The obtained results agree with the literature regarding the extent of reduction in concrete compressive strength in deteriorated concrete structures.

Keywords: modal-based model updating; Bayesian model updating; system identification; damage identification; operational health monitoring; I-girder; bridge; aging

1. Introduction

Bridges are vital components of the transportation infrastructure. The average age of in-service bridges in the United States is increasing, which necessitates methods and tools to inform decision making related to the maintenance and/or replacement of these structures [1,2]. Finite Element (FE) model updating methods have emerged as a venerable procedure for operational health monitoring and post-event structural damage identification [3–17]. In these methods, the initial/baseline FE model—developed using available as-built drawings—is updated using measured dynamic responses. During this process, uncertain model parameters—including material properties, damping parameters, boundary conditions, etc.—are calibrated/estimated. The deviation of final estimates of model parameters from their initial/baseline values reveals information regarding the location and extent of damage in the structure.

FE model updating approaches are mainly divided into two groups. The first group is *modal-based model updating*, wherein the initial FE model is updated to match the identified modal properties of the structure. In this method, the modal properties are first identified using modal identification methods (e.g., [18–22]). Then, the parameters characterizing the linear response behavior of the FE model are estimated to reduce the discrepancies between the identified and FE-predicted modal properties [3]. The accuracy of the identified modal signatures is controlled by the level of nonlinearity in the response behavior of structure, measurement noise, excitation frequency range, and sensor sparsity. The uncertainty in the identified modal properties propagates into the model updating process and is reflected in the final estimates of model parameters [23]. In addition, modal properties cannot be used to infer parameters characterizing the nonlinear material behavior. Consequently, the updated FE model may not be able to predict the dynamic response behavior of the structure correctly, especially when the structure is subjected to material nonlinearity. Although the estimated linear model parameters can be used for structural monitoring and damage identification [24], the application of modal-based model updating for damage identification of reinforced concrete bridges has been shown to be limited [25,26]. The second group of model updating approaches is referred to as *time-domain model updating* [17,27,28]. In this approach, the unknown model parameters characterizing linear and/or nonlinear response behavior of structure are updated to reduce the discrepancies between the measured and FE-predicted responses in time domain. In contrast to the modal-based model updating method, the measured dynamic responses of the structure are used directly for model updating in this approach. The direct application of dynamic responses in time-domain model updating eliminates the propagation of modal identification uncertainties into the model updating process.

Several studies in the literature (e.g., [10,23,29–35]) have focused on the system and damage identification of bridges subjected to ambient or traffic excitation (i.e., operational conditions) using modal-based model updating. The performance of these methods is mostly evaluated by comparing the identified and posterior FE-predicted modal signatures of the bridge. Studies in [36,37] showed that the accuracy of the updated model is highly sensitive to the selection of unknown model parameters. Moreover, [38] indicates that damage detection of bridges would depend on the proper simulation of boundary conditions. A two-step FE model updating process is suggested in this study to resolve modeling errors due to boundary conditions.

Using measurements other than modal properties for model updating and damage identification of bridges has attracted research interest recently. While static/pseudo-static responses (e.g., displacement measurements) have been used in previous studies for model updating [5,11,17,39–42], acceleration measurements have not been used directly for the purpose of model updating of bridges under operational conditions. It is worth noting that static/pseudo-static responses contain limited information regarding the dynamic behavior of the bridge compared to acceleration measurements. Therefore, acceleration measurements can be more informative about the uncertain model parameters compared to static/pseudo-static responses. The target in this study is to use both the modal properties and acceleration responses for model updating and damage diagnosis through a two-step FE model updating process.

Moreover, previous studies have shown that weak identifiability and mutual dependency between model parameters, modeling errors, as well as convergence of parameters to local solution may challenge the model updating process [4,43,44]. These challenges are exacerbated in a real-world application, especially in cases with large number of uncertain model parameters and/or improper selection of initial values for the uncertain model parameters.

To resolve the above-mentioned issues, this study presents a sequential combination of modal-based and time-domain model updating for operational health monitoring and damage identification of aged bridges using acceleration responses. In this procedure, first, a deterministic modal-based model updating is carried out to estimate the linear model parameters of a bridge. These model parameters are related to boundary conditions and material properties. Then, in order to refine the parameter estimation and account for the nonlinear response behavior of the bridge, a time-domain model updating is carried out. In this step, nonlinear material properties, as well as the damping energy-dissipation-related model parameters, are estimated while the linear-elastic model parameters are fixed at their final estimates obtained from the modal-based model updating. The final estimates of material properties are used to infer/quantify damage in the bridge.

In summary, the reasoning behind introducing the sequential combination of modal-based and time-domain model updating, which constitute the novelties of this work, can be summarized as follows.

- Unknown boundary conditions often challenge the application of time-domain model updating for bridges since the model parameters are often dependent on the boundary conditions. Here, the modal-based model updating is used to identify the boundary conditions first.
- The application of modal-based model updating for damage identification of bridges is limited. This is likely due to the uncertainties in the identified modal signatures that propagate through the parameter estimation process. Therefore, here, the estimation of model parameters for damage identification will be refined through a subsequent time-domain model updating.
- The dynamic measurements can provide more information about the uncertain material parameters compared to the static/pseudo-static responses. Hence, here, the acceleration measurements are used directly in the time-domain model updating.
- To improve the numerical stability and convergence of the model parameters the linear and nonlinear response behavior of the bridge are assimilated through the two-step model updating process.

To show the two-step FE model updating method and validate its applicability for damage identification in a real-world setting, a pair of full-scale precast prestressed bridge I-girders were used as testbed structures. These girders were in service from 1971 until 2009 before they were decommissioned and repurposed for research experiments [45]. A series of forced-vibration experiments were designed specifically for this study. The girders were subjected to sinusoidal force excitations, and their acceleration responses were measured at different locations. First, the collected acceleration responses are used to identify the modal signatures of the testbed structure. Then, the two-step FE model updating is carried out. In the first step, the initial FE model of girders is updated in the modal domain, and boundary conditions, including stiffness of supports and coupling beams, as well as material properties, including initial stiffness of concrete material, are estimated. The updated model is used as the prior model in the Bayesian model updating process to estimate concrete compressive strength and damping properties. Comparison between the posterior FE-predicted responses and field measurements shows a good agreement in the time domain. Moreover, the final estimates of concrete compressive strength result in a realistic damage identification/quantification for the girders. This process validates the applicability of the introduced two-step FE model updating approach for damage identification of bridge structures/components under operational conditions. While the input load used in this study varies from moving traffic load, this study proves the concept for future real-world application.

The paper is organized as follows. Section 2 is focused on test methodology and preliminary results including an introduction to the testbed structure and the experiments, modal identification, and development of the initial FE model. The two-step FE model updating method and the results are discussed in Section 3. Concluding remarks are provided in Section 4.

2. Material, Test Methodology, and Preliminary Results

In this section, first, a description of the field experiment including testbed structure, dynamic excitation system, wireless sensing network, and force-vibration tests is presented. Then, the modal identification process and corresponding results are shown in Section 2.2. Finally, the initial FE model is developed in Section 2.3.

2.1. Description of the Field Experiment

2.1.1. Testbed Structure

The testbed structure in this study includes two AASHTO precast prestressed bridge I-girders that were part of the Maryland 90 bridge. After decommissioning, the girders were salvaged and transferred to the Turner-Fairbank Highway Research Center (TFHRC) in McLean, VA, USA to be used as a research testbed [45,46]. The cross-section and elevation views of the girders are shown in Figure 1.

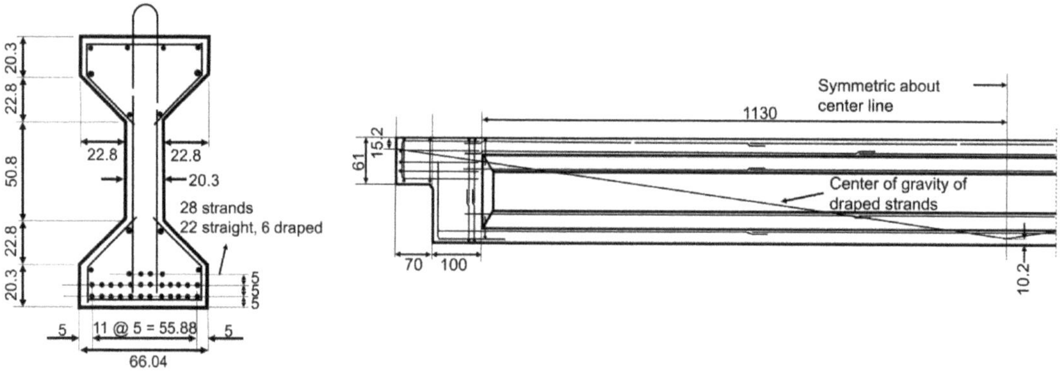

(a) (b)

Figure 1. The testbed structure: (**a**) cross-section view of each girder and (**b**) elevation view of each girder. All the dimensions are in centimeters, and all reinforcing rebars are #4 (equivalent to Φ13).

The studied girders are 1.37 m deep and 26 m long. Reinforcements and prestressing steel strands are shown in Figure 1. A side view of the testbed structure (with a 20 cm slab on top of each girder) is shown in Figure 2a. After being transferred to the TFHRC, each girder was placed on two 1×1 m^2 bearing pads on top of two $2.3 \times 1.6 \times 1.6$ m^3 geosynthetic reinforced soil (GRS) piers [47]. These can be seen in Figure 2b,c. The girders were placed parallel to each other with 2.9 m centerline spacing and were connected with four coupling beams with 0.3×1.4 m^2 cross-sectional area. Three out of four coupling beams are seeable in Figure 2d.

The studied girders were in service in a corrosive environment for almost 40 years. The environment simultaneously exposed the concrete matrix of the girders to physical and chemical deterioration processes. Concrete delamination and degradation, as well as steel corrosion, are the main damage mechanisms for concrete bridges in such environment [48]. Due to this, the girders experienced aging and deterioration in several locations, including cracking, steel reinforcement corrosion, spalling, etc. Figure 2e shows an example of the observed damage in girders.

Aside from the aging-related damage discussed above, in 2012, salt spray chambers were installed on each girder to accelerate deterioration in the girders. The chamber installed on the west girder sprayed a 15 weight percent (wt.%) NaCl solution and the chamber installed on the east girder sprayed a 3.5 wt.% NaCl solution. This was part of a study to develop protocols for non-destructive testing (NDT) methods for prestressed girder bridges [45].

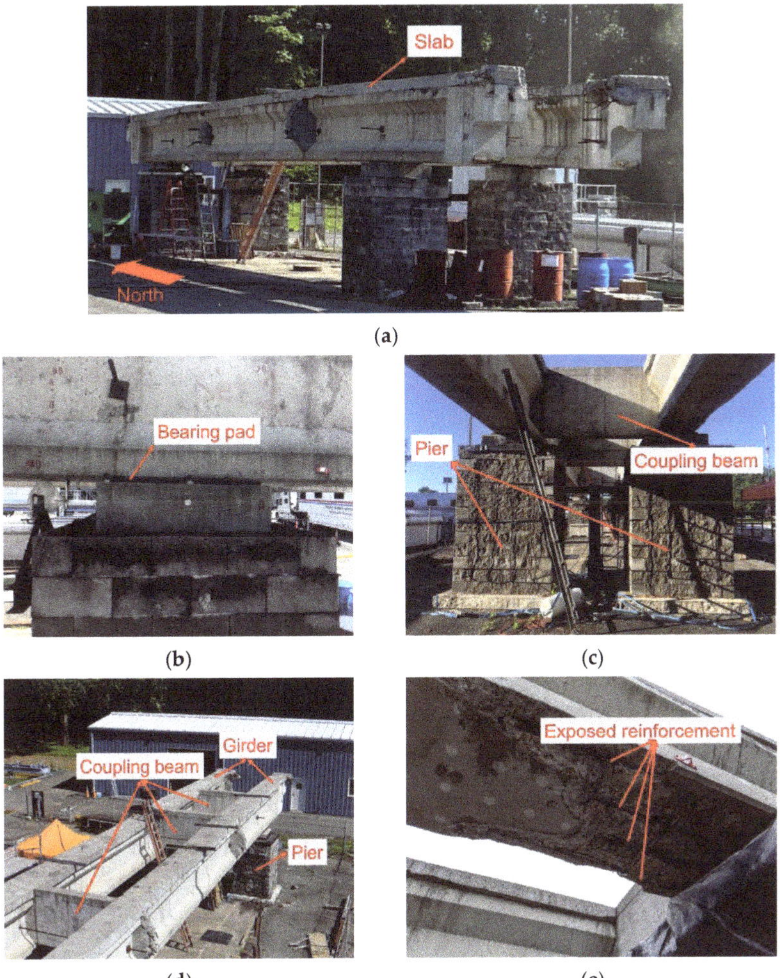

Figure 2. Testbed structure: (**a**) side view and north direction, (**b**) bearing pad, (**c**) piers and coupling beam, (**d**) top view, and (**e**) example of existing damage (west girder close to the north pier).

2.1.2. Dynamic Excitation System

The shaker used in this study was a small uniaxial hydraulic device capable of producing arbitrary displacement motions in the vertical direction. The device consisted of a hydraulic piston connected to a servo-hydraulic valve that controlled the motion of a stack of steel plates that combined to form 4450 N of moving weight. The motion was controlled by an MTS PID hydraulic controller using an LVDT sensor to provide feedback displacement. The shaker's hydraulic piston, moving weights, and steel frame of the shaker were supported on four 4450 N load cells, which were used to measure the total force generated by the shaker. The load cells were installed between the shaker plate and the clamping plate on the girder. The shaker plate with dimensions of 30 cm × 30 cm × 2 cm was located on the top center of the clamping plate with dimensions of 140 cm × 70 cm × 5 cm (see Figure 3). The shaker displacement and force time histories were collected through a LORD-Microstrain V-Link-200 wireless node [49]. The wireless sensing network is discussed further in the following section. Figure 3 shows a close view of the shaker, and

Figure 4 shows the two locations (layout 1 and layout 2) at which the shaker was installed on the testbed structure. The hydraulic power for the shaker was provided by a diesel-powered mobile pump.

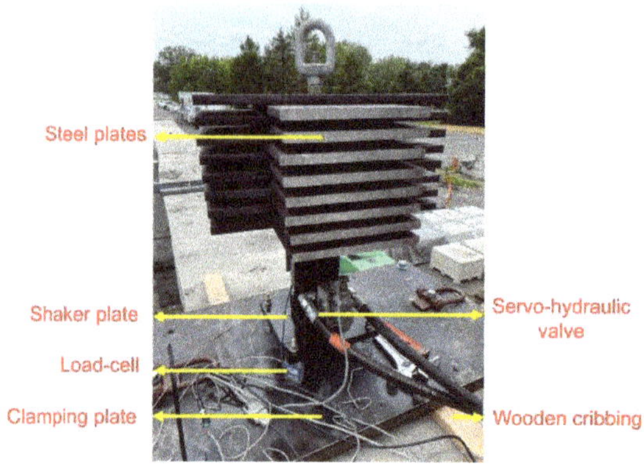

Figure 3. The shaker installed on the girder.

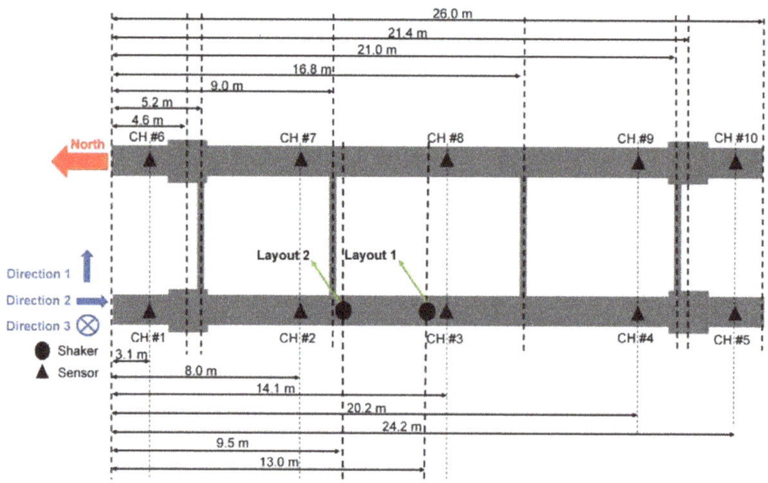

Figure 4. Test layout including the sensor and shaker locations.

2.1.3. Wireless Sensing Network

The wireless sensing network included ten battery-powered triaxial MEMS wireless accelerometers (LORD-Microstrain G-Link-200) and a V-Link-200 node [49]. Each accelerometer had an adjustable measurement range of ± 8 g and could be configured for continuous, periodic, or event-triggered sampling modes to output acceleration, tilt, or derived vibration parameters (velocity, amplitude, etc.). The measured data could be transmitted in real-time and/or be saved to the onboard memory with storage capacity up to 8×10^6 data points. The accelerometers had a noise density of 25 $\frac{\mu g}{\sqrt{Hz}}$ with a wireless range of up to 1 km and an adjustable sampling rate of up to 4 kHz. Each accelerometer had dimensions of 47 mm $\times$ 43 mm $\times$ 44 mm. To install the accelerometers, zinc-plated steel washers were glued on top surface of the girders. Then, each accelerometer was screwed to a magnetic base and attached to the washers. Figure 5 shows an installed wireless accelerometer. Data collection and coordination between the wireless nodes including the accelerometers and the V-Link-200 node, which was used to collect shaker data, were carried out through the wireless USB data acquisition gateway. The gateway used the lossless extended range synchronized (LXRS) data communication protocol and facilitated lossless data collection with node synchronization of ±50 μs. Synchronization was carried out by transmission of a continuous system-wide timing reference known as the beacon. The communication between the gateway and sensors was wireless through a license-free 2.405 GHz to 2.480 GHz radio frequency with 16 channels. The configuration of the network, data acquisition initialization, and sampling mode selection were managed through the SensorConnect software [50], which was installed on a host computer. The layout of the employed accelerometers is shown in Figure 4. In this study, the sampling rate was 128 Hz, and the acceleration data were collected in directions 1 (i.e., east-west) and 3 (i.e., up-down).

Figure 5. A wireless accelerometer installed on the girder's top surface.

2.1.4. Forced-Vibration Testing

The testbed structure was subjected to a series of designed sinusoidal force excitations through frequency sweeps with pre-defined frequency range, duration, and amplitude. Sweeps ranged between 2 Hz and 20 Hz, including 50 different frequencies increasing logarithmically with a duration of 30 s for each frequency. Moreover, the sweeps excited the girders with three different target load amplitudes equal to 445 N, 2225 N, and 4450 N. Considering two layouts (see Figure 4) and three levels of load amplitudes, girders were tested under six frequency sweeps. A general view of the testbed structure during the field experiments is shown in Figure 6. The excitation force time history and instantaneous excitation frequency—calculated using a short-time Fourier transform [51]—for each sweep are presented in Figure 7.

Figure 6. The field experiment setup corresponding to layout 2 (See Figure 4).

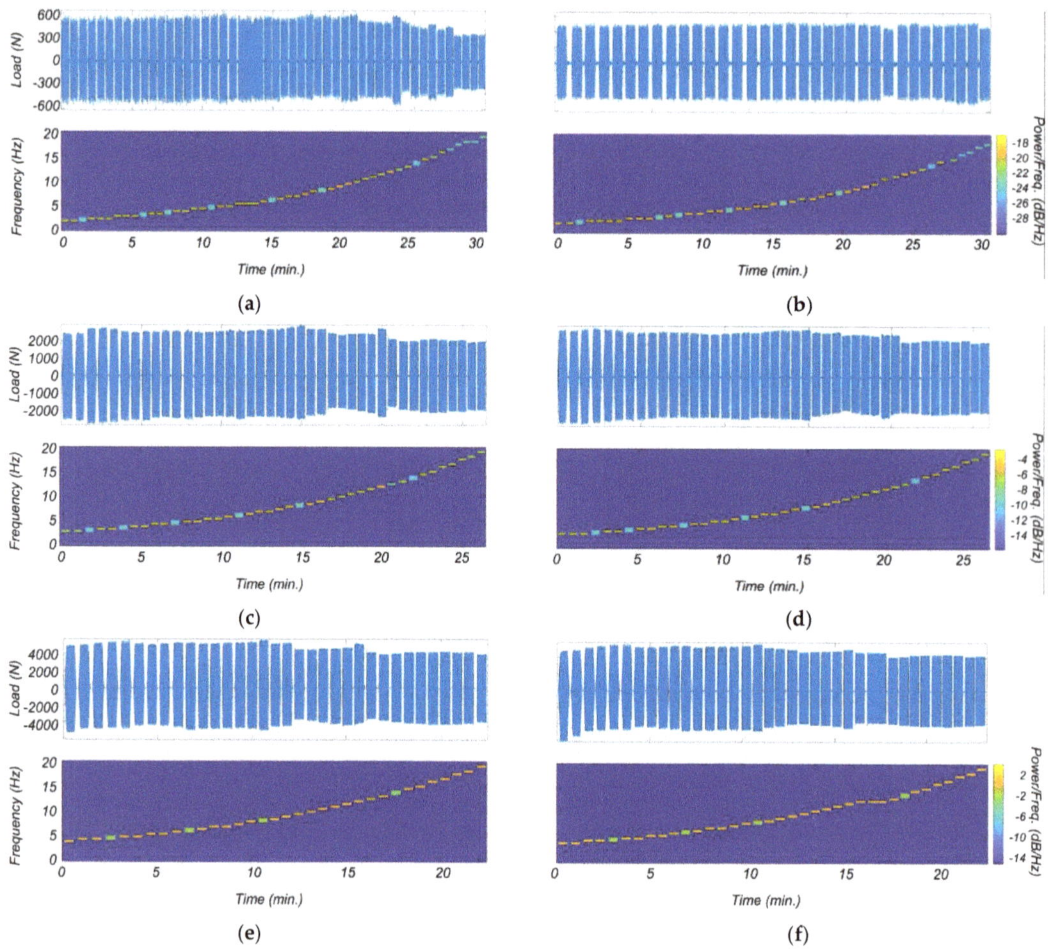

Figure 7. Excitation force time histories and instantaneous excitation frequencies: (**a**) Layout 1 with 445 N target load amplitude, (**b**) Layout 2 with 445 N target load amplitude, (**c**) Layout 1 with 2225 N target load amplitude, (**d**) Layout 2 with 2225 N target load amplitude, (**e**) Layout 1 with 4450 N target load amplitude, and (**f**) Layout 2 with 4450 N target load amplitude.

2.2. Modal Identification

In this section, the modal properties of the testbed structure are identified from the forced-vibration experimental data. For the purpose of modal identification, the collected data from the sweep with layout 1 and 445 N target load amplitude are used (see Figure 7a). These data include the acceleration of 10 channels in directions 1 and 3 (results in 20 input signals—see Figure 4) and the shaker excitation force. A brief summary of the modal identification process and the identified modal properties are presented in this section.

Various modal identification techniques are available in the literature to identify modal properties—including natural frequencies, damping ratios, and mode shapes—from experimental data [19,21,22]. In this study, due to the nature of the excitation, the empirical frequency response functions (EFRFs) [22] are calculated using applied input (shaker excitation force) and measured outputs (acceleration responses). Then, the calculated multi-output EFRFs are used to estimate a state-space model. This two-stage frequency-domain approach helps to specify the frequency band of interest in which the modal identification needs to be carried out.

The EFRF between a measurement signal $\mathbf{y}(t)$ and input signal $\mathbf{u}(t)$—depicted by $\mathbf{G}(f)$—is defined as below:

$$\mathbf{G}(f) = \frac{\mathbf{Y}(f)}{\mathbf{U}(f)} \tag{1}$$

where $\mathbf{Y}(f)$ and $\mathbf{U}(f)$ are Fourier transforms of $\mathbf{y}(t)$ and $\mathbf{u}(t)$, respectively. In a real-world setting, data are always polluted with measurement noise. To reduce the effects of the measurement noise and obtain a smooth EFRF, Welch's averaging method [52] with 12,800 data points Hamming window is used for spectral estimation. This size of window is selected to ensure covering the full length of excitation and the ambient signal before and/or after it.

Having EFRFs for all 20 signals, an n-order state-space model is estimated to fit the estimated EFRFs in the frequency band of interest. In this study, to reduce uncertainties due to low- and high-frequency noises, the frequency band of interest is selected between 2 to 25 Hz. Blue curves in Figure 8 show the calculated EFRFs using measured data.

To estimate the state-space model, the subspace state-space identification method is used [53,54]. While this method is briefly explained here, the proof of theory and more details can be found in [53]. In the subspace state-space identification method, measurements are placed in a block Hankel matrix which is divided into a past and a future part. The identification algorithm proceeds with projecting the future measurements into the past measurements, while the projection matrix can be factorized as the product of an observability matrix and a state sequence. These two matrices are identified by applying the singular value decomposition (SVD) to the projection matrix, and the order of the system is calculated as the number of non-zero singular values. By applying one block shift in separation between past and future measurements in the Hankel matrix, another projection matrix, shifted observability, and state sequence matrices can be obtained. At this point, the system matrices can be calculated from the overdetermined set of linear equations.

The numerical algorithms for system identifications are available in the Matlab *n4sid* [55] function and are used in this study. As a classical remedy, the modal identification is carried out for a range of model orders, and a stability diagram is plotted on which true modes appear as stable modes [56]. For this purpose, the stability analysis is run considering model orders from 2 to 40 with 1% and 5% error tolerances for natural frequency and damping ratios, respectively. The stability analysis showed that a model order of $n = 26$ is the lowest model order to have all stable modes within the frequency band of interest. The fits between estimated and calculated EFRFs can be improved using the prediction error minimization algorithm and nonlinear least-squares objective functions. This approach is carried out using the Matlab *ssest* function [57], which initializes the model parameters based on the previously estimated state-space model, and then updates the parameters using an iterative search to minimize the prediction errors [21]. Red curves in Figure 8 display EFRFs of the estimated state-space model at measurement points. As can be seen, the estimated state-space model is able to approximate the calculated EFRFs acceptably. Identified natural frequencies (f^{ID}) and damping ratios (ξ^{ID}) of the system are reported in Table 1. The identified mode shapes—those which will be used for modal-based model updating—are later shown in Section 3.1.

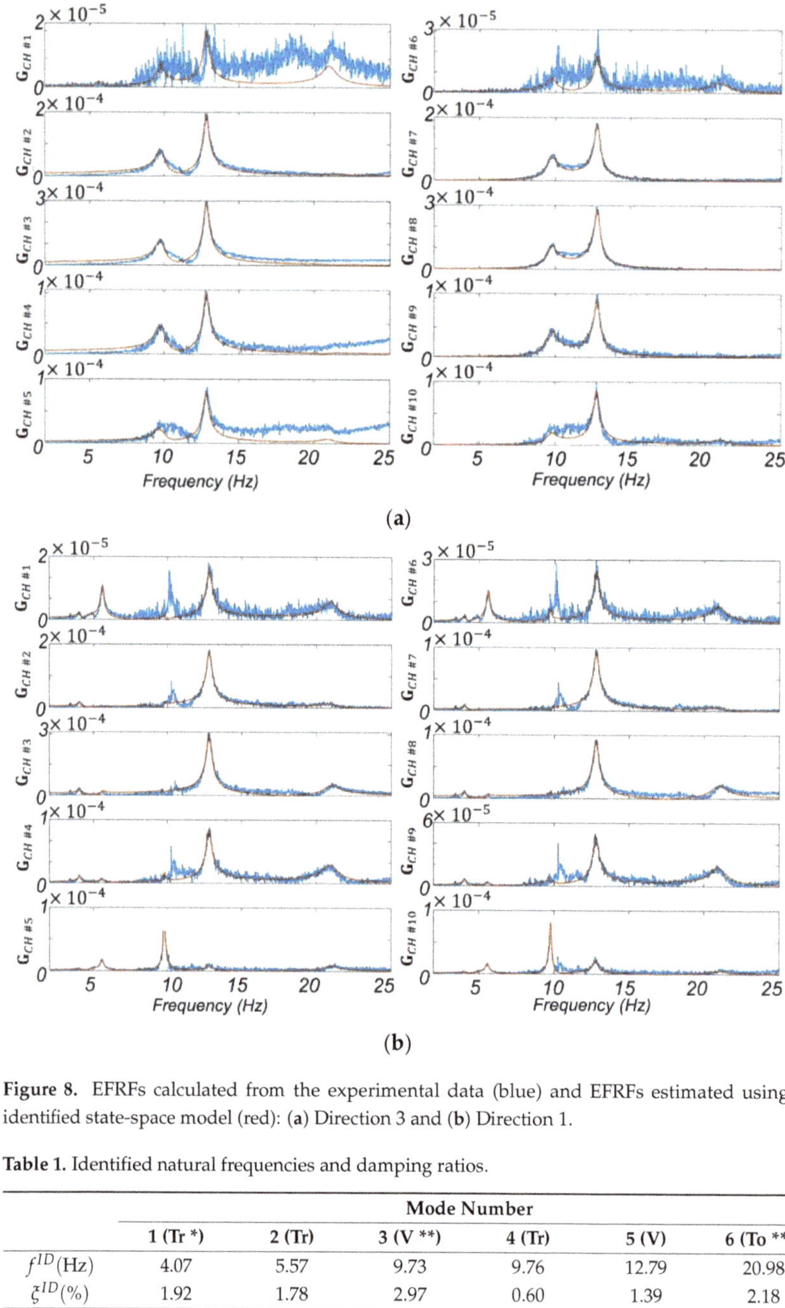

Figure 8. EFRFs calculated from the experimental data (blue) and EFRFs estimated using the identified state-space model (red): (**a**) Direction 3 and (**b**) Direction 1.

Table 1. Identified natural frequencies and damping ratios.

	Mode Number					
	1 (Tr *)	2 (Tr)	3 (V **)	4 (Tr)	5 (V)	6 (To ***)
f^{ID}(Hz)	4.07	5.57	9.73	9.76	12.79	20.98
ξ^{ID}(%)	1.92	1.78	2.97	0.60	1.39	2.18

*: Transverse mode, **: Vertical mode, ***: Torsional mode.

2.3. Finite Element Modeling of the Testbed Structure

The initial FE model of the testbed structure is developed in OpenSees [58] using the available as-built drawings. The girders are modeled using fiber-section force-based beam-column elements (*forceBeamColumn*) with approximately 30 cm length and 3 integration points. Moreover, the linear-elastic shear stiffness and torsional stiffness are aggregated to the fiber sections. The slab is also modeled using rectangular fiber sections as a part of the girders' sections. Steel reinforcement is modeled using single fibers located at the center of each bar. To model the profile of the draped strands—with a nominal cross-sectional area of 1 cm^2—the strands' depth at the integration points of each element is calculated based on Figure 1. The strand is modeled using a single fiber at its corresponding depth. Concrete material for girders and slabs is modeled with nominal compressive strength ($f'_{c,E}$ and $f'_{c,W}$ for the east and west girders) of 46 MPa, strain at maximum strength (ε_c) of 0.2%, strain at crushing strength (ε_u) of 0.5%, and zero crushing strength. As girders are already damaged, no tensile strength is assumed, and therefore, the *Concrete01* material model is employed. To model the shear and torsional stiffness of the girder sections, the shear modulus (G) is calculated based on the concrete modulus of elasticity, E (= $E_E = E_W$), and Poisson's ratio of 0.2. Moreover, the shear area for girder sections is set equal to 0.38 and 0.65 m^2 in the Y and X directions (see Figure 1), and torsional constant of girder sections is set equal to 0.03 m^4. The parameters E_E and E_w—modulus of elasticity for the east and west girders—are set equal to the initial slope in *Concrete01* model (i.e., $\frac{2f'_{c,E}}{\varepsilon_c}$ and $\frac{2f'_{c,W}}{\varepsilon_c}$) which is 46 GPa here. Steel is modeled using bilinear *steel01* material with a modulus of elasticity of 200 GPa, and yielding strength of 455 MPa and 1720 MPa for reinforcing steels and strands, respectively. Moreover, the prestressing force in strands (128.60 kN per strand) is modeled by applying its resulting initial strain to the steel material. For this purpose, the normal strain in strands' steel resulting from the prestressing force is calculated by dividing the prestressing force by strand's axial rigidity (product of the strand's modulus of elasticity and gross section area). The calculated strain (0.65%) is assigned to the strands' *steel01* material using *InitStrainMaterial*. The mass, 104×10^3 kg in total, and weight of the girders are assigned to the element nodes.

Coupling beams are modeled using *elasticBeamColumn* elements considering the gross cross-sectional area and an *Elastic* material with the modulus of elasticity equal to 30 GPa. The coupling beams are connected to the girders assuming rigid connections. The mass and weight of the coupling beams are assigned to the element nodes. To model each support, the girder nodes that are located along the bearing pad are constrained to a node at the center of the bearing pad using *rigidLink* constraint. To account for flexibility in the piers and bearing pads, supports are defined using a 6 degrees of freedom (DOFs) spring. This is modeled using *ZeroLength* elements with *Elastic* uniaxial material. Moreover, the energy dissipation in the piers and bearing pads is collectively modeled using a dashpot in the vertical direction. This is performed using *ZeroLength* elements with *Viscous* uniaxial material. The corresponding stiffness and damping parameters are initially selected based on engineering judgment [47,59] and are later updated using modal-based and time-domain model updating. In summary, the FE model consists of 214 beam-column elements, 16 *rigidLink* elements, 12 *ZeroLength* elements, and 230 joints. The list of model parameters that are later treated as unknown parameters in Section 3.1 and 3.2 and their initial values are summarized in Table 2. In this table, the equivalent values for concrete compressive strength and modulus of elasticities of girders are shown in the same row separated with '/'. The directions in this table are based on Figure 4.

Table 2. Model parameters for the initial FE model. The directions in this table are based on Figure 4.

Parameter	Parameter Description	Initial Values
$f'_{c,W}/E_W$	Concrete compressive strength/Concrete modulus of elasticity for the west girder	46 MPa/46 GPa
$f'_{c,E}/E_E$	Concrete compressive strength/Concrete modulus of elasticity for the east girder	46 MPa/46 GPa
E_c	Modulus of elasticity for coupling beams	30 GPa
K_R	Rotational stiffness of bearing pads about directions 1 and 2 and 3	9.4×10^5 kN/rad
K_V	Vertical stiffness of bearing pads in direction 3	1.2×10^5 kN/m
K_T	Transverse stiffness of bearing pads in direction 1	1.6×10^4 kN/m
K_L	Longitudinal stiffness of bearing pads in direction 2	1.6×10^4 kN/m
C_D	Damping coefficient for bearing pads in direction 3	45×10^2 kN.sec/m
ξ_1	Damping ratio for mode 1	0.02
ξ_2	Damping ratio for mode 2	0.02

The nonlinear time history analysis is performed using the Newmark average acceleration method with a constant time step size of 0.0078 s equal to the measurement sampling rate. The Newton-Raphson method is used to iteratively solve the nonlinear equilibrium equations [60]. To define energy dissipation in the structural system aside from the material nonlinearity, modal damping is modeled for the first six modes. The damping ratios, i.e., $\xi_i, \forall\, i \in \{1,2,3,4,5,6\}$, are set equal to the identified ones (see Table 1). The only exceptions are damping ratios for modes 1 and 2, which are set equal to 0.02. The reason is to have similar initial values—which is not very different from the identified values (0.0192 and 0.0178)—during time-domain model updating, which is discussed later.

3. Two-Step Model Updating: Methodology, Results, and Discussion

3.1. First Step: Modal-Based Model Updating

The modal properties of the initial FE model are different from the identified ones (Figure 9). Hence, the initial FE model needs to be updated using modal-based model updating to better fit the identified modal properties. As mentioned before, the modal-based model updating is limited to the linear response behavior of structures. Hence, only the linear model parameters are updated at this step. Modal-based model updating process is discussed next.

The modal-based model updating is a process to minimize the discrepancies between identified and FE-predicted modal frequencies and mode shapes by updating the linear parameters of the FE model [24,61]. In this process, an objective function, $g(\boldsymbol{\theta})$, is defined as shown in Equation (2). The discrepancies between modal frequencies, mode shapes, and a regularization term are respectively the first, second, and third terms in Equation (2).

$$g(\boldsymbol{\theta}) = W_f\, \mathbf{r}_f{}^T\, \mathbf{r}_f + W_M \left(N - \sum_{i=1}^{i=N} MAC_i \right) + (\boldsymbol{\theta} - \boldsymbol{\theta}_0)^T\, W_\theta\, (\boldsymbol{\theta} - \boldsymbol{\theta}_0) \tag{2}$$

In Equation (2), the parameter W_f is the weighing scalar for frequency residuals and the term $\mathbf{r}_f \in \mathbb{R}^{N\times 1}$ is the vector including the square root of normalized differences between FE-predicted and identified modal frequencies. The parameter W_M is the weighing scalar for modal assurance criteria (MAC) residuals. The term N is the total number of identified modes that are used for model updating, and the term MAC_i indicates the MAC value for mode i. The parameter W_θ is the weighing scalar for penalizing large deviation of the unknown FE model parameters from their initial values. The vector $\boldsymbol{\theta} \in \mathbb{R}^{n_{\theta M}\times 1}$ is the vector of unknown FE model parameters and $n_{\theta M}$ is the number of unknown model parameters for modal-based model updating. The vector $\boldsymbol{\theta}_0 \in \mathbb{R}^{n_{\theta M}\times 1}$ is the initial estimates of the unknown model parameters. The ith entry of the vector $\mathbf{r}_f$, denoted as $\mathbf{r}_{f,i}$, is defined as follows:

$$\mathbf{r}_{f,i} = \sqrt{\left|\frac{f_i^{ID} - f_i^{FE}}{f_i^{ID}}\right|} \quad 1,2,\ldots,N \tag{3}$$

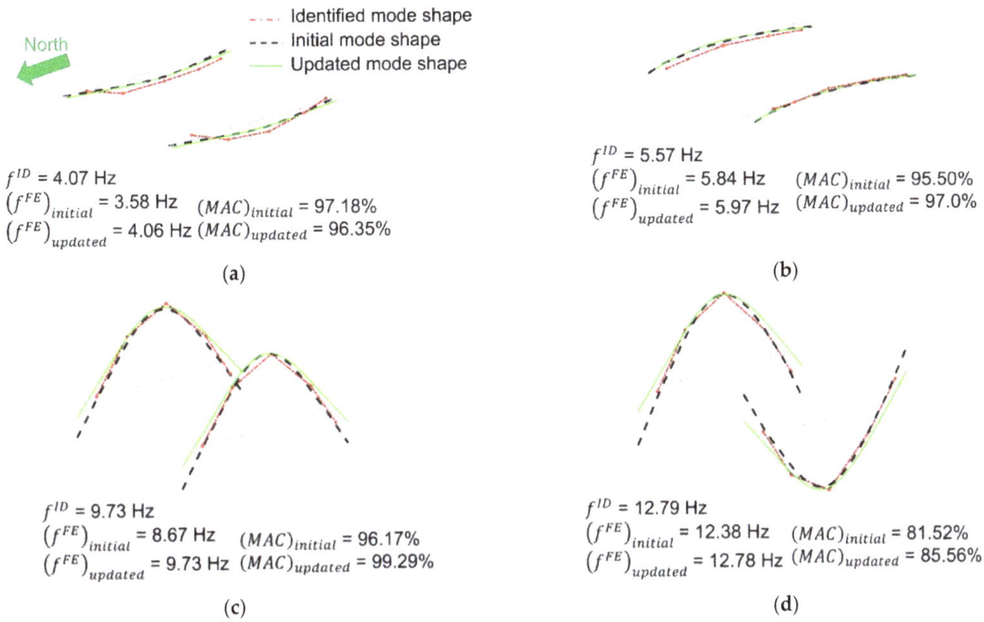

Figure 9. Comparison between identified, initial, and updated modal properties: (**a**) first lateral mode, (**b**) second lateral mode, (**c**) first vertical mode, and (**d**) second vertical mode.

In the above equation, the terms f_i^{ID} and f_i^{FE} are the modal frequencies of the ith identified and FE-predicted modes. The term MAC_i is defined as follows:

$$MAC_i = \frac{\left|\left(\psi_i^{ID}\right)^T \psi_i^{FE}\right|^2}{\left(\left(\psi_i^{ID}\right)^T \psi_i^{ID}\right)\left(\left(\psi_i^{FE}\right)^T \psi_i^{FE}\right)} \tag{4}$$

The terms ψ_i^{ID} and $\psi_i^{FE} \in \mathbb{R}^{NDOF\times 1}$ are the ith identified and FE-predicted mode shape vectors, respectively, and $NDOF$ is the number of DOFs. The superscript T denotes matrix/vector transpose operator.

To update the FE model using modal-based model updating, the stiffness-related model parameters E (= E_E = E_W), E_c, K_R, K_V and K_T are selected as unknown FE model parameters to be updated ($n_{\theta M} = 5$). Note that parameter K_L is not selected as there is no measurement in the longitudinal direction of the girders. The vector θ_0 is initiated using the initial values listed in Table 2.

The first two lateral and the first two vertical identified modes ($N = 4$) are used for the modal-based model updating. This is because the first two identified vertical modes are the only ones with frequencies less than 20 Hz (maximum excitation frequency). As can be seen in Figure 4, measurements are collected in 5 vertical (i.e., in direction 3) and 5 lateral (i.e., in direction 1) DOFs for each girder; therefore, $NDOF$ = 20. The terms W_f, W_M and W_θ are selected equal to 4, 1, and 0.01 to balance the contributions of the MAC value and frequency errors in the objective function. As a result, the difference between identified and estimated modal frequencies will be penalized more than mode shapes [61]. This is because the uncertainty in identified mode shapes is greater than frequencies. The constrained nonlinear multivariate optimization function *fmincon* within the Matlab Optimization Toolbox is used [62] with the interior-point algorithm [63]. The maximum iteration number is set to 30, and the process terminates when the relative difference between successive values of the objective function is lower than a given threshold of 10^{-4}. Moreover, the lower and upper bounds for parameter estimation are set equal to $10^{\pm 4}$ times of the initial values.

A comparison between the identified, initial, and updated modal properties is shown in Figure 9, and the updated values for the FE model parameters are shown in Table 3. It is noteworthy that as parameters $E_E = E_W$ are updated, the parameters $f'_{c,E} = f'_{c,W}$ are also updated (using $E_E = \frac{2f'_{c,E}}{\varepsilon_c}$ and $E_W = \frac{2f'_{c,W}}{\varepsilon_c}$ assuming fixed value for ε_c). These equivalent values for each girder are shown in a same column in Table 3 and are separated using '/'. A comparison between Tables 2 and 3 shows that the parameters E_E and E_W are estimated to a smaller value than their initial ones. This was expected as both girders are aged and have experienced severe damage. The parameter E_c is estimated to a smaller value than its initial one. This is probably due to the presence of cracks in the section and could also be an indication of the fact that the connections between the girders and coupling beams are not completely rigid. The parameters K_R, K_V, and K_T are estimated to have values greater than their initial ones, which indicates that the bearing pads are stiffer than what is considered in the initial model. The improvement in the modal frequencies is superior to the MAC values. This means that the modal-based model updating process compensates for the frequency match with MAC values. The maximum error in frequency is at the second lateral mode, which is less than 8% and is acceptable. The updated model is later used as the prior FE model for the time-domain model updating.

Table 3. The updated FE model parameters after modal-based model updating.

Parameter	$f'_{c,W}/E_W$	$f'_{c,E}/E_E$	E_c	K_R	K_V	K_T
Updated value	39 MPa/39 GPa	39 MPa/39 GPa	28.23 GPa	1.4×10^6 kN/rad	1.8×10^5 kN/m	2.2×10^4 kN/m

3.2. Second Step: Time-Domain Model Updating

To this point, the initial FE model of the studied girders is updated using modal-based model updating. It is noteworthy that concrete material has nonlinear response behavior even under small levels of excitation. In addition to that, presence of prestressing force pushes the concrete material across the section along its nonlinear response curve. Then, the applied excitation—shaker force here and traffic load in operational conditions—results in small loading/unloading of concrete material in nonlinear range of the response curve. The level of nonlinearity in the response behavior of concrete material increases as a function of deterioration and damage [64]. However, the nonlinear response behavior of the studied girders is not captured through modal-based model updating.

To account for the nonlinear response behavior of girders as well as refining the estimation of damage-related model parameters, time-domain model updating—here referred to as the second step of the model updating procedure—is carried out. The cumulative damage effects in a reinforced concrete section can be modeled by altering the stress-strain response behavior of the concrete material, e.g., reduction of the effective compressive strength [65–67]. Based on this, it is intended to estimate the concrete compressive strengths of girders using time-domain model updating. Moreover, the acceleration measurements contain information regarding the dynamic behavior of the testbed structure. Hence, the energy-dissipation-related model parameters can also be estimated using time-domain model updating. For this purpose, first, the most identifiable FE model parameters are selected using an information-theoretic identifiability analysis [68]. Then, these parameters are updated using the Bayesian model updating process.

3.2.1. Identifiability Analysis for Time-Domain Model Updating

The identifiability analysis is an approach to determine the most identifiable unknown model parameters using the sensitivity of FE-predicted responses with respect to the unknown model parameters. In this method, the relative information that each candidate model parameter gains from the responses and the mutual information gain between these parameters are calculated. The model parameters with high relative information gain and little dependency on other parameters are likely more identifiable than others and are selected to be updated in the model updating process.

The data used for identifiability analysis are noted as Data 1-2 in Table 4, which corresponds to layout 1 and the target load amplitude of 2225 N. It is noteworthy that no significant difference is expected in the identifiability analysis results using different layouts and target load amplitudes. However, the target load amplitude for identifiability analysis is selected high enough to have a moderate level of loading/unloading response behavior in the concrete material of the testbed structure. The data in Table 4 are later used for model updating.

Table 4. Field experimental data used for the Bayesian FE mode updating. The measured data with excitation frequencies close to mode 1 and mode 2 are lumped together.

Data I.D.	Layout	Target Load Amplitude	Excitation Frequencies	Data I.D.	Layout	Target Load Amplitude	Excitation Frequencies
1-1	1	445 N	9.73 Hz and 12.86 Hz	2-1	2	445 N	9.73 Hz and 12.86 Hz
1-2	1	2225 N	9.73 Hz and 12.86 Hz	2-2	2	2225 N	9.73 Hz and 12.86 Hz
1-3	1	4450 N	9.73 Hz and 12.86 Hz	2-3	2	4450 N	9.73 Hz and 12.86 Hz

As can be seen in Table 4, the identifiability analysis and model updating process are performed using experiments with excitation frequencies of 9.73 Hz and 12.86 Hz. These excitation frequencies are selected as they are the closest ones to the identified modal frequencies (9.73 Hz and 12.79 Hz) and are expected to provide useful information on the dynamic behavior of the testbed structure. Moreover, the measured signal-to-noise ratio in the experiments with these frequencies is higher than the similar ratio in experiments with excitation frequencies far from the identified modal frequencies. A higher measured signal-to-noise ratio results in more stable parameter estimation [69].

Each data set in Table 4 augments two experiments with excitation frequencies of 9.73 Hz and 12.86 Hz (while the layout and target load amplitude are similar). This is shown in the following equations:

$$\mathbf{y} = \begin{bmatrix} \mathbf{y}_{9.73\ \mathrm{Hz}} \\ \mathbf{y}_{12.86\ \mathrm{Hz}} \end{bmatrix} \tag{5}$$

$$\mathbf{u} = \begin{bmatrix} \mathbf{u}_{9.73\ \mathrm{Hz}} \\ \mathbf{u}_{12.86\ \mathrm{Hz}} \end{bmatrix} \tag{6}$$

The terms $\mathbf{y} \in \mathbb{R}^{20 \times t_e}$ and $\mathbf{u} \in \mathbb{R}^{2 \times t_e}$ denote the acceleration measurements and input excitations that are being used for identifiability analysis and time-domain model updating. The terms $\mathbf{y}_{9.73\ \mathrm{Hz}} \in \mathbb{R}^{10 \times t_e}$ and $\mathbf{y}_{12.86\ \mathrm{Hz}} \in \mathbb{R}^{10 \times t_e}$ refer to the collected acceleration measurements (at 10 measurement channels in direction 3) from experiments with input excitation frequencies of 9.73 Hz and 12.86 Hz. The terms $\mathbf{u}_{9.73\ \mathrm{Hz}} \in \mathbb{R}^{1 \times t_e}$ and $\mathbf{u}_{12.86\ \mathrm{Hz}} \in \mathbb{R}^{1 \times t_e}$ refer to the input excitations with frequencies of 9.73 Hz and 12.86 Hz. The term t_e is the total length of collected data.

One of the intentions of identifiability analysis/time-domain model updating is to analyze the identifiability of/update the unknown model parameters ξ_1 and ξ_2. For this purpose, it is required to implement measurements from both excitation frequencies equal to 9.73 Hz and 12.86 Hz. The augmenting process provides measurements that contain dynamic response behavior of the testbed from both its vertical modes.

Parameters $f'_{c,W}$, $f'_{c,E}$, E_c, K_R, K_V, K_L, C_D, ξ_1, and ξ_2 are selected as candidate unknown model parameters. It is intended to refine the damage state estimation of girders for the west and east girder separately. Therefore, the concrete compressive strengths of the west and east girders are modeled using two different parameters $f'_{c,W}$ and $f'_{c,E}$, respectively. While the main intention of the time-domain model updating is to update the model parameters related to the energy dissipation and nonlinear behavior of the structure, the parameters E_c, K_R, K_V, and K_L are also considered as candidate unknown parameters. As will be seen later, this is to highlight how the introduced two-step model updating helps with unidentifiability and mutual dependencies between the model parameters. Since the input load mainly excites the first two vertical modes of the girders, the modal damping ratios of these two modes are included in the list of candidates for unknown parameters. The value for parameters $f'_{c,W}$, $f'_{c,E}$, E_c, K_R, and K_V are set equal to their final estimates in modal-based model updating (see Table 3). As the input load is in the vertical direction, the parameter K_T is most probably not identifiable and is excluded from the list of candidate model parameters. The value of parameter K_L is set equal to the updated value of K_T (after modal-based model updating) as the properties of bearing pads are assumed to be the same in the transverse and longitudinal directions. The value of parameters C_D, ξ_1, and ξ_2 are set equal to their initial values (see Table 2), as these parameters were not updated using modal-based model updating.

The relative information gain of the candidate unknown model parameters and relative mutual information gain between the candidate unknown model parameter pairs are shown in Figure 10. As can be seen in this figure, although the parameters E_c, K_R, K_V, and K_L have considerable levels of relative information gain, they are highly dependent on the parameters $f'_{c,W}$ and $f'_{c,E}$. This dependency is likely because all these parameters contribute to the stiffness of the testbed structure. Based on this, parameters E_c, K_R, K_V, and K_L are most probably not identifiable together with parameters $f'_{c,W}$ and $f'_{c,E}$. Parameters E_c, K_R, K_V, and K_L reflects the linear-elastic response behavior of the testbed structure and are already estimated using modal signatures. Moreover, the parameters $f'_{c,W}$ and $f'_{c,E}$ have relatively large information gain, and their estimation is of main interest in time-domain model updating as their final estimates help to refine the damage estimation in girder level and reflect the cumulative damage status of each girder. Hence, parameters $f'_{c,W}$ and $f'_{c,E}$ are selected to be estimated using time-domain model updating while parameters E_c, K_R, K_V, and K_L are fixed at their corresponding values in Table 3 obtained from the modal-based model updating. This reduces the challenges of model updating due to the unidentifiability and/or mutual dependencies between model parameters. It is noteworthy that the final estimates of parameters $f'_{c,W}$ and $f'_{c,E}$ using time-domain model updating will inherently be dependent on the fixed values selected for parameters E_c, K_R, K_V, and K_L. Parameters ξ_1, ξ_2, and C_D have relatively moderate information gain and have a negligible dependency on other parameters. However, parameter C_D is dependent on parameters ξ_1 and ξ_2 as all of them contribute to the viscous damping energy dissipation of the structure. As the initial value for parameter C_D is selected based on judgment, it is of interest to update this parameter using the model updating process. However, the final estimates of parameters ξ_1, ξ_2, and C_D are expected to vary between different case studies and depend on each other. Therefore, the overall damping of the testbed will be calculated at the end. Based on the above discussion, parameters $f'_{c,W}$, $f'_{c,E}$, ξ_1, ξ_2, and C_D are selected to be estimated using the time-domain model updating.

3.2.2. Bayesian Inference

In the Bayesian model updating, the unknown FE model parameter vector is considered as a random vector with joint probability density function (PDF) whose mean (referred to as estimate hereafter) and covariance are updated recursively through the integration of the FE-predicted and measured responses. The unknown FE model parameters are updated using measured responses in successive overlapping windows. In this approach, known as the sequential estimation window approach [70], the a priori estimates of the unknown FE model parameters at each estimation window are updated to the a posteriori estimates. A brief review of this method is provided in the following.

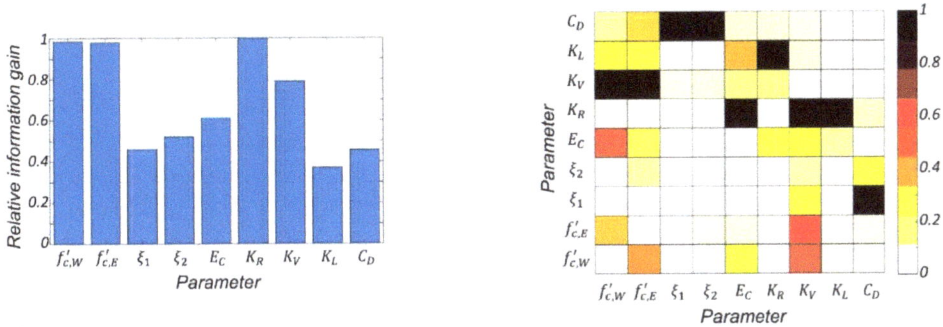

(**a**) (**b**)

Figure 10. Identifiability analysis results: (**a**) relative information gain of the candidate unknown model parameters; (**b**) relative mutual information gain between the candidate unknown model parameter pairs. Each row is normalized to its diagonal value. Then, the diagonal values are nullified.

Assuming zero-mean Gaussian white noise processes for the measurement noise ($\boldsymbol{\lambda}$) and the process noise ($\mathbf{q}$), the state-space model at each estimation window is set up as follows:

$$\boldsymbol{\vartheta}_{w+1} = \boldsymbol{\vartheta}_w + \mathbf{q}_w \tag{7}$$

$$\mathbf{y}_w = \mathbf{h}_w(\boldsymbol{\vartheta}_w, \mathbf{u}_w) + \boldsymbol{\lambda}_w \tag{8}$$

As can be seen in Equation (7), the unknown FE model parameter vector for Bayesian model updating, $\boldsymbol{\vartheta} \in \mathbb{R}^{n_{\vartheta B_{FE}} \times 1}$, evolves linearly in the state equation by a random walk process. The term $n_{\vartheta B_{FE}}$ is the number of unknown FE model parameters in the Bayesian model updating. In this study, the term $n_{\vartheta B_{FE}}$ is equal to 5 (which counts the number of parameters $f'_{c,W}$, $f'_{c,E}$, ξ_1, ξ_2, and C_D) and the initial values for the unknown model parameters are equal to those described in the previous section. In Equation (8), the term $\mathbf{y}_w \in \mathbb{R}^{(n_y \times t_w) \times 1}$ denotes the measurement vector and $\mathbf{h}_w(.) \in \mathbb{R}^{(n_y \times t_w) \times 1}$ denotes the nonlinear FE-predicted response function at the wth estimation window, which spans between time steps $t_{w,1}$ and $t_{w,2}$ with t_o time steps overlap with the previous estimation window. The parameter n_y (here 20) is the number of collected signals and t_w is number of time steps at the wth estimation window. The term $\mathbf{u}_w \in \mathbb{R}^{t_{w,2} \times 1}$ is the deterministic input vector at the wth estimation window. In this study, estimation windows have the length of 150 time steps with 50 time steps overlap, i.e., $t_{1,2} = 50$ and $t_w = 150$ and $t_{w-1,2} - t_{w,1} = 50 \ \forall \ w \geq 2$.

The measurement equation (Equation (8)) is linearized using the first-order Taylor series expansion with linearization point at the *a priori* estimate. The following equation is obtained:

$$\mathbf{y}_w \cong \mathbf{h}_w\left(\hat{\boldsymbol{\vartheta}}_w^-, \mathbf{u}_w\right) + \left.\frac{\partial \mathbf{h}_w(\boldsymbol{\vartheta}_w, \mathbf{u}_w)}{\partial \boldsymbol{\vartheta}_w}\right|_{\boldsymbol{\vartheta}_w = \hat{\boldsymbol{\vartheta}}_w^-} \left(\boldsymbol{\vartheta}_w - \hat{\boldsymbol{\vartheta}}_w^-\right) + \boldsymbol{\lambda}_w \tag{9}$$

in which the superscripts $-/+$ denote the *a priori/posteriori* estimates, and derivation of $\mathbf{h}_w(\boldsymbol{\vartheta}_w, \mathbf{u}_w)$ with respect to $\boldsymbol{\vartheta}_w$ is referred to as the sensitivity matrix and is calculated using the finite difference approach. The estimation method that is used in this study is referred to as the Extended Kalman filter (EKF) for parameter-only estimation [71]. In this method, the *a priori* estimates of the mean vector and covariance matrix of the unknown model parameters at each estimation window are considered equal to the *a posteriori* estimates at the previous estimation window. This is depicted in Equations (10) and (11). Moreover, the estimation problem is solved across each estimation window iteratively, and the subscripts 0 and k in the following equations denote the iteration number.

$$\hat{\vartheta}^{-}_{w,0} = \hat{\vartheta}^{+}_{w-1} \tag{10}$$

$$\hat{\mathbf{P}}^{-}_{\vartheta\vartheta,w,0} = \hat{\mathbf{P}}^{+}_{\vartheta\vartheta,w-1} \tag{11}$$

The term $\hat{\mathbf{P}}^{-/+}_{\vartheta\vartheta,w}$ is the *a priori/posteriori* estimate of the covariance matrix of the unknown model parameters at the wth estimation window. For the sake of brevity, the details for the derivation of the Bayesian inference based on the EKF method is not shown here, and only the recursive equations at each estimation window are presented. After completing the iteration process at each estimation window, the estimation moves to the next window, and the process is repeated. To complete the iteration process at each estimation window, the estimates of unknown model parameters need to be converged. However, to improve the efficiency of the process, the maximum number of iterations is limited. More details are available in [71].

$$\hat{\vartheta}^{-}_{w,k+1} = \hat{\vartheta}^{+}_{w,k} \tag{12}$$

$$\hat{\mathbf{P}}^{-}_{\vartheta\vartheta,w,k+1} = \hat{\mathbf{P}}^{+}_{\vartheta\vartheta,w,k} + \mathbf{Q}_w \tag{13}$$

$$\hat{\vartheta}^{+}_{w,k+1} = \hat{\vartheta}^{-}_{w,k+1} + \mathbf{K}_{w,k+1}\left(\mathbf{y}_w - \hat{\mathbf{y}}^{-}_{w,k+1}\right) \tag{14}$$

$$\hat{\mathbf{P}}^{+}_{\vartheta\vartheta,w,k+1} = (\mathbf{I} - \mathbf{K}_{w,k+1}\mathbf{C}_{w,k+1})\hat{\mathbf{P}}^{-}_{\vartheta\vartheta,w,k+1}(\mathbf{I} - \mathbf{K}_{w,k+1}\mathbf{C}_{w,k+1})^T + \mathbf{K}_{w,k+1}\mathbf{R}_w\mathbf{K}_{w,k+1}{}^T \tag{15}$$

while

$$\mathbf{K}_{w,k+1} = \hat{\mathbf{P}}^{-}_{\vartheta\mathbf{y},w,k+1}\left(\hat{\mathbf{P}}^{-}_{\mathbf{yy},w,k+1}\right)^{-1} \tag{16}$$

$$\mathbf{C}_{w,k+1} = \left.\frac{\partial\, \mathbf{h}_w(\vartheta, \mathbf{u}_w)}{\partial\, \vartheta^T}\right|_{\vartheta=\vartheta^{-}_{w,k+1}} \tag{17}$$

$$\hat{\mathbf{P}}^{-}_{\mathbf{yy},w,k+1} = \mathbf{C}_{w,k+1}\hat{\mathbf{P}}^{-}_{\vartheta\vartheta,w,k+1}\mathbf{C}_{w,k+1}{}^T + \mathbf{R}_w \tag{18}$$

$$\hat{\mathbf{P}}^{-}_{\vartheta\mathbf{y},w,k+1} = \hat{\mathbf{P}}^{-}_{\vartheta\vartheta,w,k+1}\mathbf{C}_{w,k+1}{}^T \tag{19}$$

The term $\mathbf{Q}_w \in \mathbb{R}^{n_{\theta B} \times n_{\theta B}}$ is the covariance matrix for the process noise ($\mathbf{q}_w \sim N(0, \mathbf{Q}_w)$). The matrix $\mathbf{Q}_w$ is a diagonal matrix, and its jth diagonal entry is equal to q times the jth entry in vector $\hat{\vartheta}^{-}_{w}$. The term q is set equal to 0.002. The term $\mathbf{R}_w \in \mathbb{R}^{(t_w \times n_y) \times (t_w \times n_y)}$ is the measurement noise ($\boldsymbol{\lambda}_w \sim N(0, \mathbf{R}_w)$) and is modeled as a block diagonal matrix with the simulation error covariance matrix—including measurement noise—on the diagonal blocks. In this study, the diagonal entries of the matrix $\mathbf{R}_w$ are set equal to $(0.32\%\text{g})^2$ at all measurement channels. The value 0.32%g is approximately equal to the average root-mean-square of ambient measurements. The term $\hat{\mathbf{y}}^{-}_{w,k+1}$ is the *a priori* FE-predicted response calculated at $\hat{\vartheta}^{-}_{w,k+1}$. The matrix $\mathbf{K}_{w,k+1} \in \mathbb{R}^{n_{\theta B} \times (t_w \times n_y)}$ is the Kalman-gain matrix at the $(k+1)^{th}$ iteration in the wth estimation window. The matrix $\mathbf{C}_{w,k+1} \in \mathbb{R}^{(t_w \times n_y) \times n_{\theta B}}$ is the FE response sensitivity matrix—with respect to $\hat{\vartheta}^{-}_{w,k+1}$—at the $(k+1)^{th}$ iteration in the wth estimation window. The term $\mathbf{I} \in \mathbb{R}^{n_{\theta B} \times n_{\theta B}}$ denotes the identity matrix. The matrix $\hat{\mathbf{P}}^{-}_{\mathbf{yy},w,k+1} \in \mathbb{R}^{(t_w \times n_y) \times (t_w \times n_y)}$ is *a priori* estimate of the covariance matrix of $\hat{\mathbf{y}}^{-}_{w,k+1}$, and $\hat{\mathbf{P}}^{-}_{\vartheta\mathbf{y},w,k+1} \in \mathbb{R}^{n_{\theta B} \times (t_w \times n_y)}$ is the *a priori* estimate of the cross-covariance matrix of $\hat{\vartheta}^{-}_{w,k+1}$ and $\hat{\mathbf{y}}^{-}_{w,k+1}$. The matrix $\hat{\mathbf{P}}^{+}_{\vartheta\vartheta,0}$ is initialized diagonally with diagonal entries equal to the initial variance of the initial estimate of the unknown model parameters. The jth diagonal entry in $\hat{\mathbf{P}}^{+}_{\vartheta\vartheta,0}$ is equal to the square of p_{ϑ} times the jth entry in vector $\hat{\vartheta}^{+}_{0}$. In this study p_{ϑ} is set equal to 0.1. The recursive Bayesian model updating process in each estimation window is completed after 10 iterations or meeting the following convergence criteria in the posterior estimates of unknown model parameters.

$$\left|\hat{\vartheta}^{+}_{w,k+1} - \hat{\vartheta}^{+}_{w,k}\right| < (0.02) \times \left|\hat{\vartheta}^{+}_{w,k}\right| \quad (20)$$

3.2.3. Results

This section presents the results of the second step of the introduced two-step model updating approach as well as its application for damage identification of the testbed structure. First, the Bayesian model updating is carried out, and its performance is discussed through the updating process of unknown model parameters and the fit between measurements and posterior estimates of FE-predicted responses. Subsequently, the final estimates of unknown model parameters are used to infer damage in the girders and calculate the overall damping of the testbed structure. To assess the application of Bayesian FE model updating, the data sets presented in Table 4 are used. As explained in Section 3.2.1, while the excitation frequencies close to the first two modes are lumped together, the target load amplitudes and layouts differ from one data set to another one. As mentioned before, lumping the frequencies together increases the identifiability of model parameters and the parameter estimation stability by using the dynamic response behavior of testbed structure from its two first vertical modes.

The updating process for the posterior estimates of the unknown model parameters $f'_{c,W}$, $f'_{c,E}$, ξ_1, ξ_2, and C_D using data from Table 4 is shown in Figure 11. In this figure, the estimates of the unknown model parameters are normalized to their corresponding initial values in time-domain model updating process. As can be seen in this figure, all the unknown model parameters are updated from their initial values and smoothly converged to their final estimates.

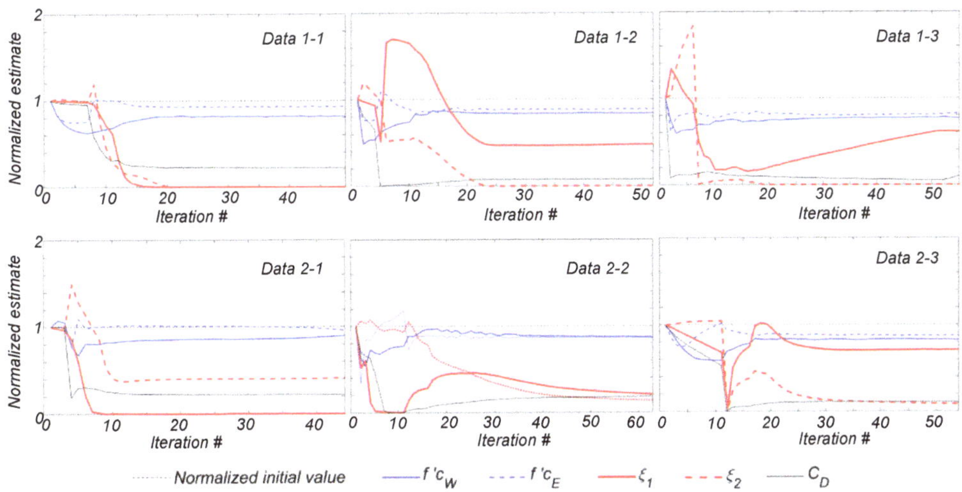

Figure 11. Updating process for the posterior estimates of the unknown model parameters using data in Table 4.

Next, the prior estimates of responses—simulated responses using the prior FE model—and the posterior ones are compared with the field measurements. For the sake of brevity, only one second of data obtained using Data 1-2 are shown in Figure 12. In this figure, the two top rows correspond to channels 1–10 and an excitation frequency of 9.73 Hz, and the two bottom rows correspond to the same channels and an excitation frequency of 12.86 Hz. For the case of excitation frequency of 9.73 Hz, the prior FE responses mostly underestimate the measurements in channels located between the supports (channels 2, 3, 4, 7, 8, and 9) and slightly overestimate the responses in channels located on the overhang parts (channels 1, 5, 6, and 10). However, the responses in all channels are initially overestimated for the excitation frequency of 12.86 Hz. This shows that although the prior FE model matches the main modal signatures of the real structure, it cannot correctly predict the measurements in the time domain. As can be seen in Figure 12, after the application of Bayesian model updating, the updated FE model better fits the measurement responses in the time domain. This improvement is noticeable in various channels and for both excitation frequencies of 9.73 Hz and 12.86 Hz.

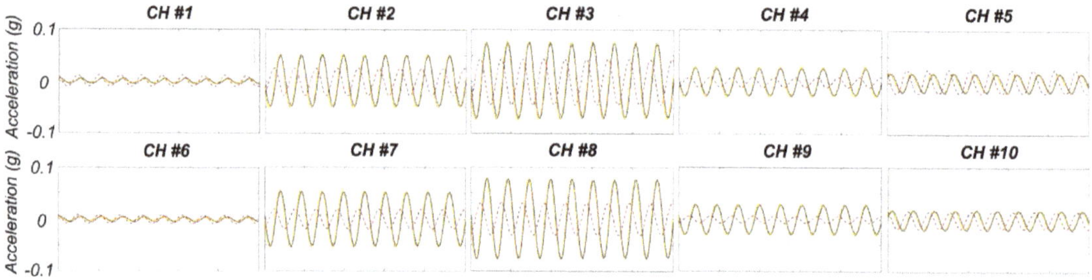

(a)

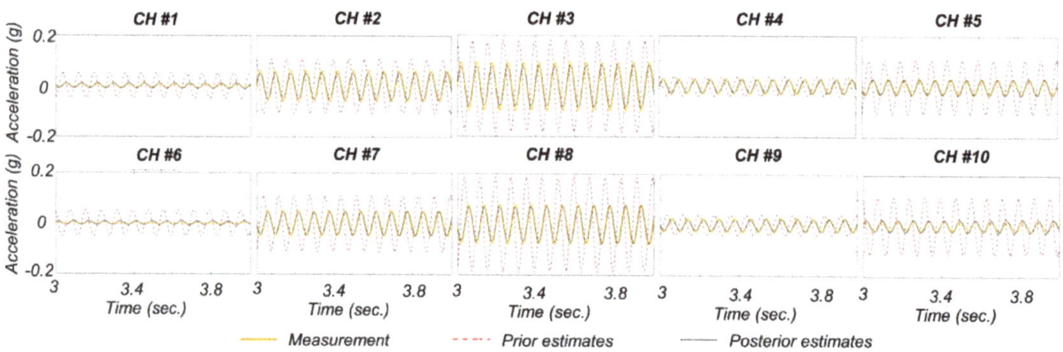

(b)

Figure 12. The measured, prior and posterior estimates of FE-predicted responses using Bayesian FE model updating based on Data 1–2: (**a**) excitation frequency of 9.73 Hz and (**b**) excitation frequency of 12.86 Hz.

To quantify the discrepancies between two signals, the relative root mean square error (RRMSE) is calculated as follows:

$$\text{RRMSE}(\%) = \frac{\sqrt{\sum_{i=t_1}^{t_n}\left(\hat{\mathbf{s}}_i - \mathbf{s}_i\right)^2}}{\sqrt{\sum_{i=t_1}^{t_n}\left(\mathbf{s}_i\right)^2}} \times 100 \tag{21}$$

In the above equation, $\mathbf{s}_i$ and $\hat{\mathbf{s}}_i$ denotes the measured and estimated responses at the ith time step. The closer the RRMSE gets to zero, the better signals $\mathbf{s}$ and $\hat{\mathbf{s}}$ match.

The RRMSEs are calculated between the measured responses and their prior/posterior FE-predicted responses and are listed in Table 5.

Table 5. RRMSEs (%) between the measurements and FE-predicted responses from the prior and posterior model.

		9.73 Hz Excitation Frequency										**12.86 Hz Excitation Frequency**									
Data I.D.	**CH #**	**1**	**2**	**3**	**4**	**5**	**6**	**7**	**8**	**9**	**10**	**1**	**2**	**3**	**4**	**5**	**6**	**7**	**8**	**9**	**10**
1-1	Prior	201	75	74	80	66	170	76	74	83	46	161	63	63	70	67	202	67	68	72	97
	Posterior	172	14	9	22	27	192	22	17	28	28	132	16	15	22	40	153	11	11	23	57
1-2	Prior	163	107	110	102	127	119	101	102	99	101	300	112	117	103	164	565	133	140	109	269
	Posterior	24	11	10	10	22	34	10	9	8	21	72	23	24	24	30	133	20	21	13	55
1-3	Prior	191	128	134	116	165	137	114	118	108	130	429	146	156	126	235	111	193	205	141	480
	Posterior	24	11	12	20	21	46	11	13	18	18	90	33	35	35	41	254	27	28	20	93
2-1	Prior	312	61	58	71	55	259	63	60	75	30	590	30	27	46	124	236	30	26	51	84
	Posterior	191	8	8	18	46	213	17	14	23	50	312	7	4	22	68	133	9	9	22	54
2-2	Prior	229	78	78	84	66	187	78	77	85	40	388	54	52	64	80	250	55	54	66	117
	Posterior	139	11	9	20	46	172	18	15	20	60	236	7	7	17	65	175	9	7	26	72
2-3	Prior	109	611	64	46	114	104	62	66	43	128	124	38	34	36	17	54	33	33	49	33
	Posterior	83	8	7	17	52	99	12	11	11	61	180	9	10	13	49	177	10	9	21	82

As can be seen in Table 5, RRMSE values are reduced from prior to posterior values for all measurement channels. This shows that the time-domain model updating reduces the discrepancies between FE-predicted and measured responses successfully. However, the highest posterior RRMSEs are due to channels 1, 5, 6, and 10. These channels are located on the overhang parts of the girders (see Figure 4), and the measurements are likely less reliable due to the low signal-to-noise ratio. It is understandable from Figure 12 and Table 5 that the FE-predicted responses at these channels are promisingly updated to match the measurements.

The final estimates of unknown model parameters using all data are shown in Figure 13. The average final estimates of parameters $f^{'}_{c,W}$ and $f^{'}_{c,E}$ are approximately 30% and 26% less than their initial values in Table 2. Previous studies [72,73] have shown more than 25% reductions in concrete compressive strength of deteriorated concrete structures—e.g., abandoned structures and structures in acidic environments. Considering that the operating environment of girders caused them to be prone to various damage mechanisms (including concrete degradation, as well as steel corrosion as discussed in Section 2.1.1), the final estimates of parameters $f^{'}_{c,W}$ and $f^{'}_{c,E}$ are reasonable. Moreover, as mentioned in Section 2.1.1, after being decommissioned, the west girder has been under more intense environmental testing conditions than the east girder [45]. This can be understood from the model updating results as the final estimates of parameter $f^{'}_{c,W}$ is smaller than $f^{'}_{c,E}$. The only exception is the case study with Data 2-1 in which the final estimate of $f^{'}_{c,E}$ is smaller than $f^{'}_{c,W}$. This is most probably an estimation error due to measurement noise, experimental error, etc. Moreover, based on Figure 13 and ignoring the case study with Data 2-1, the maximum difference in the estimation of parameter $f^{'}_{c}$ in layouts 1 and 2 is less than 7%, which shows the consistency of Bayesian model updating results regardless of the shaker location and target load amplitude.

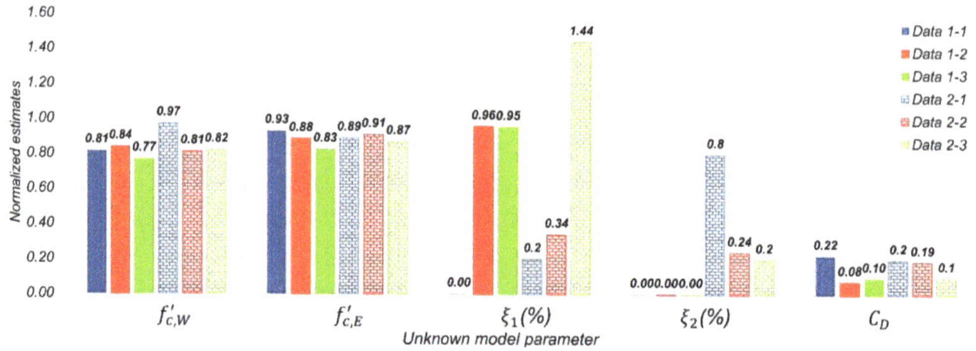

Figure 13. The final estimates of unknown model parameters using data sets in Table 4. The information on each set of data is shown in Table 4. The estimates for parameters $f'_{c,W}$, $f'_{c,E}$ and C_D are normalized to their corresponding initial values in time-domain model updating process.

The final estimates for modal damping parameters vary between different data sets. The identifiability analysis showed dependencies between parameters ξ_1, ξ_2, and C_D. This dependency is most probably a reason for the discrepancy between the final estimates of these parameters using different data sets. Aside from this, it is known that structural damping roots in various factors, including opening and closing of micro cracks, friction in structural joints, level of vibration, and other sources of energy dissipation, etc. [74,75]. These conditions could be varied from one experiment to another and result in different levels of damping. To have insight into the overall damping of the testbed structure, the modal damping ratios are calculated using the state matrix of the system. For this purpose, the stiffness, mass, and damping matrices of the system are developed for the posterior models. The stiffness (**K**) matrix is recorded as the current global system matrix (using OpenSees *printA* command) while a static analysis using *LoadControl* integrator is carried out, and damping is removed from the model. The mass matrix (**M**) is calculated based on the current global system matrix while a static analysis using *LoadControl* integrator and a transient analysis using *CentralDifference* integrator is carried out, and damping is removed from the model. Using the Central Difference formulation presented in Equation (22), the mass matrix is equal to the recorded current global system matrix ($\hat{\mathbf{K}}_{CDF}$) times Δt^2. In Equation (22), the term Δt is the size of the time step increment and is set to a very small value (here 10^{-6} s) to record the current global system matrix with high precision. The term **C** is the damping matrix.

$$\hat{\mathbf{K}}_{CDF} = \frac{\mathbf{M}}{(\Delta t)^2} + \frac{\mathbf{C}}{2\Delta t} \tag{22}$$

Including damping in the model and carrying out a transient analysis using *Newmark* integrator with $\gamma = 0.5$ and $\beta = 0.25$, the matrix $\hat{\mathbf{K}}_{Newmark}$ can be recorded using *printA*. The damping matrix is calculated as it is shown in Equation (23) using the Newmark formulation.

$$\mathbf{C} = \frac{\beta\,\Delta t}{\gamma}\left(\hat{\mathbf{K}}_{Newmark} - \mathbf{K} - \frac{\mathbf{M}}{\beta\,(\Delta t)^2}\right) \tag{23}$$

The matrices **M**, **C**, and **K** are condensed at the dynamic DOFs, and the state matrix is calculated [74]. Using the state matrix of the posterior models, the modal damping ratios are obtained for all data sets. These damping ratios are reported in Figure 14. As can be seen, the testbed structure dissipates a greater level of energy in its first mode than in the second mode, which agrees with the identified modal damping ratios (see Table 1). Moreover, the overall modal damping ratios estimated from experiments with larger target load amplitudes are smaller than those estimated from experiments with smaller target load amplitudes. Although this has been previously observed in a few system identification studies [76,77], it is in contrast with the literature in which damping increases as load amplitude increases.

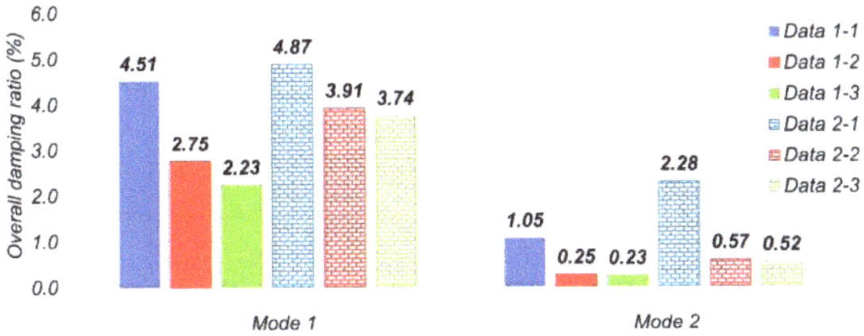

Figure 14. Overall damping ratio of the testbed structure for mode 1 and mode 2 using different data sets.

4. Conclusions

This paper proposed a two-step FE model updating process for operational health monitoring and damage identification of bridge structures. In the proposed approach, first, modal-based model updating was carried out to calibrate the initial FE model of the bridge. In this step, the stiffness-related model parameters—mainly related to boundary conditions—as well as the initial stiffness of concrete material were updated to fit the FE-predicted and identified modal signatures of the bridge. Then, to account for the nonlinear response behavior of the bridge as well as refinement of model parameter estimation, Bayesian time-domain model updating was carried out in the second step. In this step, material properties that reflect the cumulative damage in the bridge, e.g., effective concrete compressive strength, as well as damping energy-dissipation-related model parameters, were estimated. The linear-elastic boundary conditions were fixed at their final estimates obtained from the modal-based model updating. To prevent convergence of the model updating algorithm to the local solution, the initial values for concrete compressive strength were selected using the final estimates of concrete initial stiffness from modal-based model updating.

The application of the two-step model updating approach was presented using a pair of full-scale precast prestressed deteriorated bridge I-girders as the testbed structure. For this study, a series of forced-vibration experiments were planned, and the testbed structure was subjected to sinusoidal force excitations through frequency sweeps at three different amplitudesusing a small shaker. The input excitation was measured using load cells, and the acceleration responses were collected using a wireless sensing network. The findings confirm the following conclusions:

- Identifiability analysis showed significant mutual dependency between different model parameters. This mutual dependency could lead to weak identifiability of model parameters in the traditional FE model updating process. The proposed two-step model updating helped with this challenge to update the most sensitive model parameters separately using modal-based and time-domain model updating.
- Sequential application of modal-based and time-domain model updating reduced the challenges due to ill-conditioning and modeling errors.
- It was demonstrated that the updated FE model using only modal-based model updating was not capable of reflecting true response behavior of the structure in time domain.

- Concrete compressive strength was correlated with damage/deterioration in the monitored structure and could be used to assess the health condition of the structure. In this study, a 30% reduction in concrete compressive strength from its nominal value correctly showed significant deterioration in the studied girders.

It is noteworthy to mention that although the input load in this study was different from the traffic load during the operation of bridges, it provided a real-world exercise to validate the capability of the two-step model updating approach for damage identification of bridge structures.

Author Contributions: Conceptualization, F.G., H.E. and E.T.; methodology, N.M., F.G., H.E. and E.T.; software, N.M. and F.G.; validation, N.M., F.G. and H.E.; formal analysis, N.M. and F.G.; investigation, N.M., F.G. and H.E.; resources, H.E., M.B. and E.T.; data curation, N.M., M.B. and E.A.; writing—original draft preparation, N.M.; writing—review and editing, N.M., F.G., H.E., M.B. and E.T.; visualization, N.M.; supervision, H.E. and E.T.; project administration, H.E. and M.B.; funding acquisition, H.E., M.B. and E.T. All authors have read and agreed to the published version of the manuscript.

Funding: This project was funded through the United States Department of Transportation Small Business Innovative Research (SBIR) program Phase II (Contract #6913G619C100048). The project resulted from the collaboration between S.C. Solutions, Inc., the leading small business, the University of California Los Angeles (UCLA), and the University of Nevada, Reno. Opinions and findings in this study are those of the authors and do not necessarily reflect the views of the sponsor.

Data Availability Statement: Data and materials supporting the results or analyses presented in this work are available upon reasonable request from the corresponding author.

Acknowledgments: Authors would like to acknowledge Hoda Azari (FHWA NDE Center) for making available the prestressed girder testbed at the Turner Fairbank Highway Research Center and Blake Cox (FHWA NDE Center), who aided during testing.

Conflicts of Interest: The authors declare no conflict of interest.

References

1. Shahsavari, V.; Mehrkash, M.; Santini-Bell, E. Damage Detection and Decreased Load-Carrying Capacity Assessment of a Vertical-Lift Steel Truss Bridge. *J. Perform. Constr. Facil.* **2020**, *34*, 04019123. [CrossRef]
2. ASCE. *ASCE's 2021 Infrastructure Report Card*. 2021. Available online: https://infrastructurereportcard.org/cat-item/bridges/ (accessed on 1 May 2022).
3. Friswell, M.; Mottershead, J. *Finite Element Model Updating in Structural Dynamic*; Springer: Berlin/Heidelberg, Germany, 1995. [CrossRef]
4. Zapico, J.; González, M.; Friswell, M.; Taylor, C.; Crewe, A. Finite element model updating of a small scale bridge. *J. Sound Vib.* **2003**, *268*, 993–1012. [CrossRef]
5. Xiao, X.; Xu, Y.; Zhu, Q. Multiscale modeling and model updating of a cable-stayed bridge. II: Model updating using modal frequencies and influence lines. *J. Bridg. Eng.* **2015**, *20*, 04014113. [CrossRef]
6. Li, Y.; Astroza, R.; Conte, J.; Soto, P. Nonlinear FE model updating of seismic isolated bridge instrumented during the 2010 Mw 8.8 Maule-Chile Earthquake. *Procedia Eng.* **2017**, *199*, 3003–3008. [CrossRef]
7. Taciroglu, E.; Shamsabadi, A.; Abazarsa, F.; Nigbor, R.; Ghahari, S. Comparative study of model predictions and data from the Caltrans-CSMIP bridge instrumentation program: A case study on the Eureka-Samoa channel bridge. *Sacramento* **2014**. [CrossRef]
8. Ereiz, S.; Duvnjak, I.; Jiménez-Alonso, J.F. Review of finite element model updating methods for structural applications. *Structures* **2022**, *41*, 684–723. [CrossRef]
9. McCoy, R. HedegaardBrock, ShieldCarol, LindermanLauren, Updated long-term Bayesian Monitoring Strategy for time-dependent deflections of I-35W Saint Anthony Falls Bridge. In *International Conference on Structural Health Monitoring of Intelligent Infrastructure: Transferring Research into Practice*; SHMII: Porto, Portugal, 2021; pp. 1141–1147. Available online: https://experts.umn.edu/en/publications/updated-long-term-bayesian-monitoring-strategy-for-time-dependent (accessed on 1 November 2022).
10. Teughels, A.; De Roeck, G. Structural damage identification of the highway bridge Z24 by FE model updating, J. *Sound Vib.* **2004**, *278*, 589–610. [CrossRef]
11. Schlune, H.; PLoS, M.; Gylltoft, K. Improved bridge evaluation through finite element model updating using static and dynamic measurements. *Eng. Struct.* **2009**, *31*, 1477–1485. [CrossRef]
12. Costa, C.; Arêde, A.; Costa, A.; Caetano, E.; Cunha, A.; Magalhaes, F. Updating numerical models of masonry arch bridges by operational modal analysis. *Int. J. Archit. Herit.* **2015**, *9*, 760–774. [CrossRef]

13. He, L.; Reynders, E.; García-Palacios, J.; Marano, G.; Briseghella, B.; De Roeck, G. Wireless-based identification and model updating of a skewed highway bridge for structural health monitoring. *Appl. Sci.* **2020**, *10*, 2347. [CrossRef]
14. Tran-Ngoc, H.; Khatir, S.; De Roeck, G.; Bui-Tien, T.; Nguyen-Ngoc, L.; Wahab, M.A. Model updating for nam O bridge using particle swarm optimization algorithm and genetic algorithm. *Sensors* **2018**, *18*, 4131. [CrossRef]
15. Mottershead, J.; Friswell, M. Model updating in structural dynamics: A survey. *J. Sound Vib.* **1993**, *167*, 347–375. [CrossRef]
16. Ghahari, F.; Malekghaini, N.; Ebrahimian, H.; Taciroglu, E. Bridge digital twinning using an output-only Bayesian model updating method and recorded seismic measurements. *Sensors* **2022**, *22*, 1278. [CrossRef] [PubMed]
17. Abedin, M.; De Caso y Basalo, F.; Kiani, N.; Mehrabi, A.; Nanni, A. Bridge load testing and damage evaluation using model updating method. *Eng. Struct.* **2022**, *252*, 113648. [CrossRef]
18. Magalhães, F.; Cunha, Á.; Caetano, E. Online automatic identification of the modal parameters of a long span arch bridge. *Mech. Syst. Signal Process.* **2009**, *23*, 316–329. [CrossRef]
19. Juang, J.-N. *Applied System Identification*; Prentice Hall: Englewood Cliffs, NJ, USA, 1994.
20. James, G., III; Carne, T.; Lauffer, J. The Natural Excitation Technique (NExT) for Modal Parameter Extraction from Operating Wind Turbines, United States. 1993. Available online: https://www.osti.gov/biblio/10139203 (accessed on 1 November 2021).
21. Ljung, L. *System Identification Theory for the User*, 2nd ed.; Prentice Hall PTR: Upper Saddle River, NJ, USA, 1999.
22. Ewins, D.J. *Modal testing: Theory, Practice and Application*, 2nd ed.; John Wiley & Sons: Hoboken, NJ, USA, 2009.
23. Saidin, S.; Kudus, S.; Jamadin, A.; Anuar, M.; Amin, N.; Ibrahim, Z.; Zakaria, A.; Sugiura, K. Operational modal analysis and finite element model updating of ultra-high-performance concrete bridge based on ambient vibration test. *Case Stud. Constr. Mater.* **2022**, *16*, e01117. [CrossRef]
24. Moaveni, B.; He, X.; Conte, J.; Restrepo, J. Damage identification study of a seven-story full-scale building slice tested on the UCSD-NEES shake table. *Struct. Saf.* **2010**, *32*, 347–356. [CrossRef]
25. Farrar, C.; Baker, W.; Bell, T.; Cone, K.; Darling, T.; Duffey, T.; Eklund, A.; Migliori, A. *Dynamic Characterization and Damage Detection in the I-40 Bridge Over the Rio Grande*; LA-12767-M, Technical Report; U.S. Department of Energy: Los Alamos, NM, USA, 1994; pp. 1–153. [CrossRef]
26. Farrar, C.; Cornwell, P. *Structural Health Monitoring Studies of the Alamosa Canyon and I-40 Bridges*; Technical Report; U.S. Department of Energy: Los Alamos, NM, USA, 2000. [CrossRef]
27. Zheng, Y.; Wu, H.; You, X.; Xie, H. Model updating-based dynamic collapse analysis of a RC cable-stayed bridge under earthquakes. *Structures* **2022**, *43*, 1100–1113. [CrossRef]
28. Ebrahimian, H.; Astroza, R.; Conte, J.; Papadimitriou, C. Bayesian optimal estimation for output-only nonlinear system and damage identification of civil structures. *Struct. Control Heal. Monit.* **2018**, *25*, e2128. [CrossRef]
29. Zhou, X.; Kim, C.; Zhang, F.; Chang, K. Vibration-based Bayesian model updating of an actual steel truss bridge subjected to incremental damage. *Eng. Struct.* **2022**, *260*, 114226. [CrossRef]
30. Xia, Z.; Li, A.; Li, J.; Shi, H.; Duan, M.; Zhou, G. Model updating of an existing bridge with high-dimensional variables using modified particle swarm optimization and ambient excitation data. *Meas. J. Int. Meas. Confed.* **2020**, *159*, 107754. [CrossRef]
31. Fan, W.; Qiao, P. Vibration-based damage identification methods: A review and comparative study. *Struct. Heal. Monit.* **2011**, *10*, 83–111. [CrossRef]
32. Saidin, S.; Kudus, S.; Jamadin, A.; Anuar, M.; Amin, N.; Ya, A.; Sugiura, K. Vibration-based approach for structural health monitoring of ultra-high-performance concrete bridge. *Case Stud. Constr. Mater.* **2023**, *18*, e01752. [CrossRef]
33. Cheng, X.; Dong, J.; Cao, S.; Han, X.; Miao, C. Static and dynamic structural performances of a special-shaped concrete-filled steel tubular arch bridge in extreme events using a validated computational model. *Arab. J. Sci. Eng.* **2018**, *43*, 1839–1863. [CrossRef]
34. Mao, J.; Wang, H.; Li, J. Bayesian finite element model updating of a long-span suspension bridge utilizing hybrid Monte Carlo simulation and Kriging predictor. *KSCE J. Civ. Eng.* **2020**, *24*, 569–579. [CrossRef]
35. Dong, X.; Wang, Y. Finite element model. In *Computing in Civil Engineering 2019: Smart Cities, Sustainability, and Resilienc*; American Society of Civil Engineers: Reston, VA, USA, 2019; pp. 397–404.
36. Fang, C.; Liu, H.; Lam, H.; Adeagbo, M.; Peng, H. Practical model updating of the Ting Kau bridge through the MCMC-based Bayesian algorithm utilizing measured modal parameters. *Eng. Struct.* **2022**, *254*, 113839. [CrossRef]
37. Polanco, N.; May, G.; Hernandez, E. Finite element model updating of semi-composite bridge decks using operational acceleration measurements. *Eng. Struct.* **2016**, *126*, 264–277. [CrossRef]
38. Shi, Z.; Hong, Y.; Yang, S. Updating boundary conditions for bridge structures using modal parameters. *Eng. Struct.* **2019**, *196*, 109346. [CrossRef]
39. Chen, S.; Zhong, Q.; Hou, S.; Wu, G. Two-stage stochastic model updating method for highway bridges based on long-gauge strain sensing. *Structures* **2022**, *37*, 1165–1182. [CrossRef]
40. Luo, L. Finite element model updating method for continuous girder bridges using monitoring responses and traffic videos. *Struct. Control. Health Monit.* **2022**, *29*, e3062. [CrossRef]
41. Kim, S.; Kim, N.; Park, Y.; Jin, S. A sequential framework for improving identifiability of FE model updating using static and dynamic data. *Sensors* **2019**, *19*, 5099. [CrossRef]
42. Azam, S.E.; Didyk, M.M.; Linzell, D.; Rageh, A. Experimental validation and numerical investigation of virtual strain sensing methods for steel railway bridges. *J. Sound Vib.* **2022**, *537*, 117207. [CrossRef]

43. Ramancha, M.; Astroza, R.; Madarshahian, R.; Conte, J. Bayesian updating and identifiability assessment of nonlinear finite element models. *Mech. Syst. Signal Process.* **2022**, *167*, 108517. [CrossRef]
44. Yu, E.; Taciroglu, E.; Wallace, J. Parameter identification of framed structures using an improved finite element model-updating method—Part I: Formulation and verification. *Int. Assoc. Earthq. Eng.* **2007**, *36*, 619–639. [CrossRef]
45. Gucunski, N.; Lee, S.; Mazzotta, C.; Kee, S.; Pailes, B.; Fetrat, F. *Protocols for Condition Assessment of Prestressed Concrete Girders using NDE and Physical Testing*; Technical Report; Federal Highway Administrations: Washington, DC, USA, 2014.
46. U.S. Department of Transportation, Federal Highway Administration. 2020. Available online: https://highways.dot.gov/research/turner-fairbank-highway-research-center/facility-overview (accessed on 1 November 2021).
47. Adams, M.; Nicks, J.; Stabile, T. *Thermal Activity of Geosynthetic Reinforced Soil Piers*; IFCEE/ASCE: San Antonio, TX, USA, 2015. [CrossRef]
48. Liu, P.C. Damage to concrete structures in a marine environment. *Mater. Struct.* **1997**, *24*, 302–307. [CrossRef]
49. Parker Lord. 2022. Available online: https://www.microstrain.com (accessed on 1 August 2021).
50. SensorConnect. 2022. Available online: https://www.microstrain.com/software/sensorconnect (accessed on 1 August 2021).
51. Fulop, S.; Fitz, K. Algorithms for computing the time-corrected instantaneous frequency (reassigned) spectrogram, with applications. *J. Acoust. Soc. Am.* **2006**, *119*, 360–371. [CrossRef]
52. Welch, P. The use of the fast fourier transform for the estimation of power spectra: A method based on time averaging over short, modified periodograms. *IEEE Trans. Audio Electroacoust.* **1967**, *15*, 70–73. [CrossRef]
53. Overschee, P.; Moor, B. *Subspace Identification for Linear Systems Theory—Implementation—Applications*; Kluwer Academic Publishers: Dordrecht, The Netherlands, 1996. [CrossRef]
54. Peeters, B.; De Roeck, G. Reference based stochastic subspace identification in Civil Engineering. *Inverse Probl. Eng.* **2000**, *8*, 47–74. [CrossRef]
55. MathWorks, n4sid Estimate State-space Model. 2022. Available online: https://www.mathworks.com/help/ident/ref/n4sid.html (accessed on 1 April 2022).
56. Bodeux, J.; Golinval, J. Application of ARMAV models to the identification and damage detection of mechanical and civil engineering structures. *Smart Mater. Struct.* **2001**, *10*, 479–489. [CrossRef]
57. MathWorks, sset Estimate State-space Mode. 2022. Available online: https://www.mathworks.com/help/ident/ref/ssest.html (accessed on 1 April 2022).
58. McKenna, F. OpenSees: A framework for earthquake engineering simulation. *Comput. Sci. Eng.* **2011**, *13*, 58–66. [CrossRef]
59. Naeim, F.; Kelly, J. *Design of Seismic Isolated Structures: From Theory to Practice*; John Wiley & Sons: New York, NY, USA, 1999.
60. Chopra, A. *Dynamics of Structures, 4th, ed.*; Prentice Hall: Upper Saddle River, NJ, USA, 1995.
61. Nabiyan, M.; Khoshnoudian, F.; Moaveni, B.; Ebrahimian, H. Mechanics-based model updating for identification and virtual sensing of an offshore wind turbine using sparse measurements. *Struct. Control Heal. Monit.* **2021**, *28*. [CrossRef]
62. MathWorks. Matlab R2022a. 2022. Available online: https://www.mathworks.com/?s_tid=gn_logo (accessed on 1 April 2022).
63. MathWorks. fmincon Algorithms. 2022. Available online: https://www.mathworks.com/help/optim/ug/choosing-the-algorithm.html (accessed on 1 April 2022).
64. Van Den Abeele, K. Damage assessment in reinforced concrete using nonlinear vibration techniques. *AIP Conf. Proc.* **2003**, *30*, 341–344. [CrossRef]
65. Gucunski, N.; Imani, A.; Romero, F.; Nazarian, S.; Yuan, D.; Wiggenhauser, H.; Shokouhi, P.; Taffe, A.; Kutrubes, D. *Nondestructive Testing to Identify Concrete Bridge Deck Deterioration*; Transportation Research Board: Washington, DC, USA, 2013. Available online: https://nap.nationalacademies.org/catalog/22771/nondestructive-testing-to-identify-concrete-bridge-deck-deterioration (accessed on 1 November 2021).
66. Fernandez, I.; Herrador, M.; Marí, A.; Bairán, J. Structural effects of steel reinforcement corrosion on statically indeterminate reinforced concrete members. *Mater. Struct. Constr.* **2016**, *49*, 4959–4973. [CrossRef]
67. Zhu, W. Effect of Corrosion on The Mechanical Properties of the Corroded Reinforcement and the Residual Structural Performance of the Corroded Beams, Universite de Toulouse. 2015. Available online: https://tel.archives-ouvertes.fr/tel-01222175/document (accessed on 1 November 2021).
68. Ebrahimian, H.; Astroza, R.; Conte, J.; Bitmead, R. Information-theoretic approach for identifiability assessment of nonlinear structural finite-element models. *J. Eng. Mech.* **2019**, *145*, 04019039. [CrossRef]
69. Song, M.; Astroza, R.; Ebrahimian, H.; Moaveni, B.; Papadimitriou, C. Adaptive Kalman filters for nonlinear finite element model updating. *Mech. Syst. Signal Process.* **2020**, *143*, 106837. [CrossRef]
70. Ebrahimian, H.; Kohler, M.; Massari, A.; Asimaki, D. Parametric estimation of dispersive viscoelastic layered media with application to structural health monitoring. *Soil Dyn. Earthq. Eng.* **2018**, *105*, 204–223. [CrossRef]
71. Ebrahimian, H.; Astroza, R.; Conte, J. Extended Kalman filter for material parameter estimation in nonlinear structural finite element models using direct differentiation method. *Earthq. Eng. Struct. Dyn.* **2015**, *44*, 1495–1522. [CrossRef]
72. Ismail, M.; Muhammad, B.; Ismail, M. Compressive strength loss and reinforcement degradations of reinforced concrete structure due to long-term exposure. *Constr. Build. Mater.* **2010**, *24*, 898–902. [CrossRef]
73. Li, B.; Cai, L.; Wang, K.; Zhang, Y. Prediction of the residual strength for durability failure of concrete structure in acidic environments. *J. Wuhan Univ. Technol. Mater. Sci. Ed.* **2016**, *31*, 340–344. [CrossRef]

74. Rainieri, C.; Fabbrocino, G.; Cosenza, E. Some remarks on experimental estimation of damping for seismic design of civil constructions. *Shock Vib.* **2010**, *17*, 383–395. [CrossRef]
75. Liang, C.; Liu, T.; Xiao, J.; Zou, D.; Yang, Q. The damping property of recycled aggregate concrete. *Constr. Build. Mater.* **2016**, *102*, 834–842. [CrossRef]
76. Chen, G.; Omenzetter, P.; Beskhyroun, S. Modal systems identification of an eleven-span concrete motorway off-ramp bridge using various excitations. *Eng. Struct.* **2021**, *229*, 111604. [CrossRef]
77. Brownjohn, J.; Dumanoglu, A.; Severn, R.; Blakeborough, A. Ambient vibration survey of the bosporus suspension bridge. *Int. Assoc. Earthq. Eng.* **1989**, *18*, 263–283. [CrossRef]

buildings

MDPI

Article

Novel Physics-Informed Artificial Neural Network Architectures for System and Input Identification of Structural Dynamics PDEs

Sarvin Moradi [1], Burak Duran [2], Saeed Eftekhar Azam [2,*] and Massood Mofid [3]

[1] Department of Civil Engineering, University of Tehran, Tehran P.O. Box 14155-6619, Iran
[2] Department of Civil and Environmental Engineering, University of New Hampshire, Durham, NH 03824, USA
[3] Department of Civil Engineering, Sharif University of Technology, Tehran P.O. Box 11365-11155, Iran
* Correspondence: saeed.eftekharazam@unh.edu; Tel.: +1-(603)-862-0276

Abstract: Herein, two novel Physics Informed Neural Network (PINN) architectures are proposed for output-only system identification and input estimation of dynamic systems. Using merely sparse output-only measurements, the proposed PINNs architectures furnish a novel approach to input, state, and parameter estimation of linear and nonlinear systems with multiple degrees of freedom. These architectures are comprised of parallel and sequential PINNs that act upon a set of ordinary differential equations (ODEs) obtained from spatial discretization of the partial differential equation (PDE). The performance of this framework for dynamic system identification and input estimation was ascertained by extensive numerical experiments on linear and nonlinear systems. The advantage of the proposed approach, when compared with system identification, lies in its computational efficiency. When compared with traditional Artificial Neural Networks (ANNs), this approach requires substantially smaller training data and does not suffer from generalizability issues. In this regard, the states, inputs, and parameters of dynamic state-space equations of motion were estimated using simulated experiments with "noisy" data. The proposed framework for PINN showed excellent great generalizability for various types of applications. Furthermore, it was found that the proposed architectures significantly outperformed ANNs in generalizability and estimation accuracy.

Keywords: physics-informed neural networks; system identification; parameter estimation; input estimation

1. Introduction

The application of Artificial Neural Networks (ANNs) in the field of vibration-based structural health monitoring (SHM) and system identification has gained significant attention from researchers in the past decade [1,2]. Rich et al. proposed an ANN for identifying the mass of the system by acceleration measurements [3]. Xu et al. developed a framework based on ANNs to estimate the inter-story stiffness and damping coefficients of a 5-story frame [4]. In another research, eigenvalues and eigenmodes of a three-story steel frame were estimated by an ANN [5]; more recently, Liu et al. successfully identified modal properties of a cable-stayed bridge via a self-coding ANN with vibration data as input [6].

While ANNs are powerful tools for mapping nonlinear input to output, their accuracy, efficiency, and generalizability significantly depend on the size of the training data. Moreover, the interpretability of ANN results is a challenge. The former is the main drawback for ANNs application in real-life structural system identification, where collecting data is expensive, time-consuming, sometimes impractical, and often subjected to large uncertainties due to humans or equipment nature. Moreover, the common ANNs act as a "black box" and lack interpretability. Therefore, the final model of the network might not have a strong connection with the underlying physics of the system. Although some researchers have tried to find a solution to this problem, the issue of the interpretability of ANN is an open research area [7,8]. Besides, due to the complexity of determining the characteristics of dynamic systems, the ANNs require several assumptions and a significant amount of input data, such as measuring data in all DOFs or restricting the hidden layers and neurons. Khanmirza et al. suggested an ANN for parameter identification of multistory shear buildings [9]. The number of layers in their proposed neural network (NN) was equal to the system's degree of freedom, and it was assumed that displacement, velocity, and acceleration were measured at all degrees of freedom. Xie et al. proposed a discretized system and tried to define a differential model by ANNs [10]. Recently Liu et al. introduced a physically interpretable neural-fuzzy network for the identification of piecewise linear dynamical systems [11]. Wang et al. proposed an ANN using the finite element method to estimate the relationship between uncertainty in system parameters and natural frequencies [12]. Another approach is the application of more dynamic ANNs, such as dynamic neuron units [13]. Another limitation of the ANNs is their fixed architecture, the number of layers, and neurons, which makes them susceptible to large errors with minor changes in the system properties. Deng presented a hybrid NN to surpass this issue that uses a serial-parallel hybrid structure [14]. Dudek suggested introducing a data-driven randomized approach to compact the NN architecture with significant nodes [8]. However, the previous ANNs utilized for system identification lacked a rigorous and generalizable structure.

To address the shortcomings of ANNs, the Physics-Informed Neural Network (PINN) bridged the gap between ANNs and physical systems by introducing the governing equations of the system to the loss function of ANN as an additional criterion [15,16]. Since the introduction of the PINN, this approach has been applied in various research fields, including fluid mechanics [17,18] and solid mechanics [19]. The main goal of the PINN is to facilitate the modeling of phenomena with limited data and a simplified model of the system. One of the advantages of using PINNs is their generalizability in case of changes in structural properties, which is especially important in the control and diagnosis of damaged systems [20,21]. Most physical systems' behavior can be described using differential equations; therefore, introducing derivatives of variables was a challenge in the application of PINN [22]. Haghighat developed a high-level neural network Application Programming Interface (API), SciANN, for the ease of PINN application. The main advantage of the SciANN is the accessibility to define derivatives of variables in physics-informed equations [23,24]. Researchers have recently modified the PINN to extend its performance. Jagtap et al. [25] introduced the Conservative Physics-Informed Neural Networks (cPINNs) for nonlinear conservation law. Kharazmi et al. implemented the Petrov-Galerkin formulation to the PINN and presented Variational Physics-Informed Neural Networks (VPNNs) [26]. They extended their work to global approximation with local learning by developing the hp-Variational Physics-Informed Neural Networks (hp-VPNNs) [27]. The PINN was applied to the identification of parameters of materials, solid mechanics, fractures, and discrete structures such as shear-type structures, where the main focus was on finding the stiffness and viscous damping [28–31].

The existing PINN architectures do not allow for input, state, and parameter estimation using the PDEs of structural dynamics in the objective function, and this problem has not been addressed in structural dynamics for bridge-type structural systems. In this regard, two novel parallel and sequential PINN architectures were developed for seamless integration of the PDEs of structural dynamics into a Machine Learning (ML) framework within the scope of this study. Due to their inherent properties, these layouts reduce the computation time, which is essential for complex and/or real-time systems. To validate the efficacy of the proposed approaches, several dynamic systems with dynamic linear and nonlinear ODEs and PDEs are used for a comprehensive numerical investigation. Firstly, a Single-Degree-of-Freedom (SDOF) system as an ODE was studied; subsequently, a Pure Cubic Oscillator (PCO) as a nonlinear dynamic system was examined. Next, a bridge-type structural system modeled by a PDE was considered. The PDE of the structure was discretized using the Eigenfunction Expansion Method (EEM). It was shown that the two novel PINN architectures successfully estimated the parameters of the beam and the characteristics of the external load. The proposed PINN method was demonstrated to have the ability to perform successful dynamic system identification and assess moving load characteristics in more complex problems after testing it with various inputs and making a comparison to the conventional ANNs. The extensive numerical experiments presented in this study confirmed the excellent generalizability of the proposed architecture for PINN for the state, parameter, and input estimation. Additionally, it was observed that when applied to similar problems, the proposed PINN outperforms regular ANNs.

2. Theoretical Background

2.1. Artificial Neural Network (ANN)

Since the introduction of ANNs, they have been used in various fields. Adeli and Yeh were the pioneers of implementing ANN in structural analysis [32]. Multi-Layer Neural Network (MNN) is the most commonly used form of ANNs due to their simplicity and efficiency. ANNs were not very efficient before the introduction of the backpropagation algorithm and the development of fast computing; however, they have been one of the most popular black-box modeling and prediction tools [33].

In this study, Fully Connected (FC) ANNs with different hidden layers were used as conventional NNs. They were predominantly used as a benchmark to compare the performance of the proposed novel PINN architecture to a pure data-based ANN. For the ANN with multiple hidden layers and a feedforward process, each layer input is calculated based on the previous layer outputs:

$$\mathbf{a}^{s} = \sigma\left(\mathbf{W}^{s}\mathbf{a}^{s-1} + \mathbf{b}^{s}\right). \tag{1}$$

In Equation (1), the $\mathbf{W}^{s}$ is the weight matrix that connects the kth neuron in the $(s-1)^{th}$ layer to the jth neuron in the sth layer with w^{s}_{jk}. With the same notations, the $\mathbf{b}^{s}$ and $\mathbf{a}^{s}$ are the bias and activation of the sth layer, respectively. In other words, b^{s}_{j} stands for bias of the jth neuron in the sth layer, whereas a^{s}_{j} is the activation of the jth neuron in the sth layer. The $\sigma(.)$ is the activation function where $\sigma(v)_j = \sigma(v_j)$.

In this study, $\mathbf{y}$ is the state of the structural system and is approximated by ANNs:

$$\mathbf{y}(\mathbf{x}, t) \approx \hat{\mathbf{y}}(\mathbf{x}, t) = NN(\mathbf{x}, t; \mathbf{W}, \mathbf{b}). \tag{2}$$

The parameters of the ANNs, $\mathbf{W}^{s}$ and $\mathbf{b}^{s}$, are calculated through the backpropagation process. To find the optimal parameters, the following loss function is minimized:

$$\underset{\mathbf{W},\mathbf{b}}{\text{argmin}} L(\mathbf{W}, \mathbf{b}) := \|\mathbf{y}(\mathbf{x}^*, t) - \hat{\mathbf{y}}(\mathbf{x}^*, t)\| = \|\mathbf{y}(\mathbf{x}^*, t) - NN(\mathbf{x}^*, t; \mathbf{W}, \mathbf{b})\|. \tag{3}$$

where $\mathbf{x}^*$ is the training data and $\|\bullet\|$ stands for Euclidean norm of $\bullet$.

As mentioned earlier, the sensor outputs are the only input of the ANN, assuming j installed sensors on the structure, the input layer is defined with j nodes. The number of output layer nodes, k, is selected based on the number of quantities of interest to be estimated. The governing Equation of a single-layer ANN for this structure is

$$y_k = \sigma\left(\sum_{j}^{j} x_j w_{jk} + b_k\right). \tag{4}$$

where y_k is the output of the neuron, checkpoint output, w_{jk} is the corresponding weight to the input x_j, sensor output, and $\sigma(.)$ is the activation function of the ANN. Herein, the sine function was selected to be the activation function.

2.2. Physics-Informed Neural Network (PINN)

The main idea of the PINN is to guide the ANN by introducing the physical behavior of the model as an additive constraint. Therefore, the loss function consists of two components: the loss due to the difference between predicted and real training data, L_a, and the loss originates from physics constraint dissatisfaction, L_p. The data fit portion of the loss function with N_a training points are calculated as follows:

$$L_a = \frac{1}{N_a}\sum_{i=1}^{N_a}[y_i - \wp_{W,b}(x_i^*, t)]^2, \tag{5}$$

where $\wp$ is the approximate solution for the physics equation of the studied system, which is a conventional NN mapping the inputs, $\mathbf{x}, t$, to the variable $\wp$;

$$\wp : (\mathbf{x}, t) \rightarrow NN(\mathbf{x}, t; \mathbf{W}, \mathbf{b})$$

The physical constraint consists of two components; L_f, the differential equation of the system and L_b representing both the boundary and initial condition. The physics differential equation of the studied system for N_f points is introduced to the system by Equation (6) as

$$L_f = \frac{1}{N_f}\sum_{i=1}^{N_f}[\Xi(\wp) - \varsigma^*]^2, \tag{6}$$

in which the $\Xi(\wp)$ is the system's partial differential equation, and the $\Xi(.)$ is the partial differential operator; ς^* is the constant number, which here equals zero. The automated differentiation calculates the partial differential possible [34].

The initial and boundary conditions of the physical differentiation equations can be incorporated into the loss function. Hence L_b, which represents both initial and boundary conditions, is introduced as

$$L_b = \frac{1}{N_r}\sum_{i=1}^{N_r}[\hat{y}_i - g_R(x_i^*, t)]^2 + \frac{1}{N_d}\sum_{i=1}^{N_d}[\hat{y}_i - g_D(x_i^*, t)]^2, \tag{7}$$

where g_R and g_D denote the boundary and initial states of the physical equation for N_r and N_d boundary and initial points, respectively.

In conclusion, the final loss function for the PINN can be calculated in Equation (8) by

$$\underset{\mathbf{W},\mathbf{b}}{\text{argmin}}\, L(\mathbf{W}, \mathbf{b}) := \underbrace{\|\hat{y} - y^*\|}_{L_a} + \underbrace{[\overbrace{\|\Xi(\wp) - \varsigma^*\|}^{L_f} + \overbrace{\|\hat{y} - y^*\|_{\partial R} + \|\hat{y}_0 - y_0^*\|}^{L_b}]}_{L_p}. \tag{8}$$

where $\hat{\mathbf{y}}$ and $\mathbf{y}^*$ are the output of the PINN and the measured data; ∂R is the boundary domain, and subscript 0 stands for the initial condition of the physic equation. It is worth mentioning that the contribution of each phrase in Equation (8) can be adjusted regarding the accuracy of the measured data and the reliability of the physics equations. Thus, Equation (8) is rewritten in the form of Equation (9) as

$$\underset{W,b}{argmin} L(\mathbf{W}, \mathbf{b})\|\hat{y} - y^*\| + \lambda[\|\Xi(\wp) - \varsigma^*\| + \|\hat{y} - y^*\|_{\partial R} + \|\hat{y}_0 - y_0^*\|]. \tag{9}$$

λ is a tuning parameter for regulation of the proposed framework for the PINN; $\lambda = 0$ converts the framework to the ANNs, whereas $\lambda = 1$ reconciles the ANN and the physical information. The schematic algorithm for the loss function in the PINN is plotted in Figure 1. The physics equations can be introduced to the PINN as parametric equations. This makes PINNs suitable for inverse analyses. In this regard, the trained PINN can yield the unknown parameter in addition to the other quantities of interest in the system.

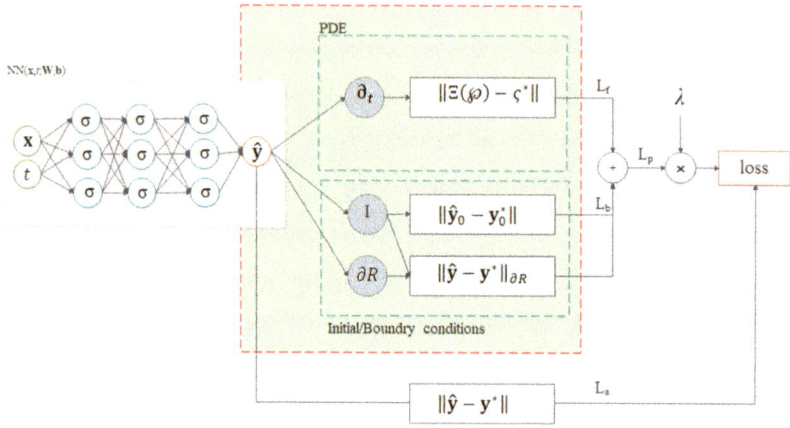

Figure 1. PINN schematic algorithm for loss function.

Although the PINNs can introduce several differential equations in the loss function, the calculation of these equations significantly increases the computational burden of the training stage. To mitigate this problem, two novel architectures are proposed that can be applied to the space-discretized PDE. In this study, the EEM was used to discretize the governing PDE. In the first layout illustrated in Figure 2, n parallel PINNs are trained simultaneously. These PINNs are connected to each other via the total loss function. The efficiency of the parallel PINNs is specifically important for complex and real-time systems in which analysis time is crucial. The proposed layout can be said to be more suitable for running it on GPUs, which significantly reduces the computational time. Besides, due to the independence of the PINNs in the parallel layout, the architecture of each PINN can be different from the rest, making the total layout of the framework more flexible. In some dynamic systems, one or two modes might be dominant and can play an important role in the system's global behavior. One can reflect this priority by introducing a weighting parameter to the output of each PINN. Another solution is to focus on the dominant modes first and later use the estimated properties in the next modes. This framework shapes n series PINNs, as demonstrated in Figure 3. In this regard, the importance of each mode is defined by the framework by order of the relative PINN in the framework. Also, one can terminate these series and control the output accuracy. Unlike the previous layout, the output of each PINN is necessary for subsequent PINNs. However, by placing the dominant modes in the early sector of the sequential framework, the subsequent PINNs benefit from previous estimates and run faster. It should be highlighted that, for generalizability purposes, one can estimate the structural properties, such as stiffness change, by introducing the proper physics-based equations and employing the PINN method. This can potentially be used for SHM of infrastructure systems subjected to unknown moving loads.

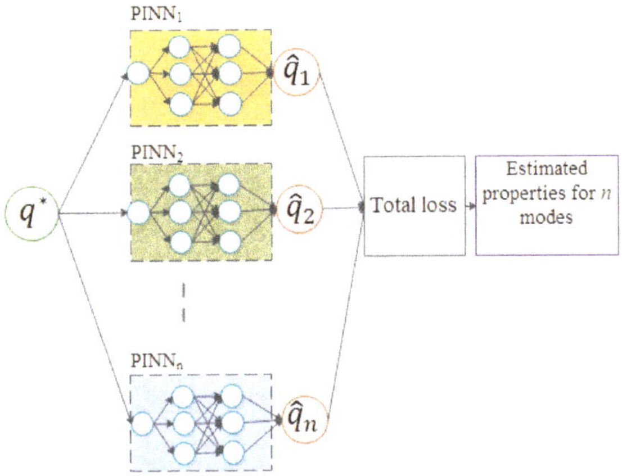

Figure 2. Schematic framework with parallel PINNs.

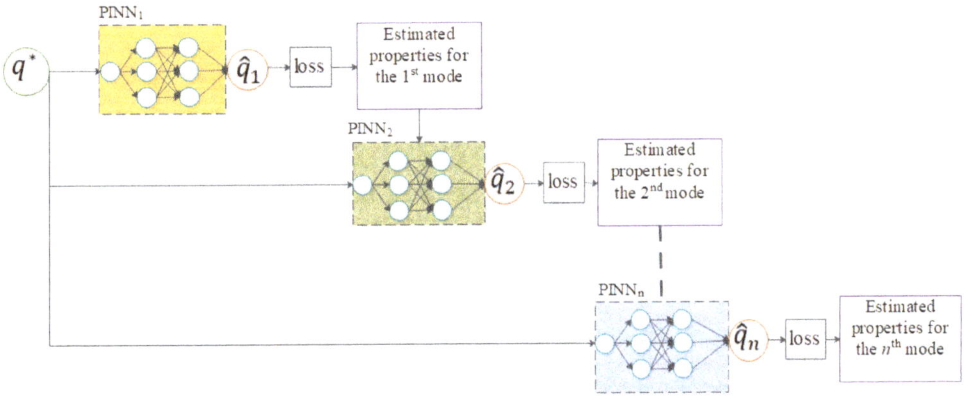

Figure 3. Schematic framework with sequential PINNs.

2.2.1. PINN for ODE

Since introducing the physical equation in the PINN is essential, the equations and loss function are briefly reviewed for ODE and PDE. The dynamic equation for an SDOF is a classic example of ODEs. Consider an SDOF dynamic system where the mass and stiffness of the system are m and k, respectively. A viscous damper, c, is added to the system to study the performance of the PINNs in tracking the vibration behavior of damping systems. The following equation represents the state-space equation for this system

$$\dot{\boldsymbol{\tau}}(t) = \mathbf{A}\boldsymbol{\tau}(t) + \mathbf{B}\mathbf{f}(t), \tag{10}$$

where

$$\boldsymbol{\tau}(t) = \begin{bmatrix} x(t) \\ \dot{x}(t) \end{bmatrix}, \mathbf{A} = \begin{bmatrix} 0 & 1 \\ -m^{-1}k & -m^{-1}c \end{bmatrix}, \mathbf{B} = \begin{bmatrix} 0 \\ m^{-1} \end{bmatrix}. \tag{11}$$

Additionally, the measurement vector $\mathbf{y}(t)$ is defined as

$$\mathbf{y}(t) = \begin{bmatrix} \mathrm{S_d} & 0 \\ 0 & \mathrm{S_a} \end{bmatrix} \begin{bmatrix} x(t) \\ \ddot{x}(t) \end{bmatrix}. \tag{12}$$

S_d and S_a correspond displacement and acceleration observation variables. Now that the physical equation and measurement vector are defined, one can simply use Equation (9) as the loss function.

2.2.2. PINN for PDE

In this section, the vibration equation for a continuous beam is studied as a PDE. Consider a simply supported beam, where ρA and EI are mass per unit length and flexural rigidity of the beam, and c is the damping of the beam. The forced vibration equation of this system can be written as follows:

$$\rho A \frac{\partial^2 w}{\partial t^2} + EI \frac{\partial^4 w}{\partial x^4} + \rho Ac \frac{\partial w}{\partial t} = f(t), \tag{13}$$

where w denotes vertical displacement. By applying the EEM, the beam response is decoupled as

$$w(x,t) = \sum_{j=1}^{\infty} \varphi_j(x) q_j(t) \approx \sum_{j=1}^{n} \varphi_j(x) q_j(t), \tag{14}$$

φ_j and q_j are the jth modal shape and coordinate of the jth mode, respectively. By substituting Equation (14) in Equation (13), the modal equation results:

$$\mathbf{M}\ddot{\mathbf{q}}(t) + \mathbf{C}\dot{\mathbf{q}}(t) + \mathbf{K}\mathbf{q}(t) = \mathbf{S_p}\mathbf{p}(t), \tag{15}$$

where,

$$\mathbf{M} = \rho A \int_0^l \psi(x)\psi^{\mathrm{T}}(x)dx, \mathbf{C} = \rho Ac \int_0^l \psi(x)\psi^{\mathrm{T}}(x)dx, \mathbf{K} = EI \int_0^l \psi(x)\psi''''^{\mathrm{T}}(x)dx. \tag{16}$$

$\mathbf{S_p}$ and $\mathbf{p}(t)$ are the influence matrix and time history of the imposed force, respectively. Subsequently, the state-space equation for a beam subjected to a moving load is

$$\dot{\boldsymbol{\tau}}(t) = \mathbf{A}\,\boldsymbol{\tau}(t) + \mathbf{H}\,\mathbf{f}(t). \tag{17}$$

where

$$\boldsymbol{\tau}^{\mathrm{T}} = \begin{bmatrix} \mathbf{q}(t) & \dot{\mathbf{q}}\ (t) \end{bmatrix}, \mathbf{A} = \begin{bmatrix} 0 & \mathbf{I} \\ -\mathbf{M}^{-1}\mathbf{K} & -\mathbf{M}^{-1}\mathbf{C} \end{bmatrix}, \mathbf{H} = \begin{bmatrix} 0 \\ \mathbf{M}^{-1} \end{bmatrix}, \mathbf{f}^{\mathrm{T}}(t) = -m_e g \boldsymbol{\Psi}(x_e)\mathbf{I}. \tag{18}$$

In this formulation, $\boldsymbol{\Psi}$ is the modal matrix, m_e and x_e are the mass and changing position of the moving loads, respectively. Also, the measured vector, $\mathbf{y}(t)$, is defined as:

$$\mathbf{y}(t) = \begin{bmatrix} \mathbf{S_d} & \mathbf{0} \\ \mathbf{0} & \mathbf{S_a} \end{bmatrix} \begin{bmatrix} \mathbf{q}(t) \\ \ddot{\mathbf{q}}(t) \end{bmatrix}, \tag{19}$$

where $\mathbf{S_d}$ and $\mathbf{S_a}$ are the correspondent displacement and acceleration observation matrix.

Since Equation (17) is composed of n independent equations, it is proposed to consider an individual PINN for each equation which can be calculated in parallel or in series. In this regard, the global architecture will be more flexible since the local architecture of each equation is independent and can be adjusted to the physical equation properly. Figure 4 demonstrates the schematic layout of the proposed framework for a beam subjected to a moving load. It can estimate the state of the system and identify the system properties. In addition, the properties of the input load, the weight, and the velocity of the moving object, which has a key role in the control of the structures, can be determined through the framework of this study. The proposed algorithm for both sequential and parallel PINNs is introduced in Algorithm 1. In these algorithms, it is supposed that the noisy measurements of s sensors are fed to the framework after discretization to n modes by the EEM. Each layout consists of n PINNs that work together.

Algorithm 1 Proposed framewoark with sequential and parallel PINNs.

Input: $\Xi_1, \Xi_2, \ldots, \Xi_n$ and Y.

Sequential layout:

for $j = 1,2,\ldots,n$
  def $\wp: (q,t) \rightarrow NN(q_i, t; \mathbf{W}_i, \mathbf{b}_i)$:
    for $k = 1,2,\ldots,s$
      calculate $\|y_s^* - \varphi_{sk}\hat{q}_k\| + \lambda\|\Xi_k - \varsigma^*\|$
      return $[\hat{q}_k; \omega_k, \zeta|\hat{p}_i]$
    end
end

Parallel layout:

for $j = 1,2,\ldots,n$
  def $\wp: (q,t) \rightarrow NN(q_i, t; \mathbf{W}_i, \mathbf{b}_i)$
end
for $k = 1,2,\ldots,s$
  calculate $\sum_{j=1}^{n}\|y_s^* - \hat{q}_k\| + \lambda\|\Xi_k - \varsigma^*\|$
  return $[\hat{q}_k; \omega_k, \zeta|\hat{p}_i]$
end

Output: [state estimation; dynamic system properties; input load].

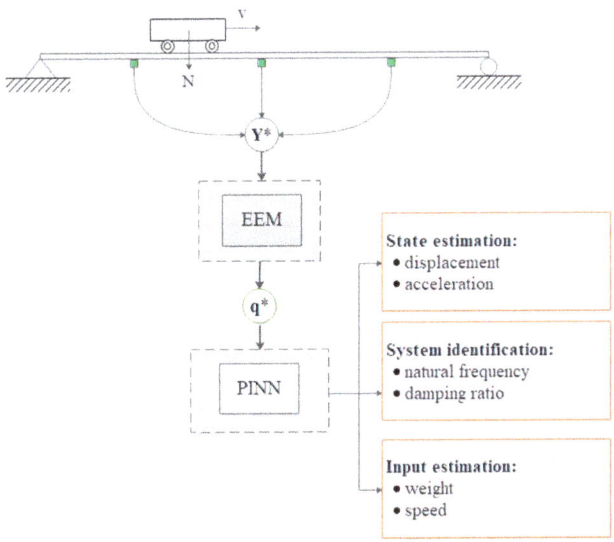

Figure 4. Input and outputs of a PINN framework for a beam subjected to moving load.

3. Demonstrative Examples

The main purpose of this section is to evaluate the performance of the PINN in dynamic structural system identification and input force estimation. To evaluate the performance of PINN for dynamic structural system identification, two types of dynamic systems were considered for verification purposes: a linear Single Degree of Freedom (SDOF) system as well as a Pure Cubic Oscillator (PCO) as ODE examples; afterward, a simply supported beam as a bridge-type structure representing PDE problems were analyzed. In the last step, the capability of the PINN in estimating input force was examined by introducing a moving load to the bridge-type system.

To evaluate the performance of the PINN in determining the dynamic structural system identification, each system was trained for a short period of time with a few displacements and/or accelerations sensor outputs. By imposing physics equations constraints, both dynamic parameters and the state of the system were estimated. The accuracy of parameter estimations was investigated with different sensor noise levels, and the PINN and ANN's capability in predicting the structure's state was compared. Finally, the competency of the PINN in estimating input force was examined. In this regard, the weight and speed of a moving load were identified for a bridge-type structure. An overview of the demonstrative examples and their results is presented in Figure 5.

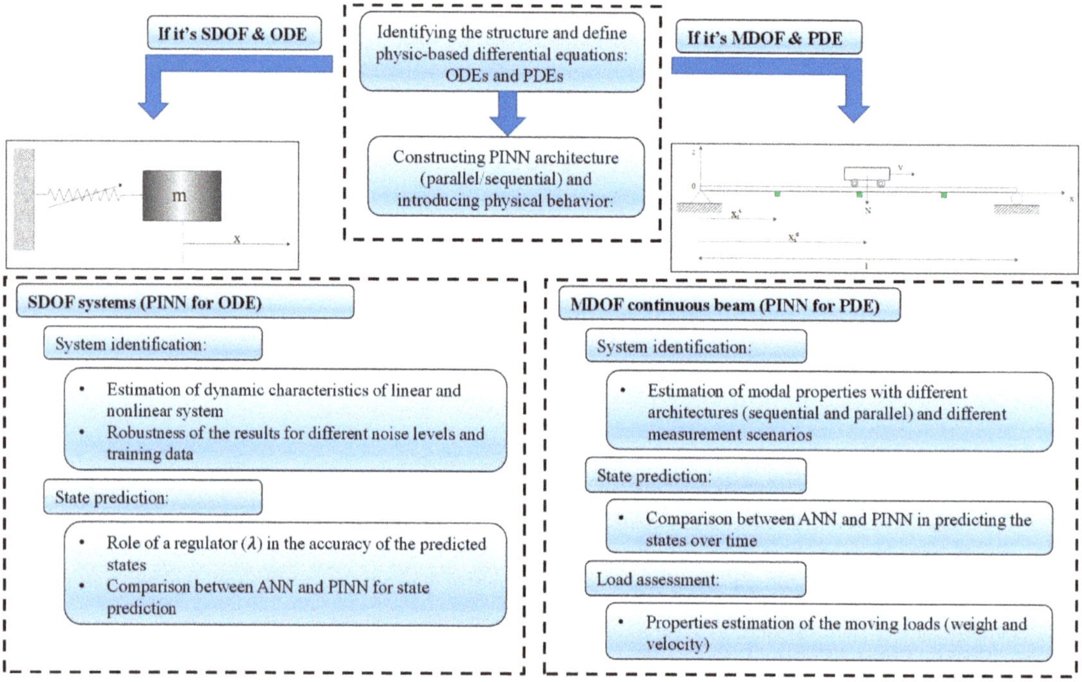

Figure 5. The workflow of the paper and demonstrative examples.

3.1. SDOF System

The vibration of an SDOF system is a classical problem in the field of structural dynamics, and an analytical solution to this system is available. By assuming circular frequency, ω, and damping ratio, ζ, as known parameters, the analytical response of the systems with $\zeta < 1$ is obtained:

$$x(t) = e^{-\zeta\omega_n t}\left[x_0 \cos(\omega_D t) + \left(\frac{\dot{x}_0 + \zeta\omega_n x_0}{\omega_D}\right)\sin(\omega_D t)\right], \tag{20}$$

where $\omega_D = \omega_n\sqrt{1-\zeta^2}$.

As discussed earlier, Equation (10) is regarded as a physics constraint for the PINN, and the outputs of the PINN are both dynamic properties of the SDOF and state prediction. In other words, the only input of the PINN is the state (displacement and/or acceleration) of the SDOF during the training period, **I**, and the PINN will estimate the SDOF's displacement and acceleration after the training period and also predict the system properties, **O**. Here below, the inputs and outputs of PINN are laid out:

$$\begin{aligned} \mathbf{I} &= \left[x(t_i); \ddot{x}(t_i)\right], & i &= 1,2,\ldots,N \\ \mathbf{O} &= \left[x(t_i), \ddot{x}(t_i); \omega, \zeta\right], & i &> N, \end{aligned}$$

where N is the number of samples in the training data, which in this case is the total time steps used for the training period. Now that the input and output of the PINN are known, the loss function is calculated:

$$
\begin{aligned}
&\underset{\mathbf{W},\mathbf{b}}{\operatorname{argmin}} L(\mathbf{W},\mathbf{b}) := \frac{1}{N}\sum_{i=1}^{N}\left(\left[x_i^* - \hat{\wp}_{W,b}(t_i)\right]^2 + \left[\ddot{x}_i^* - \ddot{\hat{\wp}}_{W,b}(t_i)\right]^2\right) \\
&+\lambda\left[\frac{1}{N}\sum_{i=1}^{N}\left[\ddot{\hat{\wp}}_{W,b}(t_i) + 2\hat{\zeta}\hat{\omega}\dot{\hat{\wp}}_{W,b}(t_i) + \hat{\omega}^2\hat{\wp}_{W,b}(t_i) - 0^*\right]^2 + \left[x_0^* - \hat{\wp}_{W,b}(0)\right]^2\right],
\end{aligned} \tag{21}
$$

$\hat{\wp}_{W,b}(t_i)$ represents the PINN estimation of the studied SDOF displacement at the ith time step; $\hat{\omega}$ and $\hat{\zeta}$ are the unknown parameter; therefore, the PINN is expected to predict these parameters after training. In order to compare the accuracy of the predicted state of the PINN and the ANN, $\lambda = 1$, and $\lambda = 0$ were introduced to the model, respectively. Aiming for an accurate comparison, an identical architecture is considered for the analysis of both PINN and ANN. Since the behavior of the SDOF is not complex, a NN with one hidden layer with 50 nodes is selected for this section. It is noteworthy that the initial condition of the system, here x_0, can also be introduced to the loss function. This criterion helps the PINN to converge numerically faster. However, since the initial condition might be unknown, it was not introduced to the presented models.

3.1.1. System Identification

In the first step, the capability of the PINN in system identification is evaluated. Since the ANN is not capable of parameter prediction, only the PINN's estimates are presented in this section. To train the PINN, the analytical displacement of the SDOF was contaminated with an additive zero-mean white Gaussian noise which its Root Mean Square (RMS) was 1% of the RMS of the signal, and it was fed into the PINN during three natural periods. The performance of the PINN in the system identification framework was evaluated for various SDOF systems. Table 1 presents the predicted dynamic characteristics of different SDOFs. With a relative error of almost less than 1.5%, the results confirmed the accuracy of the estimated parameters. The relative error provides a normalized measure of the error in the estimated value. It is expressed as a ratio of the absolute error to the true or actual value. This accuracy level reduces with increasing damping ratios and amplification of the diminution effect, although the overall estimations can be said to be accurate. Eventually, a specific SDOF with a circular natural frequency of $\sqrt{2}$ and a damping ratio of 0.005 was chosen for further studies.

Table 1. Estimation of the PINN for dynamic characteristics.

ω (rad/s)	ζ (kN.s/m)	$\hat{\omega}$ (rad/s)	Relative Error of $\hat{\omega}$ (%)	$\hat{\zeta}$ (kN.s/m)	Relative Error of ζ (%)
0.70710	0.005	0.70711	0.002	0.00495	1.050
1	0.005	1.00002	0.003	0.00501	0.281
1.414213	0.005	1.41422	0.001	0.00499	0.248
2.236067	0.005	2.23607	0.001	0.00500	0.050
3.162278	0.005	3.16221	0.002	0.00493	1.400
1.414213	0.01	1.41425	0.008	0.00998	0.156
1.414213	0.02	1.41420	0.019	0.02002	0.121
1.414213	0.05	1.41418	0.123	0.05003	0.060
1.414213	0.10	1.41459	0.009	0.00491	0.038

The efficiency of NNs relies on both the quantity and quality of the training data. In the next step, the capability of the PINN for system identification was assessed for different training periods and noise levels subjected to the various initial conditions.

To evaluate the performance of the PINN with different training periods, the accuracy of the parameter estimation after (a) half, (b) two, and (c) three natural periods of training with different initial conditions are compared in Figure 6. The PINN can estimate the system properties accurately with less than half of the natural period, and by increasing the training time, the estimations become more precise. In order to evaluate the sensitivity of the PINN to the noise levels, the analyses were repeated for different input data with noise levels ranging from 1% up to 10% RMS of the training data. Based on Figure 7, the dynamic properties of the SDOF were estimated accurately for all noise levels. Despite increasing the noise level, the accuracy of the predicted parameters decreased as expectedly. These results are valid for different initial conditions.

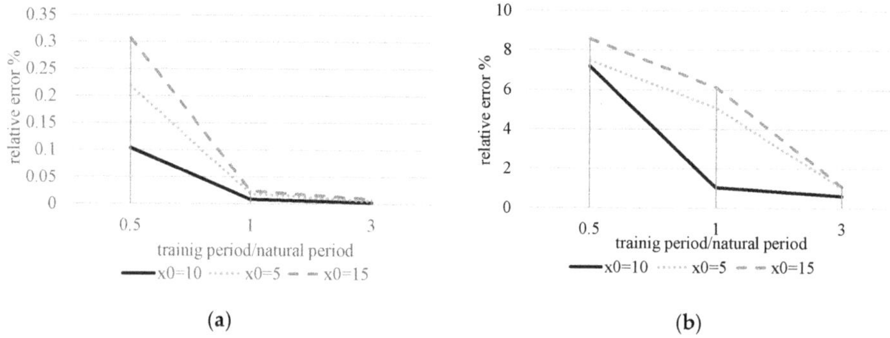

Figure 6. PINN's relative error with different training periods for (**a**) natural circular frequency; (**b**) damping ratio estimations.

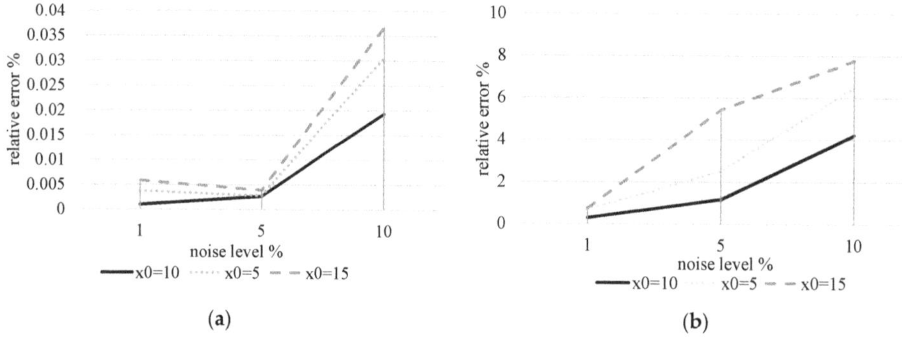

Figure 7. PINN's relative error with different noise levels for (**a**) natural circular frequency; (**b**) damping ratio estimations.

3.1.2. State Estimation

In this section, the accuracy of the predicted states during and after the training period is unveiled in detail. Figures 8 and 9 represent the state estimation of the SDOF trained by displacement and acceleration measurements during three periods, respectively. Even though both PINNs estimations are accurate for the early training steps, the estimation error increases over time. It should be noted that for the system with acceleration training, Figure 9, the PINN needs to estimate the displacement by dealing with a more complicated process, double integration, and more time-consuming compared to the system trained by the displacement, Figure 8. Unlike integration operations, the derivation operations are calculated automatically during the training process. However, the results of this system are precise for both displacement and acceleration estimates. Figure 10 shows the loss function for a PINN that was trained on both displacement and acceleration inputs. As mentioned earlier, the loss function in PINNs consists of two components: L_a, data-driven term, and L_p, differential equation term. In Figure 10, these terms, in addition to the total loss function, are presented; while the physics equation term decreases to lower loss values, the data-driven term values of the displacement and acceleration increased to 0.007 and 0.02, respectively.

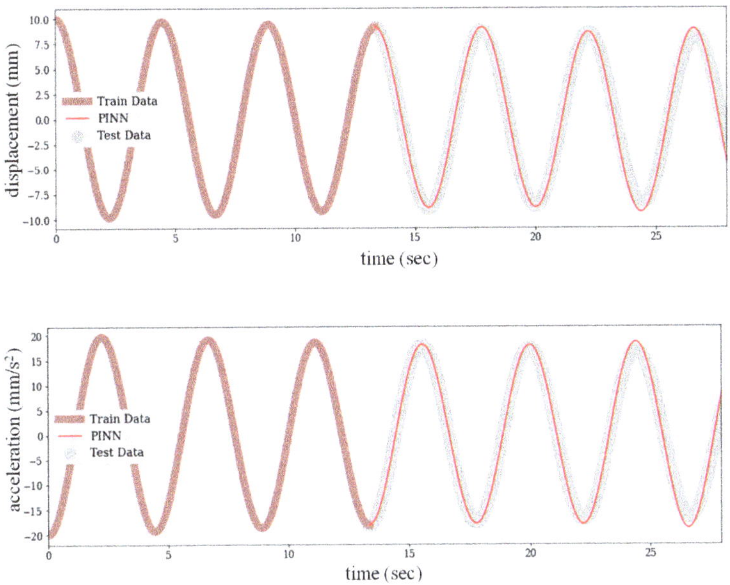

Figure 8. Displacement and acceleration estimation of the SDOF trained by displacement measurements.

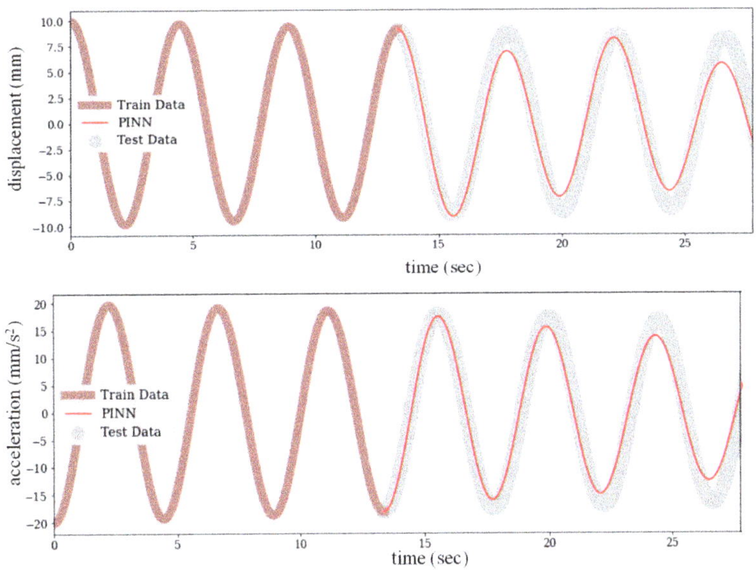

Figure 9. Displacement and acceleration estimation of the SDOF trained by acceleration measurements.

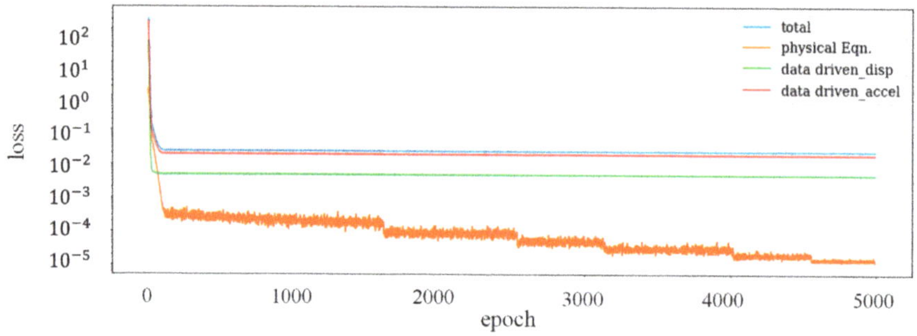

Figure 10. PINN loss function with three periods of training by displacement and acceleration input.

To be able to dig more into the performance of the PINN during the training period, Figure 11 is plotted to represent the output of the PINN for these two constraints: consistency with the training data (displacement and acceleration) and satisfaction of the physical equation during the training period. It is evident that the output of the PINN for training data is accurate, but there is an error of about 2% in the physical equation constraint that reduces over the duration of analyses.

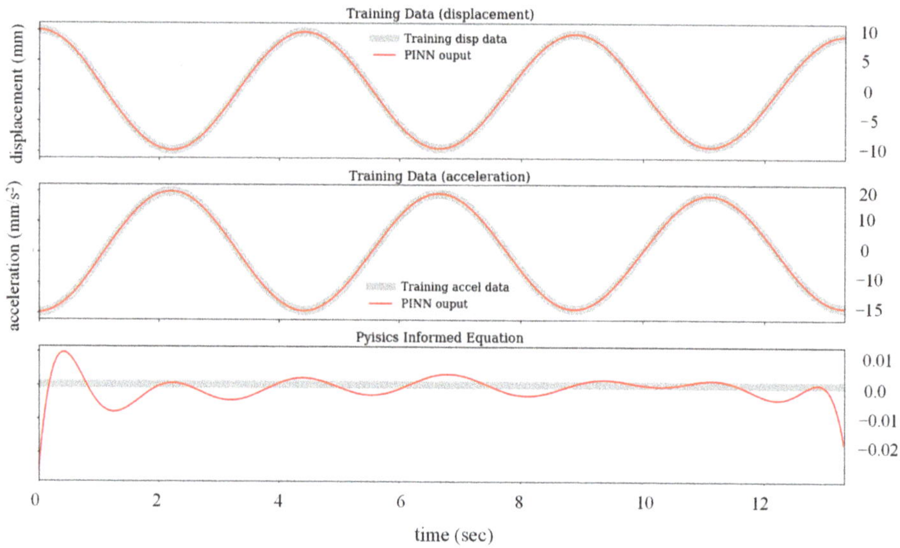

Figure 11. Constraint satisfaction of the PINN during three natural periods of training: (**top**) input data consistency with estimated displacement; (**middle**) estimated acceleration; (**bottom**) physics equation satisfaction 0*.

In the last part of this section, the role of λ as the physics-informed regulator is explained. Based on the results of the system identification, the PINN can estimate the parameters of the physics equation precisely, but according to Figure 10, the data-driven phrases decrease rapidly to a specific value. By recalling Equation (9), it is reminded that λ plays an important role in deciding on the contribution of the physics equation to the final loss function. By increasing λ, the PINN is more strictly forced to follow the physics equation, while decreasing the λ gives more credibility to the measured data. In the extreme case of $\lambda = 0$, the PINN turns into a conventional ANN, which only relies on the data during the training period. Figures 12 and 13 present the estimated displacement and acceleration of the SDOF for the test periods, which is three times longer than the training period. It is obvious from both figures that the ANN cannot represent reliable estimation for long periods, whereas the PINN estimations are capable of following the main trend in all cases. All NNs show an acceptable state estimation for a short period after the training; however, there is an increasing discrepancy over time, which is more evident in the case of using lower λ. In fact, by increasing the λ, the PINN is more restricted to satisfy the physical equation, but it should be noted that this is true when the physics equation is reliable. In other words, the λ helps the PINN to regulate its prediction based on the relative reliability of the measured data and the physics equation.

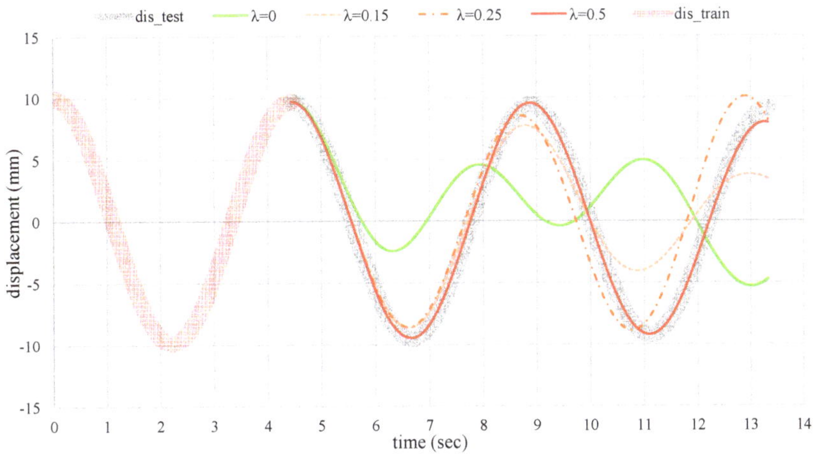

Figure 12. Displacement estimation of the PINN for various λ and the ANN with λ = 0 during the test period.

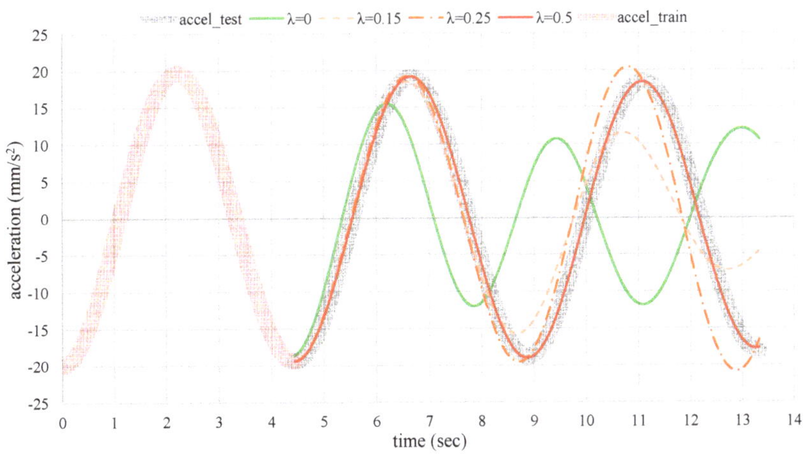

Figure 13. Acceleration estimation of the PINN for various λ and the ANN with λ = 0 during the test period.

3.2. *Pure Cubic Oscillator (PCO)*

Due to the inherent complexity of nonlinear dynamic systems, these problems are often examined through the discretization and linearization of the nonlinear system. Therefore, in most cases, numerical methods are used to find the response of the nonlinear systems. The Duffing-type oscillators are a special case with nonlinear behavior and analytical solution. These oscillators are usually implemented in the nonlinear vibration absorbers [35]. In this study, the pure cubic oscillator, as a special case of the hardening duffing oscillators, is studied. The vibration equation of the PCO plotted in Figure 14 is described as

$$m\ddot{x} + kx^3 = 0 \tag{22}$$

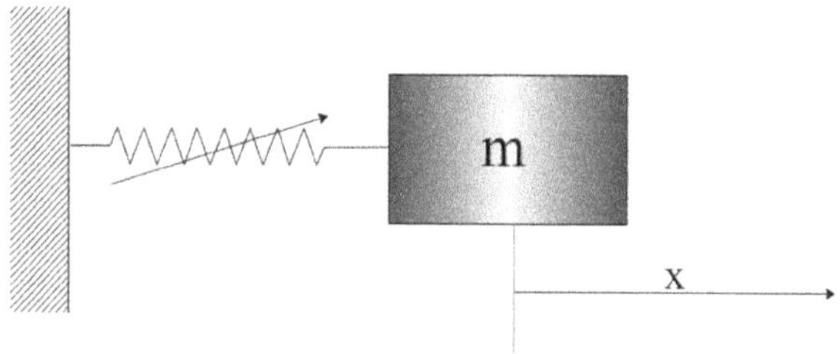

Figure 14. Schematic representation for a nonlinear dynamic system.

The analytical solution of this system is governed by the elliptical function; hence, for a system with an initial displacement of the x_0 and zero initial velocity, one can find [36]

$$x = x_0 \mathrm{cn}(\omega_{PCO}t, \sqrt{m_{PCO}}) \tag{23}$$

where cn is the Jacobi elliptic function; m_{PCO}, elliptic parameter equals 0.5, and the parameter proportional to frequency is

$$\omega_{PCO} = \sqrt{\frac{k}{m}}x_0 = \bar{k}x_0 \tag{24}$$

It is worth mentioning that, unlike linear systems, the frequency is amplitude-dependent, and its amplitude dependency makes the influence of the initial condition more critical.

To explore the performance of the NNs, the vibration of a PCO with an initial displacement was studied. The displacement of the system during the two cycles was measured and fed to the NNs. The properties of the examined PCO for different properties and initial conditions are presented in Table 2. It confirms the accuracy of the estimated parameter with less than 0.1% error. Considering the complex nature of nonlinear systems, the superiority of the PINN over ANN in state estimation is evident. According to Figure 15, the PINN estimates the states accurately while the ANN fails to deliver acceptable estimations; this incompatibility increases over time. Based on Figure 16, during the training procedure, the PINN is obligated to simultaneously comply with the dynamic equation and measure displacement, which helps it to represent more accurate predictions and better parameter estimation (Table 2). The PINN satisfies Equation (22) as the physical constraint with an accuracy of less than 1%. In fact, the added constraint for the PINN clearly improves the training performance, as displayed in Figure 17. By comparing the loss function of the PINN and ANN during the training process, which is presented in Figures 17 and 18, respectively, one can find out that, unlike the PINN, which shows a decreasing trend over epoch incrementation, the ANN shows a drastic decrease during the first epochs and stay stationary for the rest of the training. Although the fluctuations of the loss function decrease over time for both NNs, the total loss function of both NNs descends to 0.0001 at the end.

Table 2. Estimation of the PINN for the nonlinear parameter.

k°	Relative Error (%)	INITIAL Displacement (cm)
2	0.0253	1
3	0.0693	1
4	0.1072	1
2	0.0343	0.5
2	0.0882	1.5

time (sec)

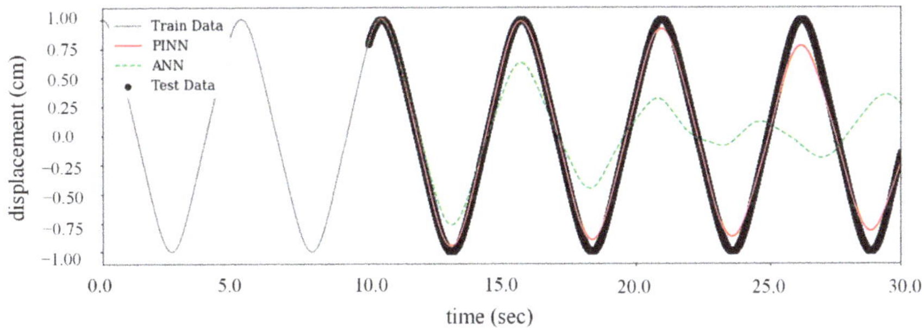

Figure 15. The PINN and ANN state estimation for the test period after two natural training periods.

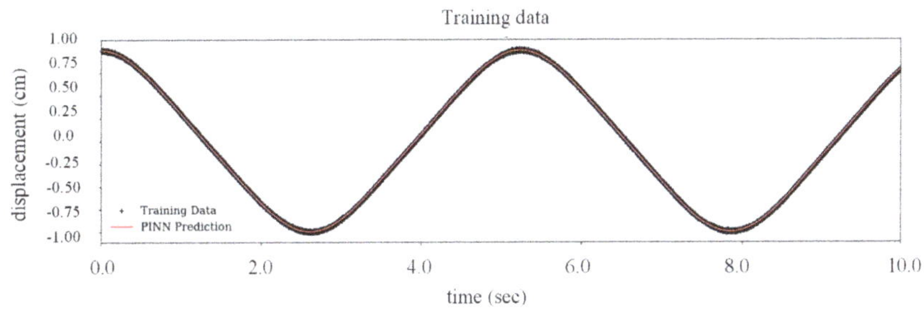

(**a**)

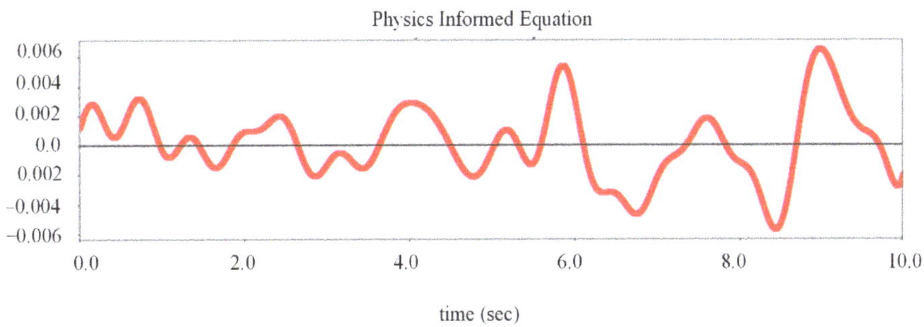

(**b**)

Figure 16. Physic equation satisfaction of the PINN during two natural training periods: (**a**) input data consistency; (**b**) physics equation satisfaction 0*.

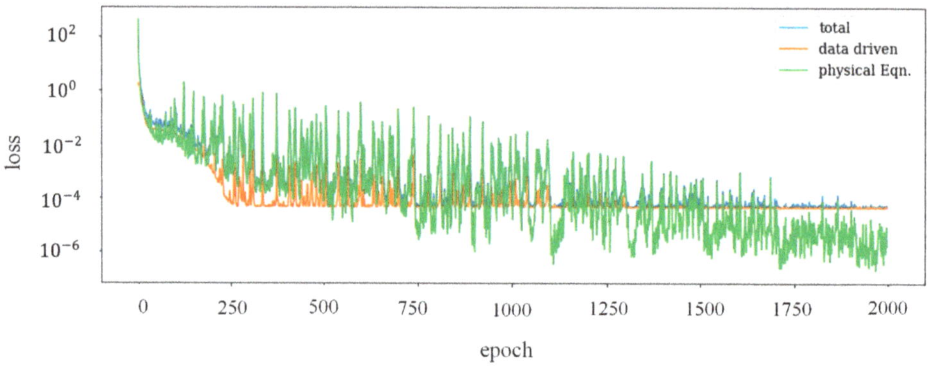

Figure 17. Loss function of the PINN.

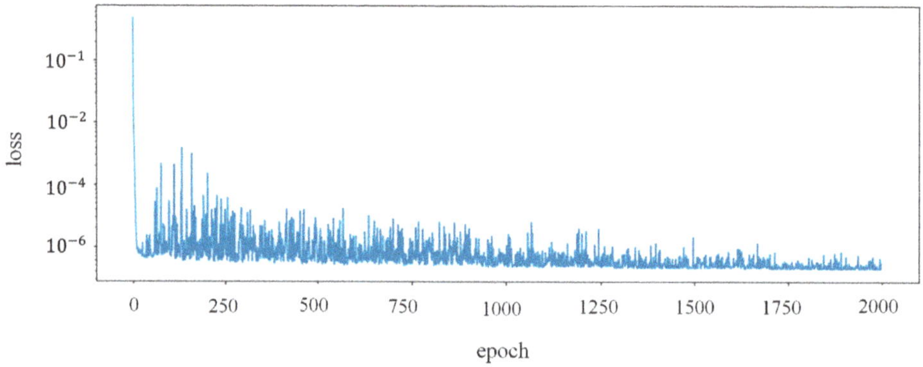

Figure 18. Loss function of the ANN.

3.3. *Simply Supported Beam*

In this section, the efficiency of the proposed method is evaluated for bridge-type structures via analysis of a simply supported beam. To investigate the performance of the PINN in system identification and input estimation of the bridge-type structures, at first, the efficiency of the parallel and sequential PINNs in both system identification and state estimation was probed. Subsequently, the effectiveness of the proposed framework in the dynamic identification of beams subjected to moving load was unveiled.

The geometrical and dynamic properties of the studied beam are length $l = 1$ m, mass per unit $\rho A = 1$ kg/m, flexural rigidity $EI = 5 \times 10^{-3}$N $\times$ m^2, and modal damping coefficient ζ = 0.005 kN.s/m. In this study, three and six modes were considered in the reduced order model for the calculation of sensor outputs and NN's analysis, respectively. It is assumed that three displacement sensors were installed at equal distances on the beam, as illustrated in Figure 19. The output of these sensors with a 1% RMS noise level is plotted in Figure 20.

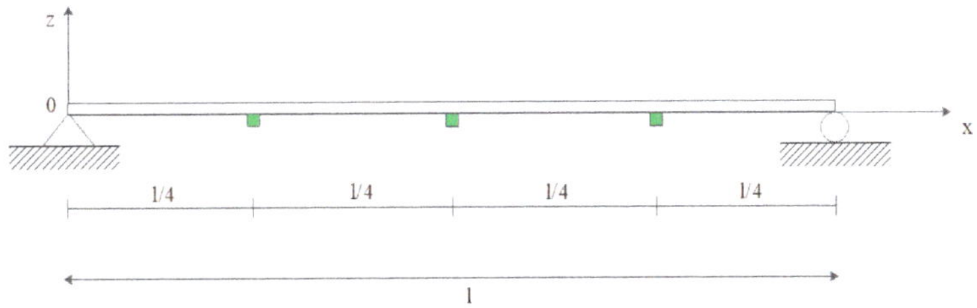

Figure 19. Simply supported beam and sensor locations.

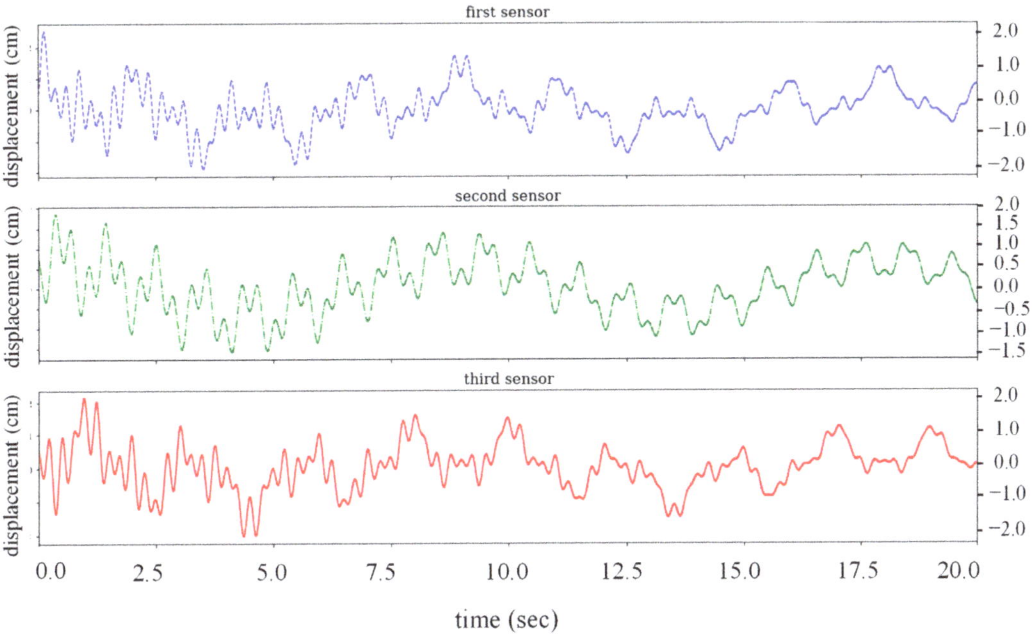

Figure 20. Sensor outputs of the beam.

To introduce the set of ODEs into the PINNs, three parallel NNs with 200 nodes and two hidden layers were considered; three sets of equations were introduced to the PINN to connect these NNs. The loss function, which was used in the training procedure of the proposed PINN framework, is presented in Equation (25).

$$\begin{matrix} j = 1,2,3 \\ k = 1,2,3 \end{matrix} \left| \underset{\mathbf{W}_i,\mathbf{b}_i}{\mathrm{argmin}} L(\mathbf{W}_i, \mathbf{b}_i) := \|\varphi_{kj}\hat{q}_j - y_k^*\| + \lambda_j[\|\hat{\ddot{q}}_j + 2\hat{\zeta}\hat{\omega}_j\hat{\dot{q}}_j + \hat{\omega}_j^2\hat{q}_j - p_j^*\|] \right. \quad (25)$$

Figure 21 plots the modal predictions for the test period, where the PINN outperforms the ANN. The ANN shows an increasing time lag in the first mode and instability in the third mode. Unlike the ANN, the PINN presents acceptable predictions, especially for the first mode. Based on Figure 22, the ANN cannot deliver the phase lag for sensors and shows inconsistency.

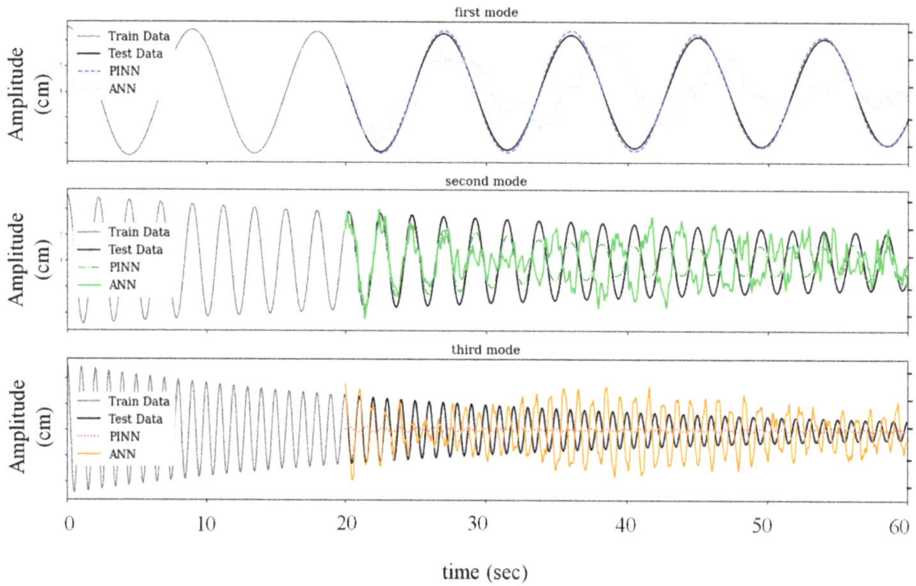

Figure 21. PINN and ANN modal estimation with two hidden layers for the test period.

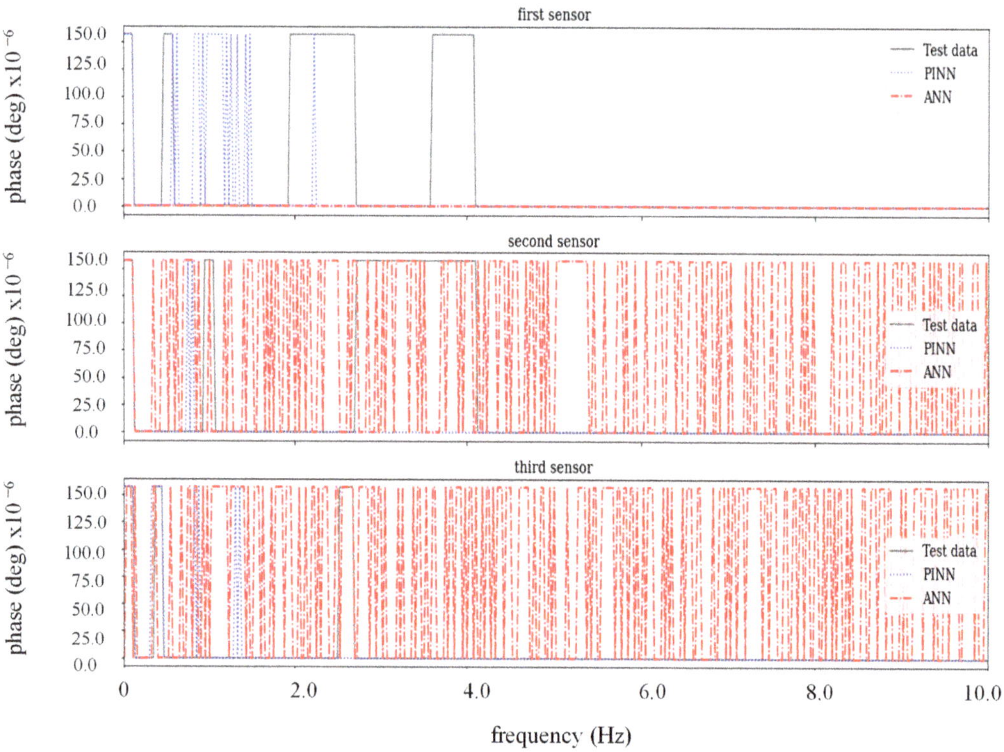

Figure 22. PINN and ANN phase lag of each sensor with two hidden layers for the test period.

To make the architecture of the NNs more efficient, three different architectures were considered for each mode based on the mode's specific attributes. Since the first mode experienced fewer oscillations due to lower vibration frequency, only one layer with 200 nodes was suggested for both NNs architectures, whereas for the next two modes with higher frequencies, three and five layers were introduced for the second and third modes, respectively. The effect of this modification is obvious in both precise circular frequency estimation in Table 3 and state estimations for the next steps in Figure 23. On the other hand, according to Figure 24, the ANN shows a more stable estimation for the second and third modes, but modal predictions deviate from their course over time. The results of the FFT analysis on both the PINN and the ANN state estimations are illustrated in Figure 25, demonstrating that, unlike the PINN, ANN cannot track the higher frequencies precisely, which results in poor modal and, thereafter, state estimations.

Table 3. Estimation of PINN for the first three modal dynamic characteristics.

ω (rad/s)	Relative Estimation Error (%)		
	One Hidden Layer	Two Hidden Layers	Modified Hidden Layers
0.6978	0.01	0.004	0.04
2.7915	50.53	0.007	0.007
6.2809	22.29	17.32	0.002

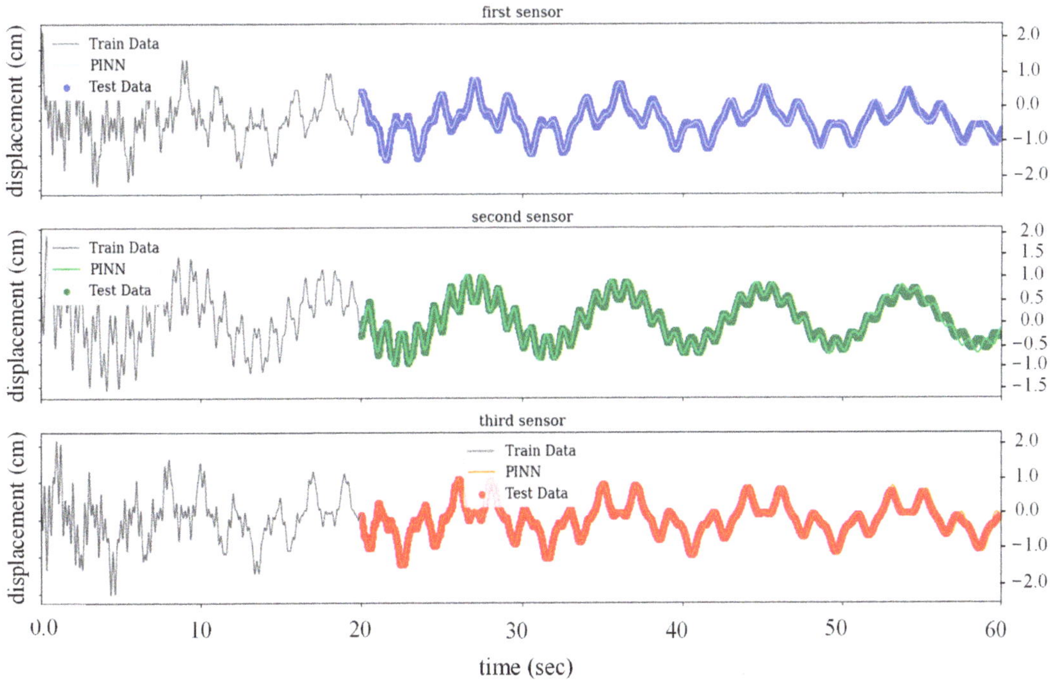

Figure 23. PINN state estimation with modified hidden layers for the test period.

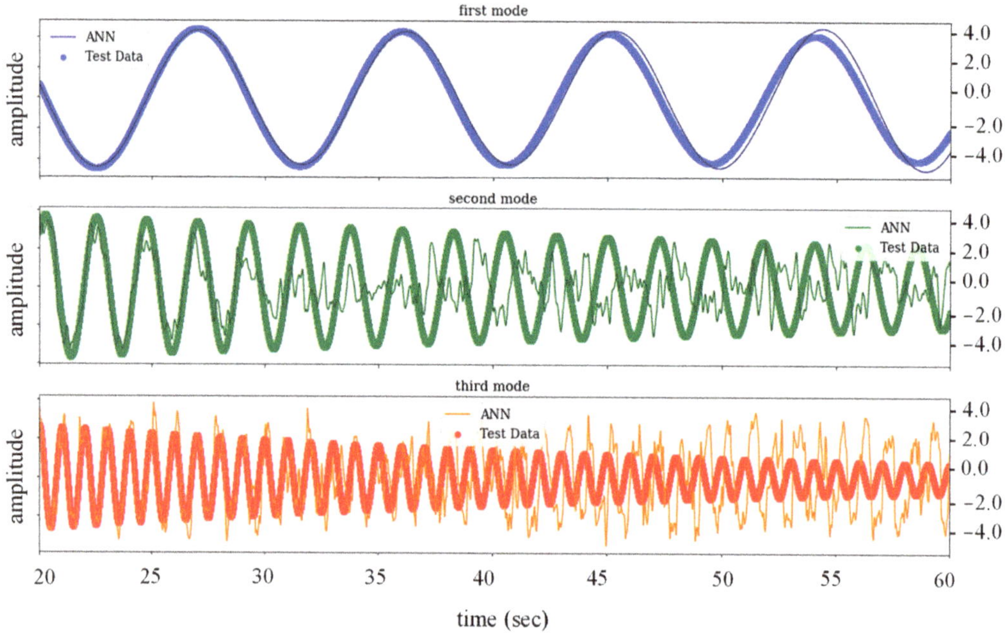

Figure 24. ANN modal estimation with modified hidden layers for the test period.

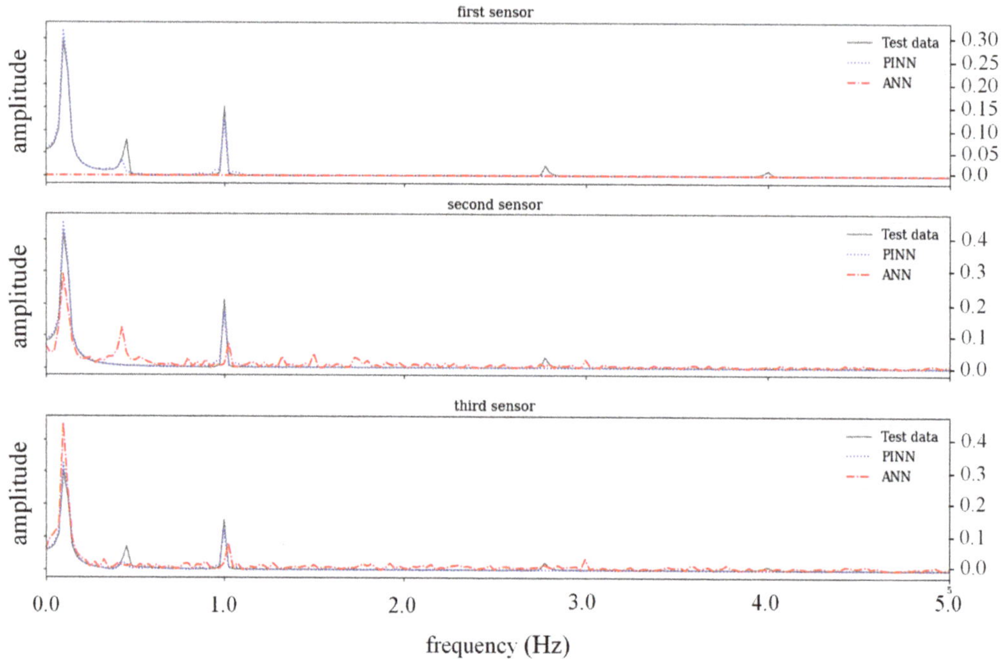

Figure 25. FFT analysis on the PINN and ANN state estimations with modified hidden layers for the test period.

Finally, to evaluate the generalizability of the NNs results on the other points of the beam, the state estimations of both PINN and ANN for two checkpoints are examined. According to Figure 26, the PINN shows accurate estimations for these two points; however, there is a minor lag at the end of the test period. This issue affects the ANN results more seriously; thus, the ANN's predictions are accurate for a much shorter time.

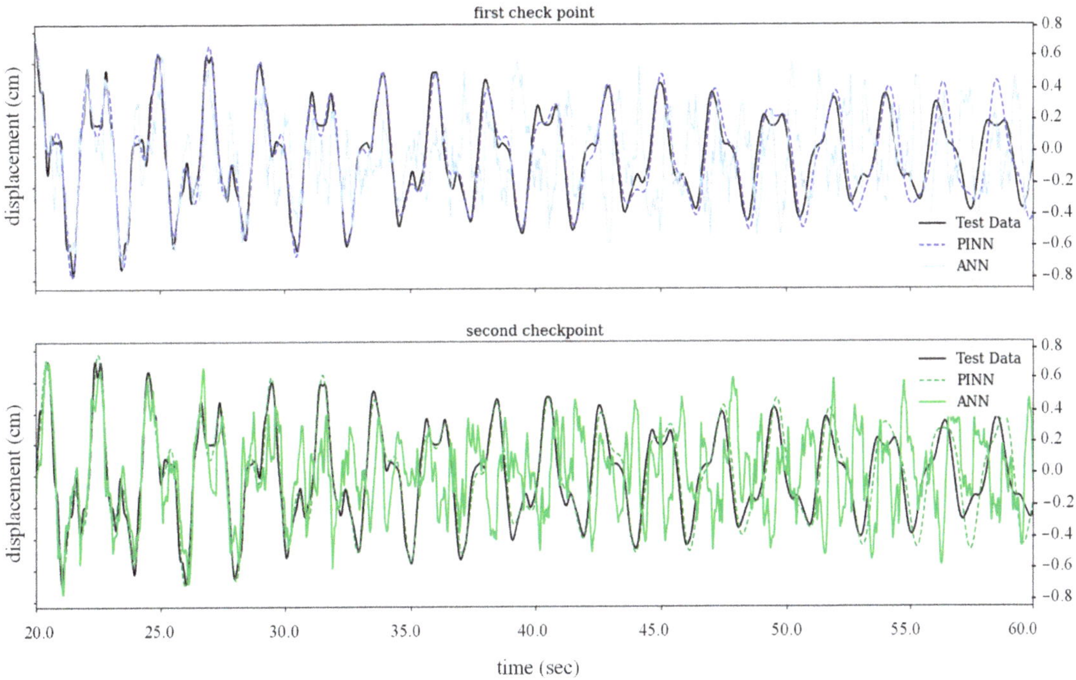

Figure 26. PINN and ANN state estimation at the checkpoints with modified hidden layers for the test period.

3.3.1. Beam Subjected to Moving Load

To examine the applicability of the proposed method for the identification of structural parameters of the beam subjected to moving loads, it is assumed that four moving loads, which weigh 0.05 kg/vehicle each, are passing the beam with equal distances, $d = 0.2$ m with a constant velocity, $v = 0.125$ m/s, which is 1.9 times higher than the critical velocity. After departing the last moving load, the beam experiences a free vibration phase, which is the desired time window for extracting the beam properties. It is assumed that three displacement/acceleration sensors, which are indicated with solid green rectangles, are installed on the beam at the middle point between the vehicles, as shown in Figure 27. The output of these sensors is contaminated with a 1% RMS noise level. Figure 28 shows the output of the third accelerometer.

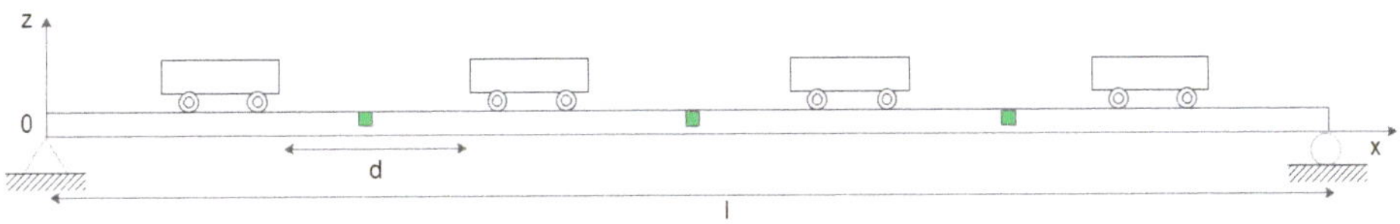

Figure 27. A simply supported beam subjected to moving inertial loads with a single span.

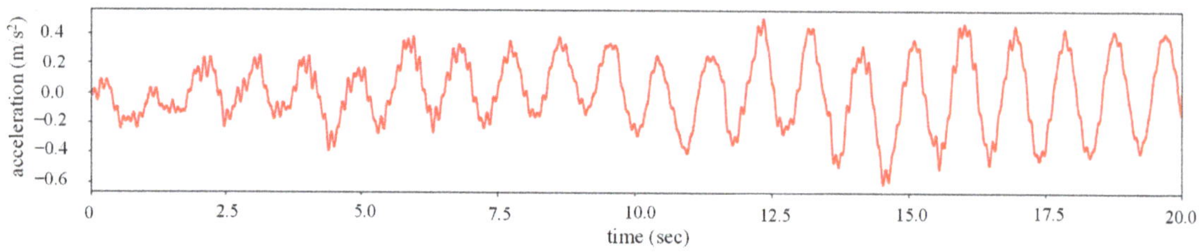

Figure 28. The output of the third accelerometer.

After the departure of the last moving load, the beam experiences a free vibration phase. During this phase, the properties of the beam are evaluated by the PINN. In the first step, NNs with three hidden layers and 200 nodes are considered. Since the first two modes are more important, two parallel NNs are considered to calculate the properties of the first two modes. This layout was presented in the previous section in Figure 2. These two NNs are connected through the modal equation of the first two modes, as explained in Equation (25). In Figure 29, the loss function of two modal equations in comparison to the displacement sensor outputs are plotted. While the first modal equation descends to lower loss values, the second modal equation experiences more fluctuations. Table 4 shows the predicted modal properties of the beam. Although the proposed architecture precisely predicts the damping ratio and the first natural frequency, the error is noticeably slight in the second natural frequency estimation. Figure 30 confirms these results.

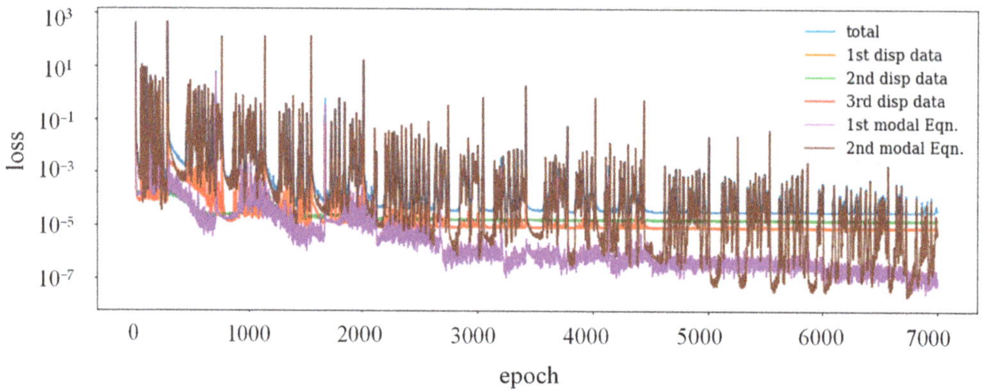

Figure 29. The loss function of the parallel PINNs with displacement input.

Table 4. The modal properties of the beam subjected to moving loads estimated by the PINN.

		ω_1 **(rad/s)**	ω_2 **(rad/s)**	ζ **(kN.s/m)**
	Exact Value	1.709	6.838	0.005
Displacement input	Parallel PINNs	1.725	6.118	0.008
	Series PINNs	1.709	6.184	0.005
Acceleration input	Parallel PINNs	1.704	6.829	0.004
	Series PINNs	1.600	6.832	0.004

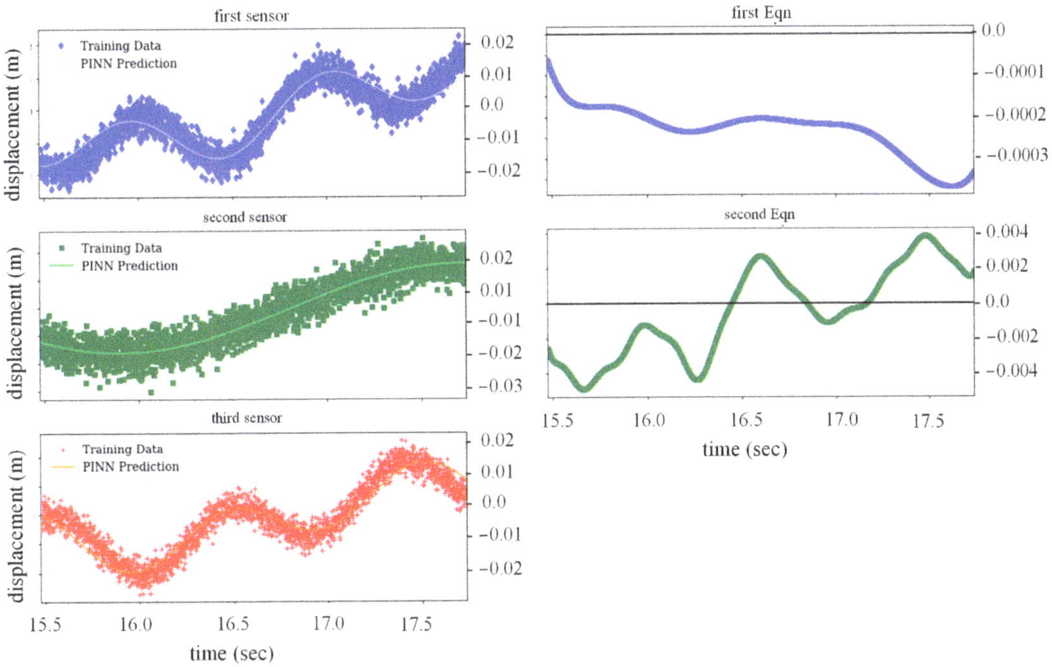

Figure 30. Parallel PINN estimations during the training period with displacement output.

In the second layout, a sequential architecture for the PINNs was considered. Since the first mode plays an important role in the displacement estimation of the beam, for the first phase, only one PINN with three hidden layers and 50 nodes is defined; therefore, only the first modal equation was considered in the loss function. In this regard, the first natural frequency, ω_1, and the damping ratio, ζ, are estimated and then fed into the next series of the PINNs. In the next step, since the properties of the first modal equations were already estimated, only the second natural frequency needs to be predicted. In a sequential layout, each step needs less calculation and therefore runs faster; it relies on the results of the previous phase. This series can also be adjusted to the desired accuracy. For instance, when only the first natural frequency plays a key role, running the next PINNs is not necessary, and the sequence can be terminated based on the desired accuracy. Once again, it is reminded that this layout is essential for situations in which specific modes contribute more to the results than the other modes. For instance, in cases with accelerometer inputs, the parallel layout is more accurate in modal frequencies and damping ratio estimations. Based on Figure 31, the sequential layout satisfies all constraints during the training period, and the parallel PINN can outperform the series layout predictions. In conclusion, both layouts represent acceptable estimations, but the input of the PINN affects the efficiency of these two layouts.

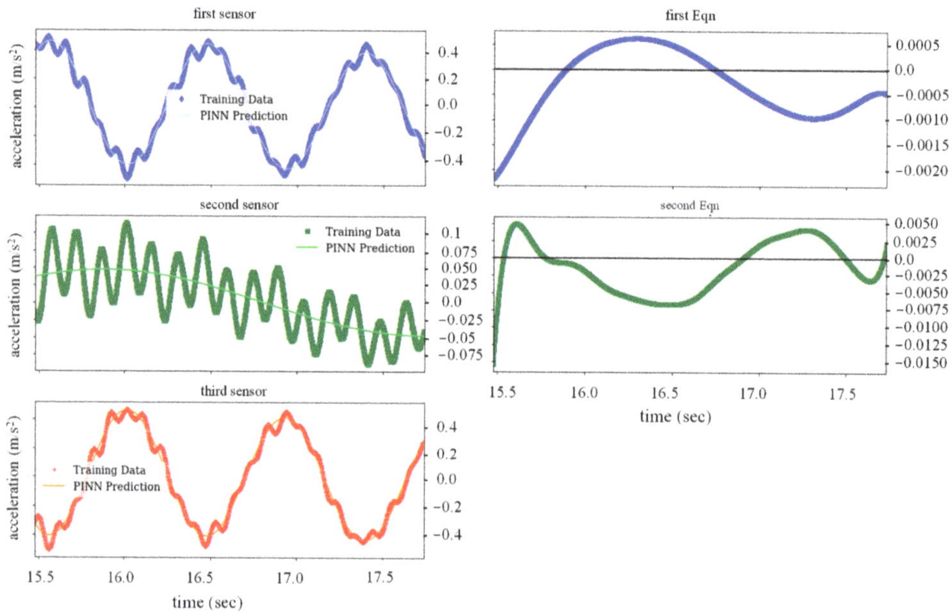

Figure 31. Sequential PINN estimations during the training period with acceleration input.

3.3.2. Force Identification of Moving Load

In the last section, the accuracy of the PINN in estimating a moving load is studied. In this regard, the properties of the beam are assumed to be known, either by the proposed algorithm from previous sections or other methods; hence, the weight and velocity of the moving load are considered unknown parameters. It should be noted that the PINN is trained by the sensor outputs during the moving load passage.

Consider a simply supported beam, the same as in the previous section that is subjected to a moving load (Figure 32). Supposing the forced movement initiates at $t = 0$ and has a constant velocity of v, the implied force is calculated as

$$p(t) = N\delta(x - x_e) = -mg\delta(x - (vt)) \tag{26}$$

where m is the mass of the moving load, g is the gravity acceleration, and δ is the Dirac delta function. By transforming the applied force to the modal force, one gets

$$P(t) = \sum_{j=1}^{n} N\varphi_j(x_e) = -\sum_{j=1}^{n} mg\varphi_j(x - vt). \tag{27}$$

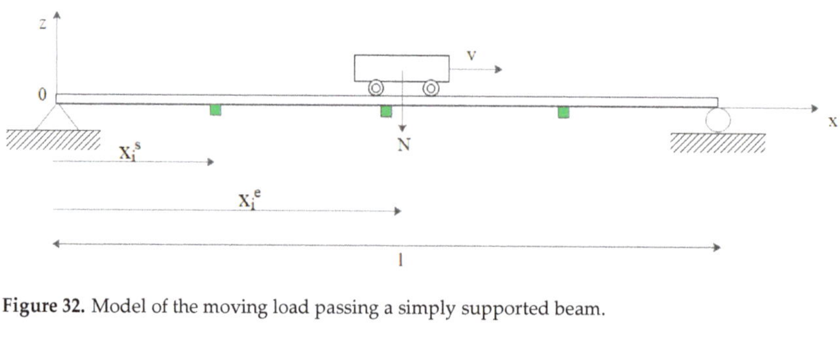

Figure 32. Model of the moving load passing a simply supported beam.

In this section, the PINN is informed by only the first two modes. In other words, the first two modal equations are introduced to the PINN. In addition, the outputs of three sensors during load passage are used for PINN training. The loss function of this system for mass and velocity identification is

$$\left.\begin{matrix} j=1,2 \\ k=1,2,3 \end{matrix}\right| \arg\min_{\mathbf{W}_1,\mathbf{b}_1} L(\mathbf{W}_1,\mathbf{b}_1) := \|\varphi_{kj}\hat{q}_j - y_k^*\| + \|\ddot{\hat{q}}_j + 2\zeta\omega_j\dot{\hat{q}}_j + \omega_j^2\hat{q}_j + \hat{m}g\varphi_j(\hat{v}t^*)\|. \quad (28)$$

To examine the performance of the PINN with different input load predictions, moving loads with various weights and velocities were applied to the proposed model. Figure 33 demonstrates the accuracy of the estimated weight and velocity for the PINN that was trained with three displacement sensors with a 1% RMS noise level. In both cases, the relative error is less than 0.1%, which affirms the PINN capability in predicting moving loads. In addition, it is inferred that by increasing the velocity of the moving load, the estimation accuracy decreases. This is because, by increasing velocity, the time that moving load passes on the beam reduces, and inevitably the time window for gathering data and training the PINN reduces. On the contrary, by increasing the weight of the moving load, the beam experiences more distinct deflections; therefore, the accuracy of the weight estimation increases by incrementing the weight of the load. Based on Figure 34, the accelerometer outputs help the PINN estimate the weight as accurately as the displacement sensors; however, the PINN with acceleration inputs fails to accurately estimate higher velocities. It is worth mentioning that, unlike the displacement sensors, the acceleration inputs make the training of the PINN more challenging and time-consuming. Once again, it is reminded that in contrast to displacement inputs, the acceleration inputs fluctuate more intensely during the passage of the moving load. Also, as mentioned in the previous sections, the process of training the PINNs with acceleration entails integration operations which, unlike the differential operation, brings more complexity to the training process.

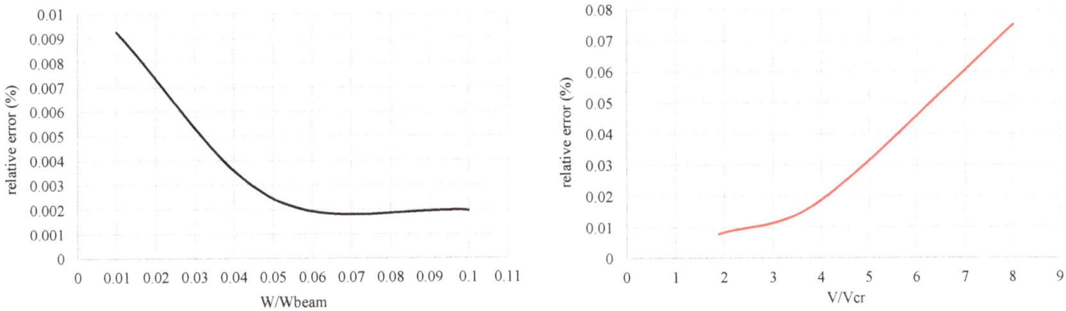

(**a**) (**b**)

Figure 33. Moving load estimations using displacement sensors: (**a**) weight; (**b**) velocity estimation of the moving load.

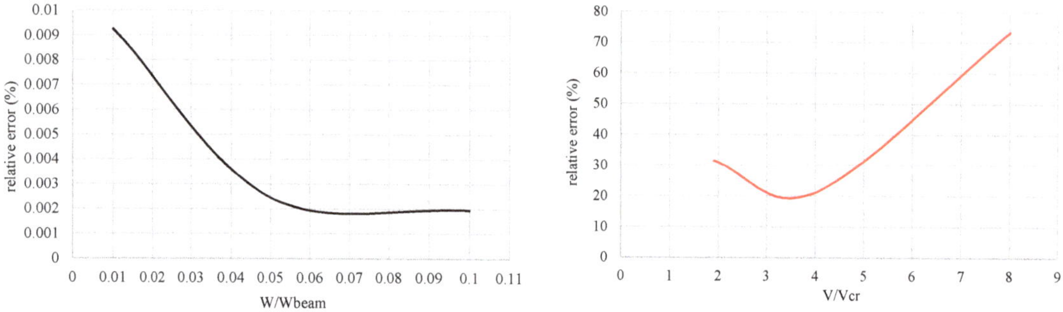

(a) (b)

Figure 34. Moving load estimations using accelerometer sensors: (**a**) weight; (**b**) velocity estimation of the moving load.

4. Conclusions

Within the scope of this study, the two novel architectures were proposed to employ the PINNs in the dynamic system identification and moving load assessment. The proposed architectures were applied to real-time structural system identification and input estimation. The presented methods were validated using simulated experiments on different dynamic systems with various levels of complexity.

In the first step, to elaborate on the details and capability of the PINN framework for system identification and state prediction, the linear and nonlinear SDOFs were studied, respectively. In this regard, various measurement scenarios with different noise levels and training periods were applied to evaluate the performance of the PINN in system identification. Based on the results, the proposed PINN framework can estimate the natural frequency and damping ratio even when the high noise level is 10% RMS. Although the PINN can predict the dynamic properties of the SDOFs with half of the natural period training, it is recommended to train the model at least for one natural period to obtain more accurate results. The capability of the PINN in state estimation was compared to the ANNs, and the superiority of the PINN frameworks was confirmed. The role of λ as the physics-informed regulator was studied, and it was shown that by increasing the λ from 0 to 0.5, the model could predict the state of the system for almost four times longer period, which highlights the significance of taking into account prior physical knowledge in the structure of the PINNs.

In the next step, the bridge-type structure was chosen to demonstrate the efficacy of the PINNs in system identification and load assessment of the beams subjected to the moving loads. The importance of the NNs architectures was assessed via various numerical analyses. It was shown that by modifying the hidden layers of the PINN, one could considerably improve the accuracy of the estimated parameters. Furthermore, the competency of the PINN frameworks for both parallel and sequential architectures was studied. The findings show that both architectures can estimate the modal properties of the beam accurately; although the parallel PINNs need less computation time, making them a better choice for more complex systems. For load assessment of the moving forces specifically important in the realm of bridge design and construction, it has been shown that using both accelerometer and displacement sensor measurements, the proposed framework delivers reliable and robust results for estimating the weight of the moving load. However, for velocity estimation of the moving load, it is recommended to use displacement measurements for training the PINNs.

The following results can be drawn based on the conducted study through extensive numerical examples;

- The PINN architectures in the published literature do not properly address the integration of structural dynamics PDEs into NN objective function. Nevertheless, the novel architecture proposed herein utilizes parallel and sequential PINNs successfully.
- The proposed parallel layout enables the PINN framework to accurately identify the properties of the continuous systems and moving loads. This is an important step toward the generalization of the PINN framework for developing an accurate model of continuous structures and, more specifically, systems subjected to moving loads, such as bridges;
- Unlike conventional ANNs, the novel PINN architectures feature excellent generalizability when applied to output-only system identification of dynamic structural systems. This makes them suitable candidates for operational system identification, where one needs to simultaneously consider input, model, and measurement uncertainties.

Author Contributions: Conceptualization, S.M. and S.E.A.; methodology, S.M. and S.E.A.; software, S.M. and B.D.; validation, S.M. and B.D.; formal analysis, S.M.; investigation, S.M., B.D., S.E.A. and M.M.; resources, S.E.A. and M.M.; visualization, S.M. and B.D.; supervision, S.E.A. and M.M. All authors have read and agreed to the published version of the manuscript.

Funding: This research received no external funding.

Data Availability Statement: Not applicable.

Conflicts of Interest: The authors declare no conflict of interest.

References

1. Hossain, M.S.; Ong, Z.C.; Ismail, Z.; Noroozi, S.; Khoo, S.Y. Artificial neural networks for vibration based inverse parametric identifications: A review. *Appl. Soft Comput. J.* **2017**, *52*, 203–219. [CrossRef]
2. De Souza, C.P.G.; Kurka, P.R.G.; Lins, R.G.; de Araujo, J.M. Performance comparison of non-adaptive and adaptive optimization algorithms for artificial neural network training applied to damage diagnosis in civil structures. *Appl. Soft Comput.* **2021**, *104*, 107254. [CrossRef]
3. Le Riche, R.; Gualandris, D.; Thomas, J.J.; Hemez, F. Neural Identification of Non-linear Dynamic Structures. *J. Sound Vib.* **2001**, *248*, 247–265. [CrossRef]
4. Xu, B.; Wu, Z.; Chen, G.; Yokoyama, K. Direct identification of structural parameters from dynamic responses with neural networks. *Eng. Appl. Artif. Intell.* **2004**, *17*, 931–943. [CrossRef]
5. Facchini, L.; Betti, M.; Biagini, P. Neural network based modal identification of structural systems through output-only measurement. *Comput. Struct.* **2014**, *138*, 183–194. [CrossRef]
6. Liu, D.; Tang, Z.; Bao, Y.; Li, H. Machine-learning-based methods for output-only structural modal identification. *Struct. Control. Health Monit.* **2021**, *28*, 1–22. [CrossRef]
7. Liang, Y.C.; Feng, D.P.; Cooper, J.E. Identification of restoring forces in non-linear vibration systems using fuzzy adaptive neural networks. *J. Sound Vib.* **2001**, *242*, 47–58. [CrossRef]
8. Dudek, G. A constructive approach to data-driven randomized learning for feedforward neural networks. *Appl. Soft Comput.* **2021**, *112*, 107797. [CrossRef]
9. Khanmirza, E.; Khaji, N.; Khanmirza, E. Identification of linear and non-linear physical parameters of multistory shear buildings using artificial neural network. *Inverse Probl. Sci. Eng.* **2015**, *23*, 670–687. [CrossRef]
10. Xie, S.L.; Zhang, Y.H.; Chen, C.H.; Zhang, X.N. Identification of nonlinear hysteretic systems by artificial neural network. *Mech. Syst. Signal Process.* **2013**, *34*, 76–87. [CrossRef]
11. Liu, Z.; Fang, H.; Xu, J. Identification of piecewise linear dynamical systems using physically-interpretable neural-fuzzy networks: Methods and applications to origami structures. *Neural Netw.* **2019**, *116*, 74–87. [CrossRef] [PubMed]
12. Wang, Q.; Wu, D.; Tin-Loi, F.; Gao, W. Machine learning aided stochastic structural free vibration analysis for functionally graded bar-type structures. *Thin-Walled Struct.* **2019**, *144*, 106315. [CrossRef]
13. Yazdizadeh, A.; Khorasani, K.; Patel, R.V. Identification of a two-link flexible manipulator using adaptive time delay neural networks. *IEEE Trans. Syst. Man Cybern. Part B* **2000**, *30*, 165–172. [CrossRef] [PubMed]
14. Deng, J. Dynamic neural networks with hybrid structures for nonlinear system identification. *Eng. Appl. Artif. Intell.* **2013**, *26*, 281–292. [CrossRef]
15. Raissi, M.; Perdikaris, P.; Karniadakis, G.E. Physics informed deep learning (Part II): Data-driven discovery of nonlinear partial differential equations. *arXiv* **2017**.
16. Raissi, M.; Perdikaris, P.; Karniadakis, G.E. Physics informed deep learning (Part I): Data-driven discovery of nonlinear partial differential equations. *arXiv* **2017**, arXiv:1711.10561.

17. Jin, X.; Cai, S.; Li, H.; Karniadakis, G.E. NSFnets (Navier-Stokes flow nets): Physics-informed neural networks for the incompressible Navier-Stokes equations. *J. Comput. Phys.* **2021**, *426*, 109951. [CrossRef]
18. Jiang, C.; Vinuesa, R.; Chen, R.; Mi, J.; Laima, S.; Li, H. An interpretable framework of data-driven turbulence modeling using deep neural networks. *Phys. Fluids* **2021**, *33*, 055133. [CrossRef]
19. Haghighat, E.; Raissi, M.; Moure, A.; Gomez, H.; Juanes, R. A physics-informed deep learning framework for inversion and surrogate modeling in solid mechanics. *Comput. Methods Appl. Mech. Eng.* **2021**, *379*, 113741. [CrossRef]
20. Azam, S.E.; Rageh, A.; Linzell, D. Damage detection in structural systems utilizing artificial neural networks and proper orthogonal decomposition. Struct. *Control Health Monit.* **2019**, *26*, 1–24. [CrossRef]
21. Jain, J.; Kundra, T. Model based online diagnosis of unbalance and transverse fatigue crack in rotor systems. *Mech. Res. Commun.* **2004**, *31*, 557–568. [CrossRef]
22. Nascimento, R.G.; Fricke, K.; Viana, F.A.C. A tutorial on solving ordinary differential equations using Python and hybrid physics-informed neural network. *Eng. Appl. Artif. Intell.* **2020**, *96*, 103996. [CrossRef]
23. Haghighat, E.; Juanes, R. SciANN: A Keras/TensorFlow wrapper for scientific computations and physics-informed deep learning using artificial neural networks. *Comput. Methods Appl. Mech. Eng.* **2021**, *373*, 113552. [CrossRef]
24. Haghighat, E.; Raissi, M.; Moure, A.; Gomez, H.; Juanes, R. A deep learning framework for solution and discovery in solid mechanics. *arXiv* **2020**, arXiv:2003.02751.
25. Jagtap, A.D.; Kharazmi, E.; Karniadakis, G.E. Conservative physics-informed neural networks on discrete domains for conservation laws: Applications to forward and inverse problems. *Comput. Methods Appl. Mech. Eng.* **2020**, *365*, 113028. [CrossRef]
26. Kharazmi, E.; Zhang, Z.; Karniadakis, G.E.M. VPINNs: Variational physics-informed neural networks for solving partial differential equations. *arXiv* **2019**, arXiv:1912.00873.
27. Kharazmi, E.; Zhang, Z.; Karniadakis, G.E.M. hp-VPINNs: Variational physics-informed neural networks with domain decomposition. *Comput. Methods Appl. Mech. Eng.* **2021**, *374*, 113547. [CrossRef]
28. Haghighat, E.; Bekar, A.C.; Madenci, E.; Juanes, R. Deep learning for solution and inversion of structural mechanics and vibrations. In *Modeling and Computation in Vibration Problems, Volume 2: Soft Computing and Uncertainty*; IOP Publishing: Bristol, UK, 2021; pp. 1–18.
29. Goswami, S.; Anitescu, C.; Chakraborty, S.; Rabczuk, T. Transfer learning enhanced physics informed neural network for phase-field modeling of fracture. *Theor. Appl. Fract. Mech.* **2020**, *106*, 102447. [CrossRef]
30. Zhang, E.; Yin, M.; Karniadakis, G.E. Physics-Informed Neural Networks for Nonhomogeneous Material Identification in Elasticity Imaging. *arXiv* **2020**, arXiv:2009.04525.
31. Lai, Z.; Mylonas, C.; Nagarajaiah, S.; Chatzi, E. Structural identification with physics-informed neural ordinary differential equations. *J. Sound Vib.* **2021**, *508*, 116196. [CrossRef]
32. Adeli, H.; Yeh, C. Perceptron Learning in Engineering Design. *Comput. Civ. Infrastruct. Eng.* **1989**, *4*, 247–256. [CrossRef]
33. Tiumentsev, Y.V.; Egorchev, M.V. Neural Network Black Box Approach to the Modeling and Control of Dynamical Systems. *Neural Netw. Model. Identif. Dyn. Syst.* **2019**, *1*, 93–129.
34. Baydin, A.G.; Pearlmutter, B.A.; Radul, A.A.; Siskind, J.M. Automatic differentiation in machine learning: A survey. *J. Mach. Learn. Res.* **2018**, *18*, 1–43.
35. Ding, H.; Chen, L.Q. Designs, analysis, and applications of nonlinear energy sinks. *Nonlinear Dyn.* **2020**, *100*, 3061–3107. [CrossRef]
36. Kovacic, I. *Nonlinear Oscillations*; Springer International Publishing: Cham, Germany, 2020.

buildings

Article

An Analytical Algorithm for Determining Optimal Thin-Walled Hollow Pier Configuration with Sunlight Temperature Differences

Lipeng An [1], Dejian Li [2], Peng Yuan [3,*] and Peng Chen [4]

1 School of Transportation, Civil Engineering & Architecture, Foshan University, Foshan 528225, China; zzdxanlipeng@163.com
2 School of Civil Engineering, Central South University, Changsha 410075, China; dejianli@vip.sina.com
3 State Key Laboratory of Internet of Things for Smart City, Department of Civil and Environmental Engineering, University of Macau, Macau SAR 999078, China
4 Shanghai Municipal Engineering Design Institute (Group) Co., Ltd., Xi'an Branch, Xi'an 710075, China; chenpeng@smedi.com
* Correspondence: pengyuan@um.edu.mo

Abstract: Formulas for computing the line shape of a thin-walled hollow pier body based on structural characteristics and measured sunlight temperature difference are derived using an analytical algorithm. In a case study of the No. 5 pier of a newly constructed continuous beam bridge on a mountainous expressway of Guizhou Province in China, the pier top's displacement calculated by the analytical algorithm, currently accepted code, and a FEM program were each compared to its measured values. Furthermore, the effects of sunlight temperature difference, pier height, and wall thickness on the line shape of the pier body were explored, and the results show that the calculation values from these formulas were closer to the measured values than the currently accepted code, with a maximum error of 0.507 mm, demonstrating that the formulas have a more dependable result, higher precision, and more specific applicability. Thus, the algorithm provides a better method for the line shape calculation and construction control of thin-walled hollow piers because it can accurately account for sunlight temperature differences and pier height.

Keywords: thin-walled hollow pier; line shape calculation; field measurement; sunlight temperature difference

1. Introduction

For thin-walled hollow high-piers in the state of the construction stage, any change in sunlight temperature leads to an uneven temperature field on a pier that typically presents with a nonlinear distribution [1,2]. This can lead to a large nonlinear temperature difference in the pier structure that can cause structural deformation, the temperature difference is affected by atmospheric temperature, sunlight radiation, section position, and other factors [3–5]. With the recent development of highway construction in China, the amount of high-pier bridge construction has been correspondingly increasing, and the height of piers has also been gradually increasing. However, it is common that pier top displacements exceed the allowable value (25 mm, H/3000, and $\leq$30 mm (H is pier height)) during the bridge construction stage [6]. The optimal configuration is that the linear of the pier after construction has no offset, and the offset value is 0 mm. In construction, when the linear offset of pier meets the requirements of the design specification, it can also be accepted. Nonlinear temperature changes not only affect a pier's alignment but also produce a large eccentric bending moment at the bottom of the pier that affects the bearing capacity of the pier [7–9]. Hence, studies of the effects of sunlight temperature differences on the alignment of such high piers [10–13] are urgently needed.

The research on the effect of temperature on concrete structures began in the 1950s, when scholars from various countries initially studied the temperature field of concrete structures through a series of theoretical analyses and a large number of field experiments and successfully obtained some temperature field distribution laws for concrete structures [14–20]. However, the research on the effect of a sunlight temperature difference mostly focused on the superstructure of the constructed bridge [21–29] rather than the substructure under construction [30–33]. Research on the influence of sunlight temperature differences on the shape of thin-walled hollow piers under construction was even more scant. In recent years, though, with the wide application of hollow, thin-walled high piers in engineering, people have begun to pay more and more attention to the effects of temperature in their analyses of bridge substructures, especially high the piers [34,35].

Based on the theory of sunlight temperature distribution, Zhang [30] analyzed the temperature effect of a double-limbed rectangular hollow pier under a sunlight temperature load using the secondary development function of ANSYS and analyzed the influence of section shape on the pier's temperature field. Dai [31] et al. used the generalized Pareto distribution and Centennial return period additive model to analyze the temperature–time change in the direction of pier thickness and predicted the most extreme temperature values during the return period, although they did so without experimental verification. Similarly, Liu [32] analyzed the influence of sunlight radiation and air temperature on thin-walled hollow piers in high-altitude areas according specifically focusing on low temperatures, the large temperature differences between day and night, and stronger solar radiation at higher altitudes. To study of the effects of sunlight temperature differences on the shape of thin-walled hollow piers, Lin [33] analyzed the effect of temperature with both three-cavity sections and double-limb sections. The results showed that the displacement of the pier top caused by the sunlight temperature difference was closely related to the change in air temperature, and the lateral displacement of the pier top was the largest when the external temperature was the highest. Likewise, Bi [3] established a three-dimensional simulation model of a thin-walled hollow pier using ANSYS software and analyzed the displacement of the pier top under the fabrication constraints. Based on ANSYS analysis and railway specification formula [36], Tang [37] calculated the top displacement of a thin-walled hollow high pier under the action of sunlight temperature difference, and then put forward an active control method for high pier body alignment.

However, there are still few studies on the effects of sunlight temperature differences on the substructure of bridges under construction, and research results on the effects of sunlight temperature differences on the alignment of high piers are still scarce. The only useful references have come in the Chinese Code TB 10052-1997 railway flexible pier bridge technical specifications [36] for the calculation formula of pier top displacement, or from the use of finite element software. However, the displacement calculation formula in the railway code is meant for railway flexible piers, and it has a long history of successful applicability, whereas the thin-walled hollow structure commonly used in high pier structures is not well suited for that formula as it fails to reflect the piers' structural characteristics.

To solve the above-mentioned disadvantages, the finite element methods are used to tackle the problem of optimal high pier construction including modeling, boundary condition simulation, calculation, and result analysis, and an optimal high pier construction configuration is obtained in this study. Additionally, the field environment is complex and ever-changing, so that question of how best to simulate the real-world situation of these structures is also difficult. Therefore, this article uses the function integral method together with the equivalent linearization method to derive an analytical formula for thin-walled hollow pier top displacement caused by sunlight temperature differences according to this type of pier's particular structural characteristics. Furthermore, it takes the No.5 thin-walled hollow high pier under construction of a continuous beam bridge in the mountainous area of Guizhou Province in China as a field example and use monitored temperature effects to compare the analytical formula to existing finite element calculations. These comparison results show that the analytical calculation formula has very high precision and thus can provide a new means for better taking sunlight temperature differences into account. Using the above formula and pre-bias method, it can provide a method for the linear construction control of the hollow pier.

2. Derivation of the Analytical Formula

2.1. Fundamental Assumptions

Consider a theoretical thin-walled hollow pier. It is assumed that the temperature distribution along the height direction of the pier is uniform, and the temperature distribution difference caused by local changes is ignored. The concrete materials are assumed to be homogeneous and isotropic, in line with the law of elastic deformation before cracks occur, and the Bernoulli plane deformation assumption is also applied. Deformation loads are thus calculated as unidirectional temperature loads and then superimposed to form the basis for multidirectional temperature load calculations.

2.2. Derivation

At present, the research on the influence of sunlight temperature differences on the alignment of bridge piers usually begins from the one-dimensional problem. First, the temperature field effects in the one-dimensional direction of the structural plane are analyzed, as shown in Figure 1, and then the temperature field effects in another dimension are analyzed. The total offset value is then obtained using a coupling calculation, and this analysis method has achieved the expected results in several practical engineering projects [31,32]. Referring to the relevant literature and specifications [37], under the action of a sunlight temperature difference between the AB-side and the CD-side, as shown in Figure 1, the temperature gradient pattern of box piers along the wall thickness is distributed exponentially [38] as follows:

$$T(x) = T_0 e^{-ax} \tag{1}$$

where $T(x)$ is the temperature difference between the calculation point and the back surface (°C); T_0 is the temperature difference between the front and back of the pier (°C); a is the exponential coefficient, generally taken to be 7; and x is the distance from the calculated point to the heated surface (m).

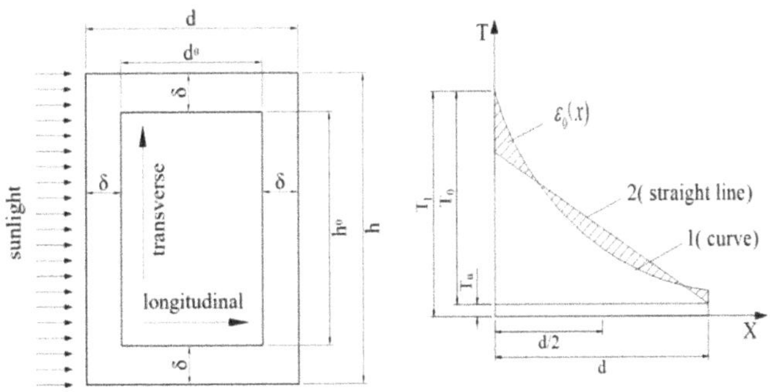

Figure 1. Deformation of the pier section under a sunlight temperature difference.

Figure 1 shows the deformation of the thin-walled hollow pier cross-section using a sunlight temperature difference diagram. The right side of the deformation diagram can be expressed in the X direction along the bridge but can also be expressed in the Y direction transverse bridge. Under the action of the nonlinear temperature gradient from sunlight, the longitudinal (transverse) fiber on the cross-section of the bridge pier swells and shrinks, and in the free state the free deformation curve of the fiber is similar to the shape of the sunlight temperature gradient curve (an exponential curve), as shown in Figure 1. However, since the longitudinal (transverse) fiber deformation on the cross-section needs to maintain the straight-line distribution assumed by the plane cross-section, the actual cross-section line shape is given by line 2 in Figure 1, and the self-constrained deformation between the longitudinal (transverse) fibers of the cross-section are shown as the shaded line in the figure. Since there is a free end, the thermally induced stresses should be 0 in the vertical direction of the pier.

Figure 1 shows that the self-constrained temperature stress in the cross-section is proportional to the free strain of the fiber and that the strain difference is retained after plane deformation. The calculation is as follows:

$$\varepsilon = \alpha T_0 e^{-ax} \tag{2}$$

$$T'_x = T_1 - (T_1 - T_n)\frac{x}{d} \tag{3}$$

$$\varepsilon' = \alpha T'_x \tag{4}$$

$$\varepsilon_0(x) = \varepsilon - \varepsilon' = \alpha\left[T_0 e^{-ax} - T_1 + (T_1 - T_n)\frac{x}{d}\right] \tag{5}$$

$$\sigma_0(x) = E\varepsilon_0(x) = E\alpha\left[T_0 e^{-ax} - T_1 + (T_1 - T_n)\frac{x}{d}\right] \tag{6}$$

In the above formulas: α is the material linear expansion coefficient ($\alpha = 1.0 \times 10^{-5}/°\text{C}$); d is the width of a single pier; ε is the free strain of the fiber; ε' is the actual deformation of the section; T_1 is the surface temperature of the sunny side when the section is deformed in the plane (°C); T_n is the surface temperature of the dorsal side when the section experiences planar deformation (°C); T_0 is the temperature difference between the front and back of the pier (°C); T'_x is the equivalent temperature at the position x when the cross-section is plane deformed (°C); $\varepsilon_0(x)$ is the strain difference between fiber-free expansion and section plane deformation (°C); $\sigma_0(x)$ is self-constrained temperature stress; and E is the elastic modulus of concrete (MPa).

When the temperature field is stable, the deformation of the structure is related to the linear expansion coefficient of the material. In a typical pier structure, the volume of the steel bars is only 1% to 2% of the total volume of the structure, so the thermal conductivity of the pier structure is mostly affected by the thermal conductivity of the concrete. Therefore, α takes $1.0 \times 10^{-5}/°\text{C}$. During construction the pier during construction is a statically determinate structure, so the temperature stress caused by sunlight temperature difference is only temperature self-stress $\sigma_0(x)$. According to the principle of temperature stress self-balance on a section, when the bar is in a free state, the bending moment on the section is $\sum M = 0$, i.e., $\int_0^d \sigma_0(x)\left(\frac{d}{2} - x\right)h_0 dx = 0$. From this formula, it can be obtained:

$$\begin{aligned} &h\left[T_0\left(\frac{1}{a^2}e^{-ad} + \frac{de^{-ad}}{2a} + \frac{d}{2a} - \frac{1}{a^2}\right) - \frac{1}{12}(T_1 - T_n)d^2\right] - \\ &h_0\left[T_0\left(\frac{1}{a^2}e^{-a(d-\delta)} + \frac{(d-2\delta)e^{-a(d-\delta)}}{2a} + \frac{(d-2\delta)e^{-a\delta}}{2a} - \frac{e^{-a\delta}}{a^2}\right)\right] \\ &+\left[h_0\left(\frac{1}{12d}(T_1 - T_n)(d-2\delta)^3\right)\right] = 0 \end{aligned} \tag{7}$$

The temperature gradient η along the wall thickness can be obtained from Equation (7) by equivalent linearization of the actual temperature difference curve:

$$\begin{aligned} \eta &= \frac{T_1 - T_n}{d} \\ &= \frac{6T_0}{ahd^3 - ah_0 d_0^3}\left\{\begin{array}{c} hd\left[\left(\frac{2}{ad} + 1\right)\left(e^{-ad} - 1\right) + 2\right] - \\ h_0 d_0\left[\left(\frac{2}{ad_0} + 1\right)\left(e^{-ad_0} - 1\right) + 2\right]e^{-a\delta} \end{array}\right\} \end{aligned} \tag{8}$$

where d is the longitudinal width of the pier section; h is the transverse width of the pier section; d_0 is the longitudinal width of the hollow part of the pier section; h_0 is the transverse length of the hollow section of the pier; and δ is thin-walled hollow high pier wall thickness. The rotation angle for the pier micro-segment dy is given by $\alpha\eta dy$, so pier top displacement ΔS can be expressed as:

$$\Delta S = \int_0^H \alpha\eta y dy \tag{9}$$

Now, it assumes that $A = h \times d$ and $A_0 = h_0 \times d_0$ so that when d is constant, η changes with the temperature difference. Substituting Formula (8) into Formula (9), thus gives:

$$\begin{aligned}\Delta S &= \frac{\alpha\eta H\Delta H}{2} \\ &= \frac{3\alpha H\Delta H A T_0\left[\left(\frac{2}{ad}+1\right)\left(e^{-ad}-1\right)+2\right]}{ahd^3-ah_0d_0^3} \\ &- \frac{3\alpha H\Delta H A_0 T_0 e^{-a\delta}\left[\left(\frac{2}{ad_0}+1\right)\left(e^{-ad_0}-1\right)+2\right]}{ahd^3-ah_0d_0^3}\end{aligned} \tag{10}$$

where H is the vertical length of the pier (m), and ΔH is the distance from pier center to pier top (m). Since $e^{-a\delta}$ is small, on the order of 10^{-3} to 10^{-6}, the second term of the molecule is much smaller than the first term, and it can be ignored here. Thus,

$$\Delta S = \frac{3\alpha H\Delta H dh T_0\left[\left(\frac{2}{ad}+1\right)\left(e^{-ad}-1\right)+2\right]}{ahd^3-ah_0d_0^3} \tag{11}$$

If the sunlight temperature differences are different for n different sections of the pier, the pier needs to be divided formally into these n sections to calculate the displacement of the pier top caused by the sunlight temperature difference in each section ΔS_i [39] as follows:

$$\Delta S_i = \frac{3\alpha H_i\Delta H_i d_i h_i T_i\left[\left(\frac{2}{ad_i}+1\right)\left(e^{-ad_i}-1\right)+2\right]}{ah_i{d_i}^3-ah_{i0}d_{i0}^3} \tag{12}$$

where ΔS_i is the free displacement of section i in the sunlight (m); α is the material linear expansion coefficient (α = 1.0 × 10^{-5}/°C); H_i is the length of segment i (m); ΔH_i is the distance from the center of section i to the pier top (m); d_i is the width of the pier along the bridge in section i (m); h_i is the transverse width of pier i (m); T_i is the temperature difference between the sunny side and the back-sun side along the bridge in section i (°C); d_{i0} is the longitudinal width of the hollow part of the pier body in section i (m); h_{i0} is the transverse width of the hollow part of the pier body in section i (m); and n is the calculated number of segments. Superimposing the pier top displacement caused by each segment, the total pier top displacement is thus:

$$\Delta S = \sum_{i=1}^{n}\Delta S_i = \sum_{i=1}^{n}\frac{3\alpha H_i\Delta H_i d_i h_i T_i\left[\left(\frac{2}{ad_i}+1\right)\left(e^{-ad_i}-1\right)+2\right]}{ah_i{d_i}^3-ah_{i0}d_{i0}^3} \tag{13}$$

The above analytical formula applies to both the calculation of pier top displacement along the bridge and the calculation of pier top displacement across the bridge under the measured sunlight temperature difference. When calculating the transverse displacement of the pier top, it is only necessary to exchange the position of the along-bridge parameter d_i (d_{i0}) and the transverse bridge parameter h_i (h_{i0}) in Formula (13). Furthermore, after calculating the longitudinal and transverse displacements of the pier top, the total displacement S of the pier top can be obtained by coupling these two displacements [40,41].

$$S = \sqrt{\Delta S_{longitudinal}^2 + \Delta S_{transverse}^2} \quad (14)$$

where $\Delta S_{longitudinal}$ and $\Delta S_{transverse}$ are the displacement of the pier top along the bridge direction and transverse bridge directions, respectively.

3. Engineering Profile and Case Study Selection

3.1. Engineering Profile

For the Case study, taking a newly constructed continuous beam bridge on a mountainous expressway of Guizhou Province in China with a span arrangement of (3 × 40 + 3 × 40 + 3 × 40) m as an example, and whose superstructure uses a prestressed concrete (post-tensioned) continuous beam. The substructure abutment is U-shaped, and the No. 4 and No. 6 piers have a rectangular solid pier construction. The No. 5 pier, however, is a rectangular thin-walled hollow pier, and the rest of the piers are double-limbed cylindrical solid piers whose abutments pile foundations.

The No. 5 pier is 75 m high, making it the highest pier of this bridge, and the areas 1 m from the upper and lower ends of the No. 5 pier are a solid section, with the others being thin-walled hollow equal sections. Each section size is 3 m (along the bridge direction) × 6 m (transverse bridge direction), and the wall thickness is 0.55 m, as shown in Figure 2. A diaphragm plate with a thickness of 0.5 m is also set every 18.15 m in the vertical direction of the pier A generalized construction diagram of this pier is shown in Figure 2.

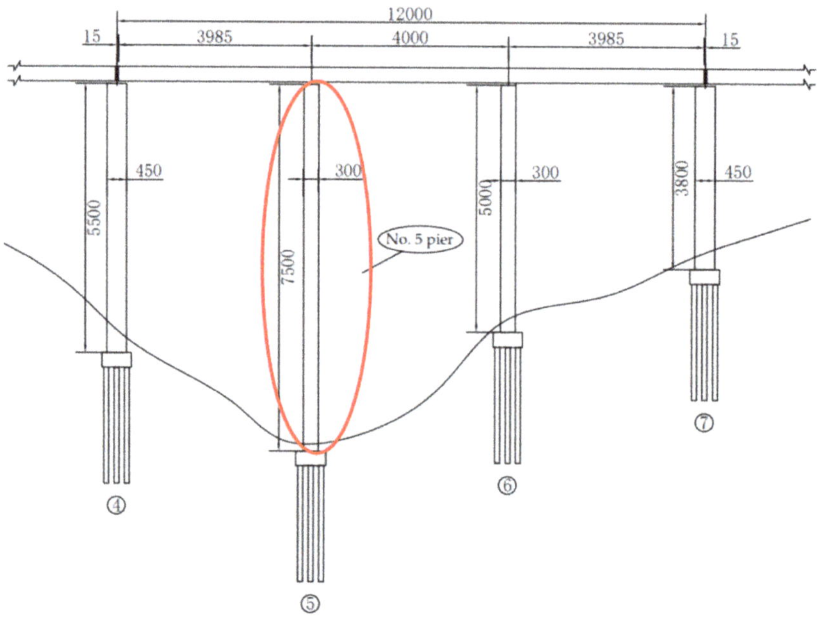

(**a**) Lateral plane (④ ~ ⑦ is the pier number, unit: cm)

Figure 2. *Cont.*

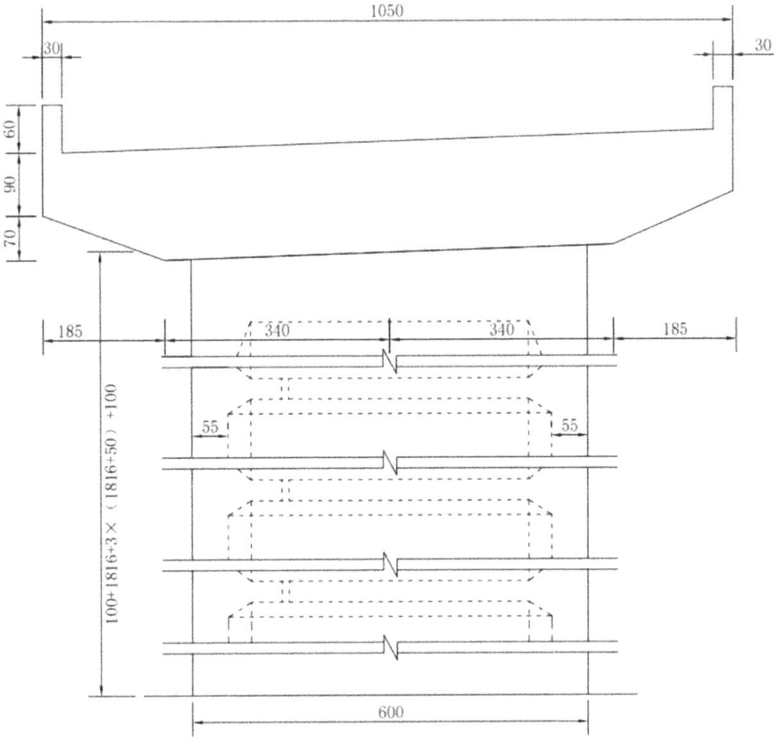

(**b**) Frontal plane (unit: cm)

Figure 2. Pier No. 5 structural design drawing (frontal plane) (unit: cm).

3.2. Temperature Monitoring Scheme

To analyze the sunlight temperature effect on the No. 5 pier, embedded thermometers were placed in its vertical 1/4, 1/2, and 3/4 sections. The layout scheme is shown in Figure 3. When pier construction height is low, the influence of sunlight temperature difference is small, so the thermometer measuring points were arranged starting from the vertical 1/4 section. Externally embedded thermometers were placed close to the outer side, and internally embedded thermometers were placed close to the inner side.

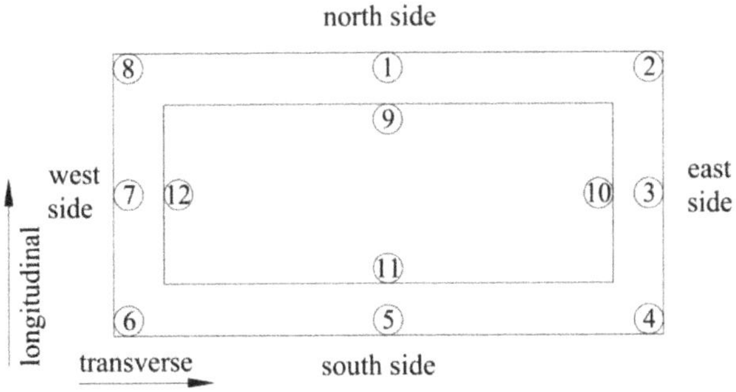

Figure 3. The arrangement of measuring points on the pier No. 5's vertical 1/4 cross-section(①–⑫ is the number of the measuring points).

Figure 3 shows the layout of the measuring points at the vertical 1/4 section of the No. 5 pier, and the vertical 1/2 and 3/4 section measuring points were arranged in the same manner. The middle measuring point on the north side was the starting point, and clockwise from the outside to the inside, these were numbered 13 to 24 and 25 to 36, respectively. When installed, embedded thermometer is close to the inner or outer side of the pier, each thermometer was first fixed on the steel bar of the pier body, and then buried in the concrete after moisture and mechanical damage prevention steps were taken. In addition, a signal cable of embedded thermometer needed to be drawn out of the concrete's surface.

The specific test instrument was an embedded thermometer (model: BGK-3700-0.2) provided by Geokon Instruments (Beijing) Co., Ltd. As shown in Figure 4, the core component of embedded thermometer adopts semiconductor thermistor sensor, and the measurement range was −30 °C ~ +70 °C, with a test accuracy of 0.2 °C. This instrument included a BGK-MICRO-40 automatic data acquisition instrument provided by Geokon Instruments (Beijing) Co., Ltd., and the supporting BGK-Logger software system was based on the WINDOWS working platform, which can read, save, and process the measurement data quickly and conveniently, as shown in Figure 5. The acquisition device also supported wireless transmission. The thermometer's data signal was sent using BGK-187V3 professional cable transmission because this type of cable has superior waterproof performance, good data signal stability, and high precision.

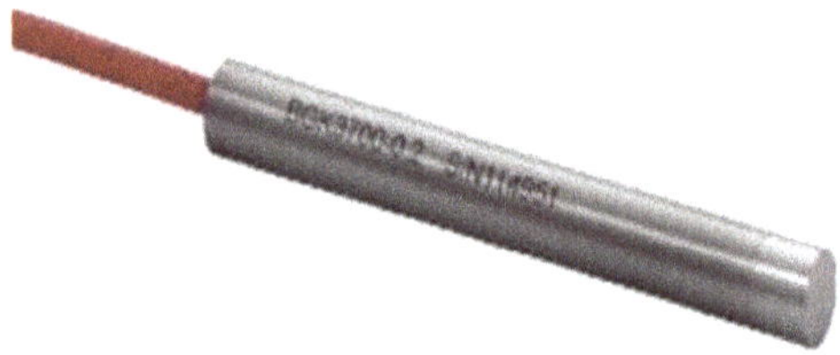

Figure 4. An embedded thermometer.

Figure 5. The BGK-MICRO-40 automatic data acquisition instrument.

3.3. *Temperature Field Analysis of the Pier Section*

Nearly three months of on-site temperature testing from the time the thermometers were installed in May, 2019 until the construction of the piers was completed in July, 2019. During the hot summer months, a large amount of temperature data were collected. To select the appropriate calculation parameters for the sunlight temperature differences, the temperature field changes of the pier section on a sunny day in May, June, and July were selected for analysis: May 16 (when the pier had been constructed to the 1/2 cross-section),

June 16 (when the pier had been constructed to the 3/4 cross-section), and July 15 (when the pier had been completed). The three selected time points all reflect the period immediately after the completion of pier construction and maintenance, and the temperature of these measuring points is influenced almost entirely by sunlight. Temperature changes at each measuring point on the vertical 1/4 section of the pier on May 16 are shown in Figure 6.

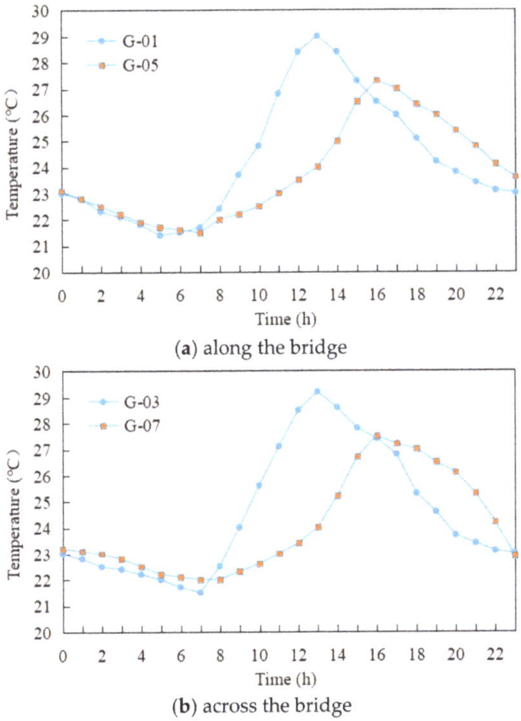

Figure 6. Temperature changes at each measuring point at the 1/4 section of the pier on May 16.

From the above figures, it was considered that the temperature changes at the measuring points at similar positions on the pier in the longitudinal and transverse directions (both positive and negative sides) were the same. From early morning to about 7 a.m., the temperature of each side measuring point cools. In the morning, the east side and the north side were located on the positive side, and the temperature of the measuring points increased rapidly, however. By 13:00, the highest temperatures were 29.2 and 29 °C, respectively, before these sides came under shadow and began to cool. The measuring points on the west side and the south side were located in shadow in the morning, and thus they were subject to a similar (the peak temperatures were both slightly lower than the other two sides) yet time-shifted cycle compared to the east and north sides. At about noon, the temperature difference between the two sides of the pier along the bridge and across the bridge reached their maxima, which were 6 °C (along the bridge) and 5 °C (across the bridge), respectively. On June 16, the pier had been constructed to the vertical 3/4 section. The temperature changes in each measuring point of the vertical 1/4 section and 1/2 section of the pier on June 16 are listed below.

As shown in Figure 7, the temperature changes at the measuring points at the 1/4 cross-section and the 1/2 cross-section of the pier were the same at similar positions both along and across the bridge (both positive and negative sides). However, the maximum temperatures at each site in Figure 7 were higher than that in Figure 6 because the average daily temperature in June was higher than in May. In June, the sunlight time became longer as well. Starting at 6 a.m., the sides of the pier were already in a state of warming. The temperature rose until noon, at which point the highest temperature was reached, and the temperature at the measuring points began to trend downward. At about noon on the same day, the temperature differences between the two sides of the 1/2 section in the longitudinal and transverse directions were the largest, at 7.5 °C and 7 °C, respectively. On July 15, the pier had been completed, the temperature changes in each measuring point of the vertical 1/4 section, 1/2 section, and 3/4 section of the pier on July 15 are listed below. For brevity, only the temperature changes in along the bridge on the vertical 1/4 section and 1/2 section are listed.

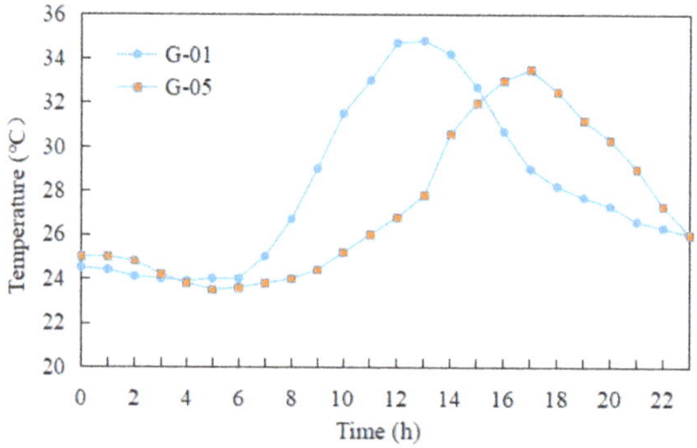

(**a**) along the bridge (1/4 section)

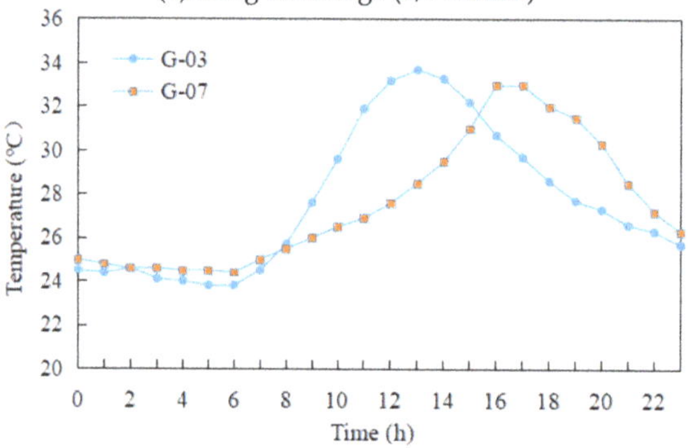

(**b**) across the bridge (1/4 section)

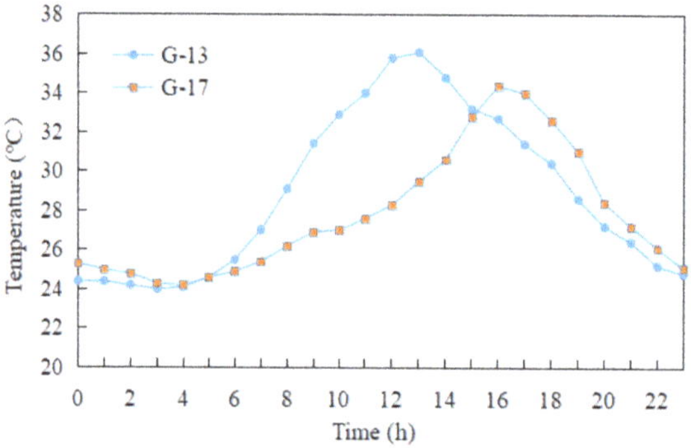

(**c**) along the bridge (1/2 section)

Figure 7. *Cont.*

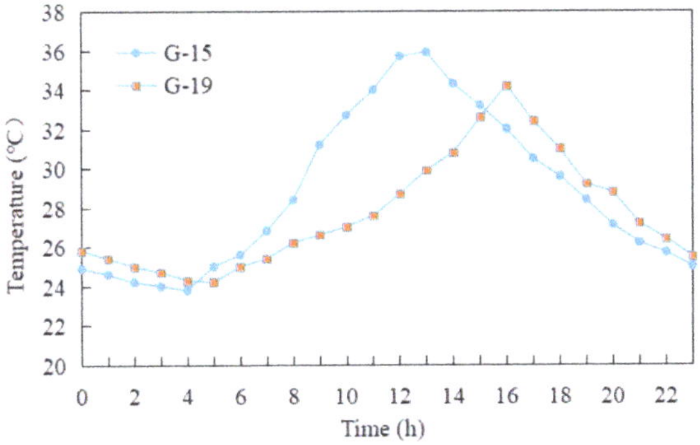

(**d**) across the bridge (1/2 section)

Figure 7. Temperature changes at each measuring point at the 1/4 section and the 1/2 section of the pier on June 16.

Figure 8 shows that the temperature changes at the measuring points at similar positions of each section of the pier along both the bridge direction and the transverse bridge direction (both on the positive and negative side) were roughly the similar. In the morning, the temperature of the measuring point on the positive side of the 3/4 section rose rapidly and reached its highest temperature of 40.5 °C (north side) and 41 °C (east side) at 12:00 and 13:00, respectively. In the afternoon, the two sides came under shade, and the temperature of the measuring point cooled until 0:00 at night. The temperature of the 3/4 section of the pier was higher than the other sections as well, at up to 41 °C, but this was mainly due to the fact that July had much hotter weather than June or May as well as an earlier sunrise and higher sunlight intensity. Moreover, the higher the height of the pier section, the longer the sunlight radiation time. At about 12:00, the temperature difference between the two sides of the pier in both the longitudinal and transverse directions reached their maxima, which were 8.5 °C and 7 °C, respectively.

3.4. Selection of Calculation Conditions for Sunlight Temperature Differences

Since the transverse moment of inertia of the thin-walled hollow pier is larger than that along the bridge, the temperature difference along the bridge is more likely to cause linear deviation of the pier. Thus, the temperature condition of the maximum temperature difference along the bridge is selected for calculation.

Condition 1: On May 16, the sunny, maximum temperature difference between the front and back sides occurred at noon along the vertical 1/4 section of the pier. At this time, the external temperature of the north side along the bridge was 29 °C, the external temperature of the south side was 23 °C, the temperature difference was 6 °C, and the No. 5 pier of the bridge had been constructed to the vertical 1/2 section height. As shown in Table 1.

Condition 2: On June 16, the sunny, maximum temperature difference between the front and back sides again occurred at noon along the vertical 1/2 section of the pier. At this time, the external temperature of the north side along the bridge was 34.5 °C, the external temperature of the south side was 27 °C, the temperature difference was 7.5 °C, and the No. 5 pier of the bridge had been constructed to the vertical 3/4 section height. The maximum temperature difference conditions of 1/4 section and 1/2 section along the bridge are shown in Table 1.

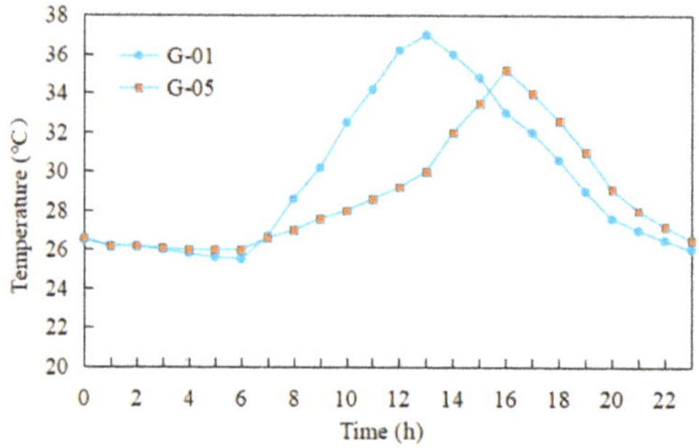

(**a**) along the bridge (1/4 section)

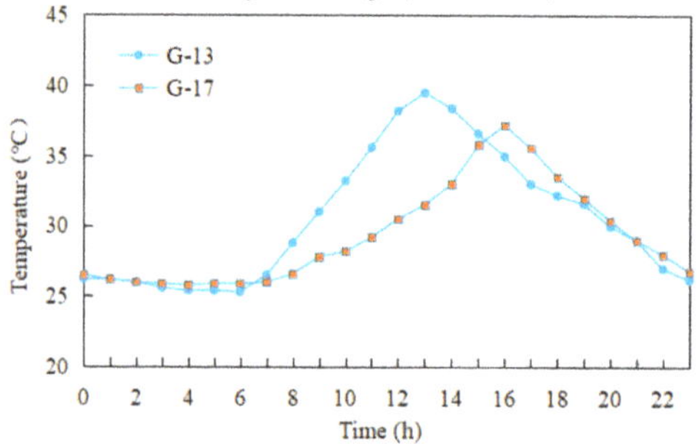

(**b**) along the bridge (1/2 section)

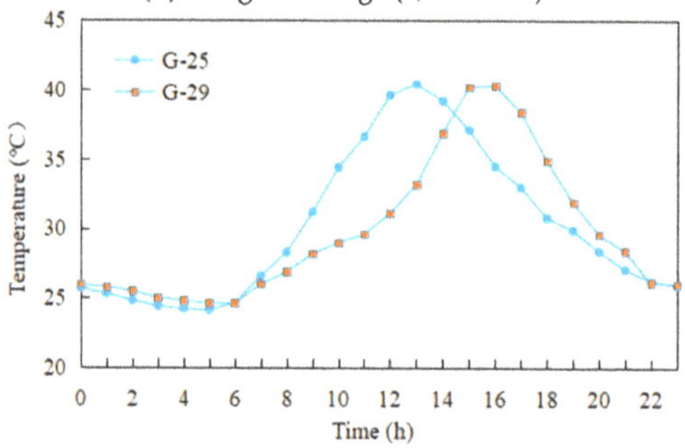

(**c**) along the bridge (3/4 section)

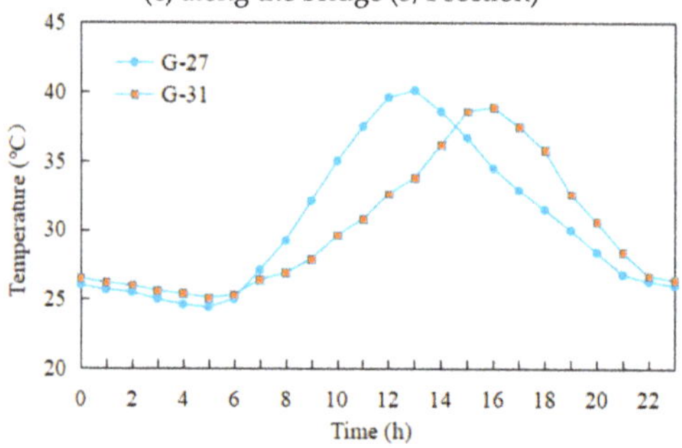

(**d**) across the bridge (3/4 section)

Figure 8. Temperature changes at each measuring point at 1/4 section, 1/2 section, and 3/4 section of the pier on July 15.

Table 1. The measured temperatures of the pier on each side in the longitudinal direction (unit: °C).

Condition	Location	Northside	Southside	Temperature Difference
Condition 1	1/4 section	29	23	6
Condition 2	1/4 section	33	26	7
	1/2 section	34.5	27	7.5
Condition 3	1/4 section	37	30	7
	1/2 section	39.5	31.5	8
	3/4 section	40.5	32	8.5

Condition 3: On July 15, the sunny, maximum temperature difference between the front and back sides also occurred at noon along the vertical 3/4 section of the pier. At this time, the external temperature of the north side along the bridge was 40.5 °C, the external temperature of the south side was 32 °C, the temperature difference was 8.5 °C, and the No. 5 pier of the bridge had been completed. The maximum temperature difference conditions of each section along the bridge are shown in Table 1.

Since the No. 5 pier is located in a valley, the wind speed was small, and the duration of exposure to direct sunlight of each part of the pier is different. Thus, the lower the pier section height, the smaller the temperature difference.

4. Basic Theory and Finite Element Methods in Temperature Effect Research

4.1. Basic Theory of Temperature Effect Research

Beginning with a homogeneous isotropic material, a micro-element is shown in Figure 9.

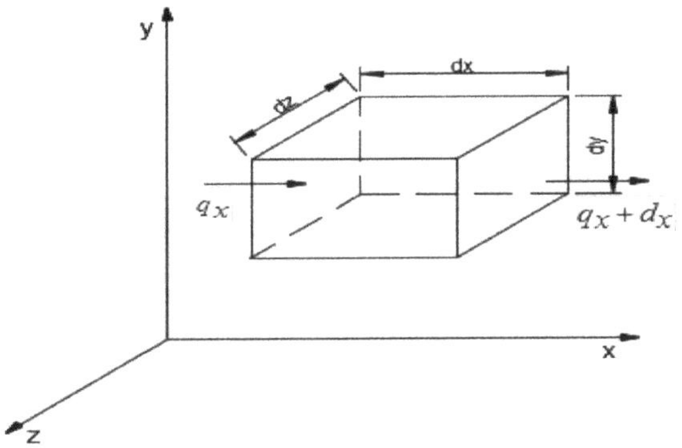

Figure 9. Microunit thermal conductivity schematic.

In the Figure 9, *dx*, *dy*, and *dz* are the length of the three coordinate directions of the micro-element; *q*x is the heat flow into the micro-element, *q*x + *d*x is the heat flux of the derived micro-element.

According to the law of conservation of energy to analyze the thermal energy balance of the micro-element. The difference between heat entering and leaving the micro-element for a change *dt* in time is equal to the heat of the heat source and the thermodynamic energy, as shown in Figure 9. The heat conduction equation can be expressed as [42]:

$$\frac{\partial T}{\partial t} = \lambda\left(\frac{\partial^2 T}{\partial x^2} + \frac{\partial^2 T}{\partial y^2} + \frac{\partial^2 T}{\partial z^2}\right) + \frac{\partial \theta}{\partial t} \quad (15)$$

where λ is the thermal conductivity coefficient (m^2/h); θ is adiabatic heating (°C); T is the temperature (°C); and t is time (h).

If there is no internal heat source, the above formula becomes

$$\frac{\partial T}{\partial t} = \lambda\left(\frac{\partial^2 T}{\partial x^2} + \frac{\partial^2 T}{\partial y^2} + \frac{\partial^2 T}{\partial z^2}\right) \tag{16}$$

When $\frac{\partial T}{\partial t} = 0$, there is a steady temperature field, and the field is unsteady otherwise. In addition, there exists a unique solution to the above equation for every set of initial conditions for the temperature problem.

4.2. Finite Element Methods

In ANSYS thermal analysis software, five kinds of thermal loads (boundary conditions) can be directly applied to solid or element models, namely constant temperature, heat flux, convection, and heat generation rate. There are three main means of heat exchange between bridge piers and the external environment: solar radiation, radiation heat transfer, and convection heat transfer. Compared to solar radiation and convection heat transfer, radiation heat transfer has little effect on the temperature distribution of the structure and can be ignored. Thus, only the temperature effect caused by solar radiation and convective heat transfer are analyzed, so that the final thermal load applied to the boundary nodes of the pier finite element model is the combination of the external comprehensive temperature and the comprehensive heat transfer coefficient [42]. The comprehensive temperature is the sum of air temperature and solar radiation equivalent temperature. it can be calculated as follows:

$$T_{sa} = T_\alpha + a_t I / h \tag{17}$$

where T_{sa} is the combined temperature (°C); T_α is the measured external temperature (°C) of the bridge pier; I is the solar radiation intensity (W/m^2), (refer to the Chinese national standard Specifications for design of heating ventilation and air conditioning for details [43]); h is the comprehensive heat exchange coefficient (W/(m^2·°C)), which is the sum of the convective heat exchange coefficient hc and radiative heat exchange coefficient hr; and a_t is the absorption rate, generally taken to be 0.65 [37]. Since the heat radiation exchange coefficient is negligible compared to the convective heat exchange coefficient, the overall heat exchange coefficient can be taken as the convective heat transfer coefficient hc, which is given by: hc = 5.6 + 4.0 v, where v is wind speed (*m/s*).

4.3. A finite Element Model of the No. 5 Pier

Both the three-dimensional solid finite element method and the space bar finite element method can be used in the finite element analysis of the displacement of the pier top under the action of sunlight temperature difference of the No. 5 pier. The space bar finite element method is also called the space truss displacement method and is suitable for the calculation of grid structures. Although it is simple to model and fast to calculate, the error is slightly larger than that of the former because the model is not as fine as the three-dimensional solid finite element model [44]. For this reason, the more accurate solid finite element models are elected to study the pier according to theoretical construct.

First, the finite element analysis models of bridge piers under three working conditions were established by ANSYS finite element analysis software: finite element models of vertical 1/2 section height, vertical 3/4 section height, and bridge pier construction completion. The temperature field unit of the ANSYS finite element model was the SOLID87 three-dimensional solid thermal analysis unit, and the structural field unit was the SOLID187 three-dimensional solid structural analysis unit. When the thermal coupling analysis of the temperature field was carried out, the temperature load could only be applied to the thermal analysis unit SOLID87. After thermal analysis, conversion between the thermal unit SOLID87 and the structural unit SOLID187 was achieved through ANSYS interface operations and command streams in order to apply the structural stress load generated by thermal analysis and then complete the thermal coupling analysis of the structural temperature field. The three ANSYS finite element models of the No. 5 pier are shown in Figure 10.

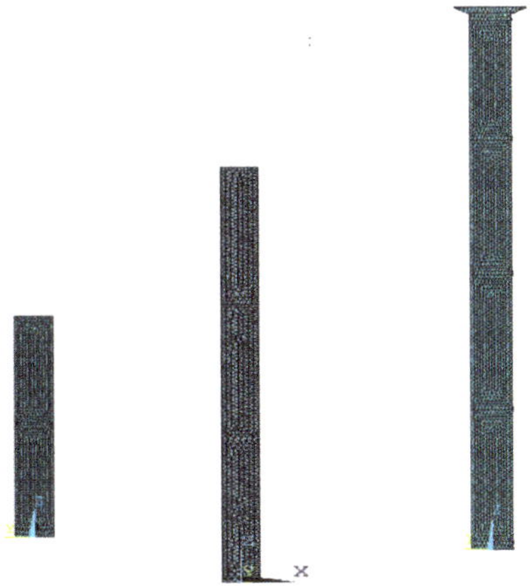

Figure 10. The finite element models of the No. 5 pier.

5. Comparative Analysis of Algorithm Results

According to the measured meteorological data during the construction of the No. 5 pier, the average wind speed of the bridge site was 2.0 *m/s*. Measurement of the average wind speed is needed to calculate the comprehensive heat transfer coefficient. the reference temperature was set to 0 °C according to the relevant literature [45] in order to facilitate the application of temperature load in the ANSYS calculations. The deviation of the pier along the bridge caused by the sunlight temperature difference was more obvious than in any other direction, so the deviation of the pier top along the bridge in this direction is selected for calculation and analysis.

5.1. Finite Element Calculated Values

The displacement calculation under the action of a temperature field was carried out by ANSYS, and the displacement of the pier top of the No. 5 pier three sunlight temperature difference conditions was 2.157 mm (Condition 1), 5.801 mm (Condition 2), and 10.935 mm (Condition 3). As shown in Figures 11–13.

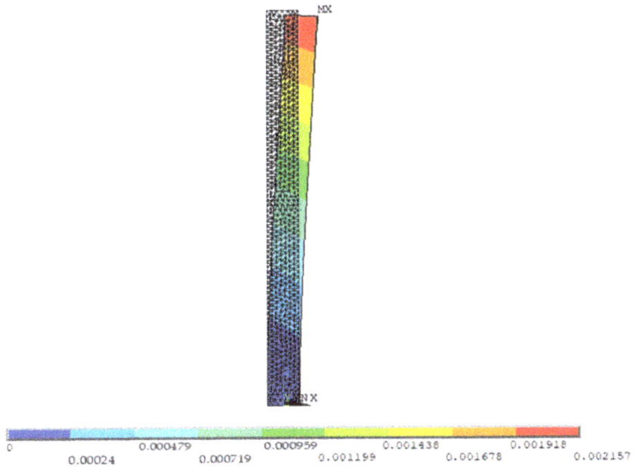

Figure 11. The calculated longitudinal displacement result for the top of the No. 5 pier under Condition 1 (unit: m).

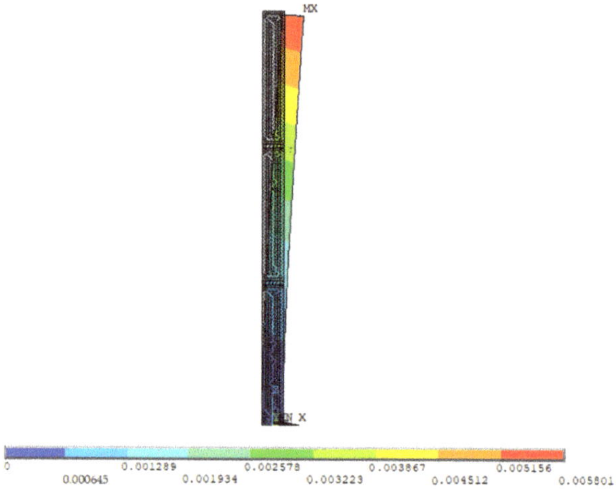

Figure 12. The calculated longitudinal displacement result for the top of the No. 5 pier under Condition 2 (unit: m).

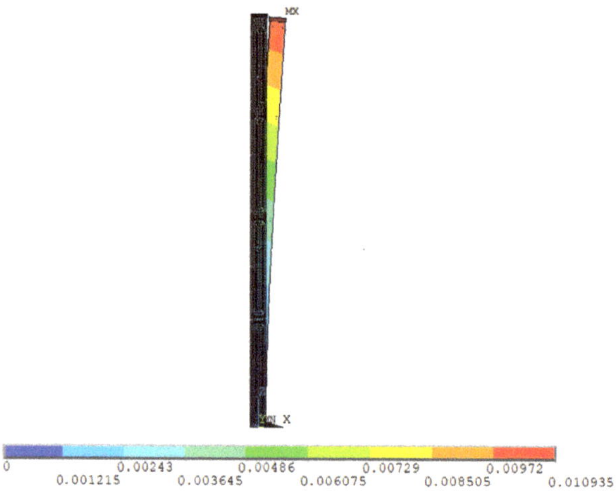

Figure 13. The calculated longitudinal displacement result for the top of the No. 5 pier under Condition 3 (unit: m).

5.2. Analytic Algorithm Results

The data from Section 3.4 were used to inform analytical calculations of the pier top displacement under a sunlight temperature difference effect along the bridge direction. The results are given in Table 2.

Table 2. The pier top calculated displacement results from the analytical algorithm (unit: mm).

Condition	Location	Longitudinal Direction
Condition 1	total	2.293
Condition 2	0 ~ 1/2 segment	5.351
	1/2 ~ 3/4 segment	0.717
	total	6.068
Condition 3	0 ~ 1/2 segment	8.027
	1/2 ~ 3/4 segment	2.293
	3/4 ~ top segment	0.807
	total	11.127

5.3. Specification Calculation Results

Referring to the Chinese national standard TB10052-1997 Technical specifications for flexible pier railway bridges, Article 2.3.6 [15,36], the formula for pier top displacement caused by sunlight temperature differences along the bridge is:

$$\Delta_S = \sum_{i=1}^{n} \frac{6\alpha T_i y_i \Delta y_i (ad_i - 2)}{a^2 d_i^3} \tag{18}$$

where, Δ_S is the pier top displacement caused by sunlight (m); α is the linear expansion coefficient of concrete, taken to be 10^{-5} (1/°C); T_i is the temperature difference between the sunny side and the back of the bridge pier (°C); a is the coefficient, taken to be 7 (1/m); n is the number of segments; Δy_i is the length of each calculated segment (m); y_i is the distance from the center of segment i to the top of the pier (m); d_i is the width of section i of the pier along the bridge (m). The results for the calculated pier top displacement along the bridge under three sunlight temperature difference conditions are given in Table 3.

Table 3. The pier top calculated displacements from the Chinese reference formula (unit: mm).

Condition	Location	Longitudinal Direction
Condition 1	total	3.635
Condition 2	0 ~ 1/2 segment	6.946
	1/2 ~ 3/4 segment	1.136
	total	8.082
Condition 3	0 ~ 1/2 segment	9.756
	1/2 ~ 3/4 segment	3.18
	1/2 ~ top segment	1.281
	total	14.217

5.4. Comparative Analysis

The analytic algorithm calculation values, standard calculation values, finite element calculation values, and field-measured pier top displacement values for the No. 5 pier under all three sunlight temperature difference conditions are now compared. To measure the actual displacement in the field, the TC1800 high-precision total station from Leica, Germany was used, which has a minimum angle measurement accuracy of 1″ and a minimum reading of 0.01 mm. The comparison results are shown in Table 4.

Table 4. The results comparison of the pier top displacement calculations in the longitudinal direction (unit: mm).

Calculated Values	Condition 1	Condition 2	Condition 3
analytical algorithm	2.293	6.068	11.127
railway code	3.635	8.082	14.217
finite element	2.157	5.801	10.935
measured value	2.8	6.48	11.36
analysis-measured	0.507	0.412	0.233
code-measured	0.835	1.602	2.857
digital-measured	0.643	0.679	0.425

It can be seen from the table that the four groups of results were generally close, but the calculation results of the analytical algorithm were closer to the measured values of the pier than of the railway specification, and the maximum difference was only 0.507 mm. This shows that the calculation results of the analytical algorithm are reliable and have considerable accuracy.

The measured value of the pier top displacement was larger than the calculated value from the analytical algorithm, however, because the measured pier top offset was not only affected by the sunlight temperature difference but also by wind load and other factors. Under each of the three kinds of sunlight temperature difference conditions, the finite element calculation value was also consistent with the measured value of the pier top, and the maximum difference was 0.679 mm. The finite element calculated values were slightly smaller than the measured values because the ANSYS finite element method only considered the sunlight temperature load.

Among the three calculation methods, the difference between the calculated value from the Chinese reference and the measured value of the pier top was the largest, up to 2.857 mm (Condition 3). This is mainly because the reference is somewhat outdated, and it is mainly aimed at the calculation of the displacement of pier tops of flexible railway piers under sunlight temperature differences. its applicability is not directly transferrable to a thin-walled hollow pier.

5.5. Parametric Analysis

The No.5 pier of the Bridge is located in a V-shaped valley area, and due to the influence of this topography, the sunlight temperature difference was not very large. However, this is not the case in many different regions across China, and the linear changes in piers in these regions can also be quite different from those of the case study. Figure 14 shows the displacement results of the pier top calculated by analytical algorithm under different sunlight temperature difference conditions for the No.5 pier. It is assumed that the pier had been constructed to a height of 75 m at this time and that the pier consisted of only one segment. The remaining parameters were unchanged.

As Figure 14 shows, with an increase in sunlight temperature difference, the displacement of the pier top along and across the bridge also gradually increases. Specifically, when the sunlight temperature difference is 10 °C, the longitudinal bridge is 15.289 mm and the transverse bridge is 4.868 mm. Furthermore, the greater the temperature difference, the greater the displacement of the pier top, and the greater the displacement along the bridge compared to across the bridge. This is because the thin-walled hollow pier section transverse moment of inertia is larger than that along the bridge.

Figure 15 presents a graph of the changes in displacement for different pier heights. Because this graph is only for the height of the pier, it is assumed that the temperature difference between the longitudinal and transverse directions of the bridge was 5 °C and that the pier only consisted of one section. The other parameters remained unchanged. From the figure we see that the greater the height of the thin-walled hollow pier, the greater the offset of the pier top for a given sunlight temperature difference. Furthermore, the greater the sunlight temperature difference and pier height, the more significant the influence the sunlight temperature difference on pier alignment.

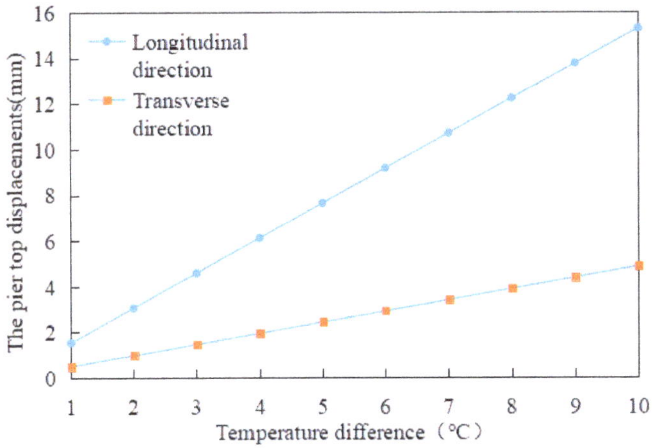

Figure 14. The pier top displacements caused by different temperature differences.

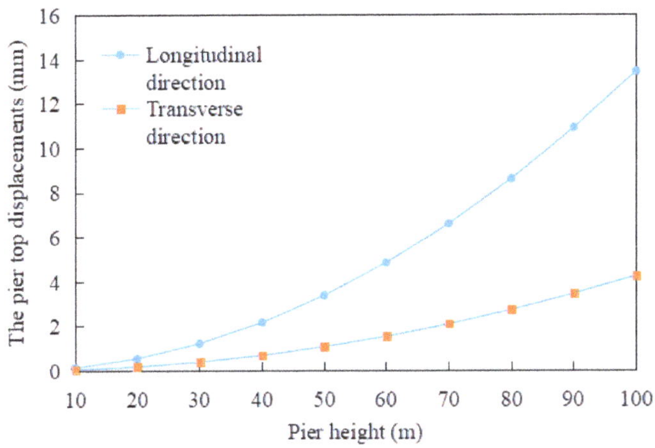

Figure 15. The pier top displacement caused by different pier heights.

Figure 16 shows the displacements of the pier top when the wall thickness of the pier changes. If the wall thickness is too small, it affects structural safety, so the minimum wall thickness was set to be 0.5 m. Assuming the temperature difference of the pier to be 5 °C and that the pier only consisted of one segment, as above, other parameters remain unchanged. It can be seen from the Figure 16 that the offset of the pier top decreases with an increase in pier wall thickness, but when the wall thickness reaches 0.75 m, the offset of the pier top does not change much with further increases. This shows that for thin-walled hollow piers, increasing wall thickness has little significance after a point in limiting the displacement of the pier top.

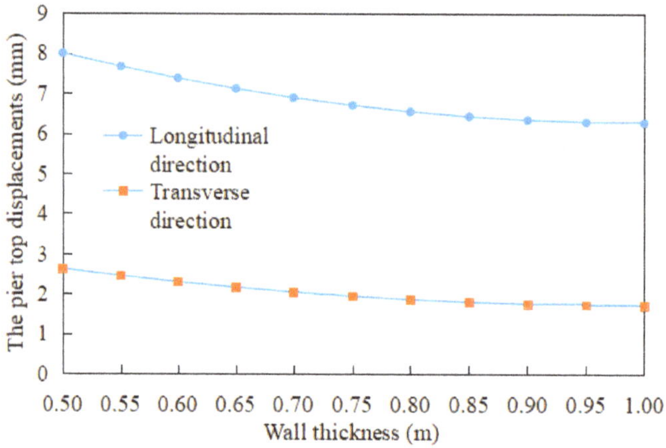

Figure 16. The pier top displacements caused by different pier wall thicknesses.

6. Conclusions

Based on the structural characteristics of thin-walled hollow concrete piers, an analytical formula for the displacement of pier top caused by sunshine temperature difference was derived. The calculated values from this analytical formula are compared with the reference calculation values, finite element calculation values, and field measured values and analyzed the influence of sunlight temperature difference, pier height, and wall thickness on pier alignment. The conclusions are enumerated below.

(1) The analytical calculation formula derived in this paper has quite a high accuracy and thus can provide a new means to calculate and control the alignment of thin-walled hollow piers.
(2) The analytical formula is more suitable for the linear calculation of thin-walled hollow piers under the influence of sunlight temperature differences than the formula in the current Chinese railway code. The formula has more specificity to its application, and was also closer to the measured values than the reference formula, with a maximum difference of only 0.507 mm.
(3) The temperature difference due to sunlight has a great influence on the alignment of thin-walled hollow piers, and the greater the temperature difference and height, the more significant the influence on the alignment. When the temperature difference was 10 °C, the displacement of the pier top (75 m high) along the bridge was 1.5 cm. However, increasing pier wall thickness did not significantly limit the pier top displacement.

Author Contributions: Writing—original draft preparation, Investigation, L.A. and P.C.; Formal analysis, L.A. and P.Y.; Supervision, Project administration, D.L. and P.Y.; Writing, Review & editing, L.A. and P.Y.; Conceptualization, Funding acquisition, D.L. All authors have read and agreed to the published version of the manuscript.

Funding: This project received funding from the National Natural Science Foundation of China (51378504) and the Joint Fund for Basic Research of High Speed Railways of China (U1334203).

Data Availability Statement: Not applicable.

Acknowledgments: The authors thank AiMi Academic Services (www.aimieditor.com) for English language editing and review services, accessed on 24 March 2023.

Conflicts of Interest: The authors declare no conflict of interest.

References

1. Hu, A.-Q.; Ren, D.-H. Effect of concrete hollow pier wall thickness on the temperature difference distribution of sunlight. *J. Shijiazhuang Tiedao Univ.* **2021**, *34*, 21–27.
2. Yuan, P.; Cai, C.S.; Li, D.; Xu, G.; Li, C. A novel method for determining the spatial responses of a cable-stayed bridge with four cable-planes. *Eng. Struct.* **2019**, *180*, 223–233. [CrossRef]
3. Bi, Z.-G.; Yan, X.-J.; Yu, Y.-J.; Liang, B. Analysis of temperature field and temperature effect of the thin-walled hollow high pier. *Henan Sci.* **2022**, *40*, 46–53.
4. Zhou, L.-Y.; Liu, Z. Investigation of the bucking behavior of thin-walled hollow concrete piers. *Struct. Concr.* **2020**, *22*, 1363–1377.
5. Tayeh, B.; Marijana, H.-N.; Magdy, Y.-R.-R.; Radwa, D.-A.-H. Behavior of Ultra-High-Performance concrete with Hybrid synthetic fiber waste exposed to elevated temperatures. *Buildings* **2023**, *13*, 129. [CrossRef]
6. *JTG F80/1-2017*; Inspection and Evaluation Quality Standards for Highway Engineering-Section 1 Civil Engineering. China communications press Co., Ltd.: Beijing, China, 2017.
7. Zheng, G.; Gu, X.-Y.; Zhang, T.-Q.; Sun, J.-B.; Zheng, W.-H. Random forest method-based prediction and control of bridge pier displacements during construction of two overlapped EPBM tunnels. *Eur. J. Environ. Civ. Eng.* **2022**, *26*, 2273–2293. [CrossRef]
8. Suman, R.; Ikwulono, D.-U.; Andrew, D.-S. Reliability assessment and sensitivity analysis of vehicle impacted reinforced concrete circular bridge piers. *Structures* **2022**, *37*, 600–612.
9. Shim, C.-S.; Dang, S.-N.; Park, S. Three-Dimensional information delivery for design and construction of prefabricated bridge piers. *Struct. Eng. Int.* **2018**, *28*, 6–12. [CrossRef]
10. Editorial Department of China Journal of Highway and Transport. Review on China's Bridge Engineering Research: 2021. *China J. Highw. Transp.* **2021**, *34*, 1–97.
11. Liu, Y.-J.; Liu, J.; Zhang, N. Review on solar thermal actions of bridge structures. *China Civ. Eng. J.* **2019**, *52*, 59–78.
12. Sun, Z.-G.; Wang, D.-S.; Wang, T.; Wu, S.-W.; Guo, X. Investigation on seismic behavior of bridge piers with thin-walled rectangular hollow section using quasi-static cyclic tests. *Eng. Struct.* **2019**, *200*, 109708. [CrossRef]
13. Chen, H.-H.; Xu, B.; Liu, Q.; Gu, J.-F. Study on failure performance of the thin-walled steel-reinforced concrete pier under low cyclic loading. *Buildings* **2022**, *12*, 1412. [CrossRef]
14. Choi, S.; Won, M.-C. Thermal strain and drying shrinkage of concrete structures in the field. *ACI Mater. J.* **2010**, *107*, 498–507.
15. Rodriguez, L.-E.; Barr, P.-J.; Halling, M.-W. Temperature effects on a box-girder integral-abutment bridge. *J. Perform. Constr. Facil.* **2014**, *28*, 583–591. [CrossRef]
16. Li, Y.; He, S.-H.; Liu, P. Effect of solar temperature field on a sea-crossing cable-stayed bridge tower. *Adv. Struct. Eng.* **2019**, *22*, 1867–1877. [CrossRef]
17. Suzuki, J.; Ohba, Y.; Uchikawa, Y.; Hoshikawa, K.; Kimura, K. Monitoring temperatures on a real box-girder bridge and energy budget analysis for basic information on bridge cooling and surface freezing. *J. Bridge Eng.* **2014**, *12*, 45–52. [CrossRef]
18. Zhao, J.-T.; Cui, C.-Y.; Zhang, P.; Wang, K.-P.; Zhao, M. Parameter sensitivity analysis of the seismic response of a piled wharf structure. *Buildings* **2023**, *13*, 349. [CrossRef]
19. Tang, Z.-Z.; Xue, H.-Y.; Liu, H.; Zhang, W. Prediction of ultralow cycle fatigue damage of thin-walled steel bridge piers. *Adv. Steel Constr.* **2021**, *17*, 403–411.
20. Jelena, P.; Nina, S.; Radenko, P. Damage assessment of road bridge caused by extreme streamflow in Montenegro: Reconstruction and structural upgrading. *Buildings* **2022**, *12*, 810.
21. Hagedorn, R.; Marti-Vargas, J.R.; Dang, C.N.; Hale, W.M.; Floyd, R.W. Temperature gradients in bridge concrete I-girders under a heat wave. *J. Bridge Eng.* **2019**, *24*, 04019077. [CrossRef]
22. Lawson, L.; Ryan, K.L.; Buckle, I.G. Bridge temperature profiles revisited: Thermal analyses based on recent meteorological data from Nevada. *J. Bridge Eng.* **2020**, *25*, 11–24. [CrossRef]
23. Abid, S.R.; Taysi, N.; Ozakça, M. Experimental analysis of temperature gradients in the concrete box–girders. *Constr. Build. Mater.* **2016**, *106*, 523–532. [CrossRef]
24. Abid, S.R.; Taysi, N.; Ozakça, M.; Xue, J.; Briseghella, B. Finite element thermomechanical analysis concrete box-girders. *Structures* **2021**, *33*, 2424–2444. [CrossRef]
25. Deng, Y.; Li, A.; Liu, Y.; Chen, S. Investigation of temperature actions on flat steel box girders of long-span bridges with temperature monitoring data. *Adv. Struct. Eng.* **2018**, *21*, 2099–2113. [CrossRef]
26. Tao, T.; Wang, H.; Zhu, Q.; Zou, Z.; Li, J.; Wang, L. Long-term temperature field of the steel-box girder of a long-span bridge: Measurement and simulation. *Eng. Struct.* **2021**, *236*, 111924. [CrossRef]
27. Sheng, X.-W.; Zheng, Y.-H.; Zheng, W.-Q.; Zhu, Z.-H. Vertical temperature gradient model of concrete box girders based on real-time shadow technology. *J. South. China Univ. Technol.* **2020**, *48*, 40–47.
28. Wang, Z.-W.; Zhang, W.-M.; Zhang, Y.-F.; Liu, Z. Temperature prediction of flat steel box girders of Long-Span bridges utilizing in situ environmental parameters and machine learning. *J. Bridge Eng.* **2022**, *27*, 04022004. [CrossRef]
29. Phuong, H.-H.; Hoang, N.-P. Kriging Metamodel-based seismic fragility analysis of single-bent reinforced concrete highway bridge. *Buildings* **2021**, *11*, 238.
30. Zhang, Y.-B.; Cai, T.-T.; Liu, Y. Effects analysis of the cross-sectional shape on the double-limb thin-walled piers temperature field. *J. Railw. Eng. Soc.* **2015**, *198*, 41–45.

31. Dai, G.-L.; Tang, Y.; Liang, J.-B. Study of temperature load extreme value of a bridge pier base on generalized pares to distribution. *Bridge Constr.* **2017**, *47*, 48–53.
32. Liu, C.-C.; Huang, Y.-J.; Zhang, G.-S.; Liu, W.-H.; Liu, C.; Zhang, Y.-S. Sunshine temperature effect of the thin-wall hollow high pier in high altitude area. *J. China Foreign Highw.* **2021**, *41*, 76–79.
33. Lin, G.-T.; Su, B. Numerical analysis of sunshine temperature effect on a super high pier of mountain bridge. *J. Civ. Eng. Manag.* **2018**, *35*, 112–116.
34. Andrei, K.; Ivar, T. Computationally efficient method for steel colum bucking in fire. *Buidings* **2023**, *13*, 407.
35. Faham, T.; Zhang, L.-D.; Park, S.-W.; Sepasgozar, S. Numerically evaluation of FRP-Strengthened members under dynamic impact loading. *Buildings* **2021**, *11*, 14.
36. *TB10052-97*; Technical Specification for Flexible Pier Railway Bridge. China railway publishing house Co., Ltd.: Beijing, China, 1997.
37. Tang, Y. *Active Control of Verticality of Thin-Wall High Pier Based on the Measured Temperature Field*; Harbin Institute of Technology: Harbin, China, 2017.
38. Liu, X.F. *Analysis of Thermal Stress Force Concrete Structure*; China Communications Press: Beijing, China, 1991.
39. Tang, F.; Li, D.-J.; An, L.-P. Research of solar radiation influence on pier's line shape and alignment monitoring of high-pier bridge. *J. Railw. Sci. Eng.* **2016**, *13*, 1970–1976.
40. Yuan, P.; Dong, Y. High-efficient decoupling method for coupling systems with multiple subdomains and time steps. *Mech. Syst. Signal. Process.* **2022**, *163*, 15. [CrossRef]
41. Yuan, P.; Li, D.-J.; Cai, C.-S.; Xu, G.-J. A Novel Decoupling Dynamic Method with Third-order Accuracy and Controllable Dissipation. *Comput. Struct.* **2021**, *249*, 44–56. [CrossRef]
42. Wang, R.F.; Chen, G.R. *Temperature Field and Thermal Stress*; Science Press: Beijing, China, 2005.
43. *GB50019 2003*; Specifications for design of heating ventilation and air conditioning. China nonferrous metal industry's foreign engineering and construction Co., Ltd.: Beijing, China, 2003.
44. Zhang, Z.-M.; Sun, C.; Vahid, J. Structural damage identification of off shore wind turbines:A two-step strategy via FE model updating. *Struct. Control. Health Monit.* **2021**, *29*, 1–21.
45. Song, M.-M.; Silas, C.; Babak, M.; Eric, M.-H. Joint parameter-input estimation for virtual sensing on an offshore platform using output-only measurements. *Renew. Energy* **2023**, *202*, 1032–1045. [CrossRef]

buildings

MDPI

Article

Long Short-Term Memory Network for Predicting Wind-Induced Vibration Response of Lightning Rod Structures

Guifeng Zhao [1], Kaifeng Xing [1], Yang Wang [2], Hui Qian [1] and Meng Zhang [1,*]

[1] School of Water Resources and Civil Engineering, Zhengzhou University, Zhengzhou 450001, China; gfzhao@zzu.edu.cn (G.Z.)

[2] Institute of Civil Engineering, Huzhou Vocational & Technical College, Huzhou 313000, China

* Correspondence: zhangmeng@zzu.edu.cn

Abstract: Lightning rod structures are susceptible to wind loads due to their high slenderness ratio, high flexibility, and light weight. The wind-induced dynamic response of a lightning rod is critical for structural safety and reliability. The traditional methods for this response, including observation and simulation, focus on structural health monitoring (SHM), wind tunnel tests (WTTs), or fluid–structure interaction (FSI) simulations. However, all these approaches require considerable financial or computational investment. Additionally, problems such as data loss or data anomalies in the sensor monitoring process often occur during SHM or WTTs. This paper proposes an algorithm based on a long short-term memory (LSTM) network to predict the wind-induced dynamic response and to solve the problem of data link fracture caused by abnormal sensor data transmission or wind-induced damage to lightning rod structures under different wind speeds. The effectiveness and applicability of the proposed framework are demonstrated using actual monitoring data. Root-mean-squared error (RMSE), determination of coefficient (R^2), variance accounted for (VAF), and the refined Willmott index (RWI) are employed as performance assessment indices for the proposed network model. At the same time, the random forest algorithm is adopted to analyze the correlation between the data of the different measurement points on the lightning rod structure. The results show that the LSTM method proposed in this paper has a high accuracy for the prediction of "missing" strain data during lightning rod strain monitoring under wind speeds of 15.81~31.62 m/s. Even under the extreme wind speed of 31.62 m/s, the values of RMSE, MAE, R^2, RWI and VAF are 0.24053, 0.18213, 0.94539, 0.88172 and 0.94444, respectively, which are within the acceptable range. Using the data feature importance analysis function, it is found that the predicted strain data of the measurement point on the top part of the lightning rod structure are closely related to the test strain data of the two adjacent sections of the structure, and the effect of the test strain data of the measurement points that are far from the predicted measurement point can be ignored.

Keywords: lightning rod structure; long short-term memory (LSTM) network; structural health monitoring (SHM); wind-induced response prediction; data augmentation

1. Introduction

Substations are key places in the power grid system to receive, transform and distribute electric energy, and it is very important to ensure their safe operation. During the service period of a substation, lightning strikes, as a common natural disaster, seriously threaten the safety of electrical equipment and transmission lines in the substation. To prevent lightning damage, lightning rods are usually installed on the substation frame to form a frame–lightning rod structure [1], as shown in Figure 1. Lightning rods are usually thin and long, with a height of 10 m~30 m. They are typically towering structures and are very sensitive to wind loads. In the past ten years, lightning rod destruction accidents in substations caused by wind-induced vibration have occurred from time to time. For example, in December 2014, the lightning rod of the 220 kV outlet side of a 500 kV substation was broken [2]. In March 2015, the lightning rod of the 330 kV incoming line frame of a 750 kV substation was fractured [3]. In September 2015, the lightning rod of the frame on the incoming side of the main transformer of a 750 kV substation collapsed [4] and caused regional power outages, resulting in the interruption of people's normal production and living order and great economic losses. Therefore, it is of great practical significance to carry out fine analysis of the wind-induced vibration response of a frame lightning rod structure using finite element numerical (FEM) simulations, wind tunnel tests (WTTs), or structural health monitoring (SHM) techniques [5–7]. This helps to accurately grasp the bearing performance of the structure and then take reasonable maintenance measures to ensure its work safety and reliability. However, FEM simulations, WTTs and SHM require high analysis costs and support for a large amount of data. In recent years, with the development of smart sensor technology, wireless sensor networks have been gradually used in laboratory experiments and online monitoring of large-scale building structures and infrastructures [8–10]. However, in the process of testing or monitoring, transmission interruption caused by the unstable state of individual sensors and the destruction of environmental factors is unavoidable, resulting in frequent data loss and drift, which has a great adverse effect [11] on the accuracy and reliability of testing and monitoring results. Therefore, it is very important to address these abnormal data chains [12].

(a) (b)

Figure 1. Substation structure and lightning rod structure. (**a**) A substation. (**b**) Frame–lightning rod structure.

With the advent of the fourth scientific and technological revolution, the growing digital resources of computer technology and deep learning have opened up many new possibilities for the refinement experiments of engineering structures and the processing of sensor data in the field of SHM [13]. Yuen and Kuok [14] conducted a monitoring study on a 22-story building structure, determined the correlation relationship between environmental conditions and structural modal frequencies, and found that normalizing the sensor system data can reduce the influence of environmental noise and sensor failures on the results. Cross et al. [15] extracted damage-sensitive features using principal component analysis (PCA) techniques to remove operational and environmental effects. Sarmadi and Karamodin [16] proposed a method based on the Mahalanobis squared distance that combines a class of k-nearest neighbor (kNN) rules and an adaptive distance metric, which can eliminate the influence of different environmental conditions on the anomaly detection process. Padil et al. [17] proposed a structural damage identification method combining the RFR function and principal component analysis, which can reduce the amount of input data, minimize the influence of uncertain and abnormal data on model disturbance, and thus reduce the model error. Bao et al. [18] proposed a method for detecting abnormal data using a deep neural network (DNN), which can detect abnormal data by converting data signals into image signals for computer visualization. Tang et al. [19] developed a structural damage identification system that can convert real-world anomaly detection acceleration data into dual-channel time–frequency images with an overall average correctly identified accuracy of 93.5%. Avci et al. [20] proposed a decentralized 1D-CNN system for structural damage identification. In this system, each sensor uses a 1D-CNN for local damage identification. Through the classification network for each sensor, damage identification and localization are realized, which effectively reduces the need for data transfer and aggregation.

However, the above research mainly focuses on the health monitoring of traditional civil engineering structures such as bridges, and there is still a lack of necessary research on wind tunnel tests or online monitoring data processing of the wind-induced vibration response of power system structures such as frame–lightning rod structures. Therefore, in this paper, a method based on a long short-term memory (LSTM) network is proposed to predict the "missing" data of wind-induced dynamic response in the process of testing or monitoring lightning rod structures under different wind speeds. Using the existing test data from some measurement points, the predicted value of another part of the measurement points can be obtained and compared with the measured data. On this basis, the random forest algorithm analysis function is used to determine the parameter correlation proportion between the different sections of the lightning rod. It is expected to provide necessary data support for the analysis of the cause of frame–lightning rod structure fracture accidents and to provide a basis for the wind resistance design and daily monitoring and maintenance of similar lightning rod structures.

2. Long Short-Term Memory Network

The long short-term memory (LSTM) [21] network originally evolved from the recurrent neural network (RNN) model. By introducing "gate" units such as forget gates, input gates and output gates (as shown in Figure 2), the LSTM model can effectively predict long sequence problems with long-term dependencies, which cannot be reasonably solved by the traditional RNN model because of problems such as gradient disappearance and explosion [22–24].

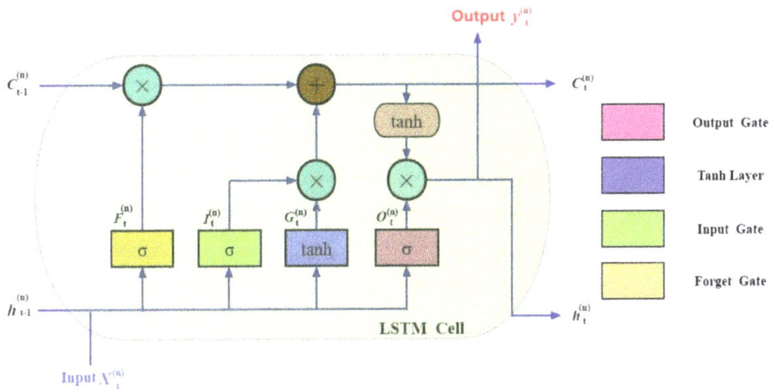

Figure 2. Structure diagram of the nth LSTM cell hidden layer at time t.

The LSTM method uses the same activation function σ for the three "gate" structures and introduces the unit memory cell C_t through them. Overall, the hidden state h_t and cell state C_t simultaneously flow over time. This well-designed "gate" structure enables the LSTM method to have the function of data memory and forgetting [25,26].

2.1. Forget Gate

The forget gate can selectively forget the last few sets of states and correct the parameters [27], which determines what information the LSTM unit needs to forget from the cell state C_t and what information it needs to retain. The forget gate checks the output vector C_{t-1} from the previous LSTM unit, combines the parameter h_{t-1} passed from the previous time step with the input value X_t of the current time step, and outputs the number from 0 to 1 [28] via the activation function σ (i.e., sigmoid function), where 0 means forget it completely, and 1 means keep it completely. Then, by combining the product with C_{t-1}, information screening is achieved [29,30]. The calculation formula of the activation vector F_t is as follows:

$$F_t = \sigma(W_{xf}X_t + h_{t-1}W_{hf} + b_f) \quad (1)$$

In Formula (1), X_t is the input data of the current time step, h_{t-1} is the hidden state of the previous time step, W_{xf} and W_{hf} are the weight parameters, and b_f are the bias vectors, which are learned from the training samples during the training process. The activation function σ is used to convert multiple linear inputs into nonlinear relationships in the neural network to realize the linear to nonlinear mapping function. The calculation formula of σ is as follows:

$$\sigma(x) = \frac{e^x}{e^x - 1} \quad (2)$$

2.2. Input Gate

The input gate contains two activation functions: the σ function and *tanh* function. Combining the input X_t at the current time step and the hidden state h_{t-1} transmitted from the previous time step, the two activation vectors I_t and G_t can be obtained through the above two activation functions. Here, G_t is also called a candidate memory cell, and its information is added to the cell state C_t medium [31,32]. The calculation formulas of activation vector I_t and G_t are as follows:

$$I_t = \sigma(W_{xi}X_t + h_{t-1}W_{hi} + b_i) \quad (3)$$

$$G_t = \tanh\left(W_{xg}X_t + h_{t-1}W_{hg} + b_c\right) \quad (4)$$

In Formulas (3) and (4), W_{xi}, W_{hi}, W_{xg}, and W_{hg} are weight coefficients, and b_c and b_i are bias vectors. Here, the tanh function is the hyperbolic tangent function, which is another activation function in the model. Its calculation formula is as follows:

$$\tanh(x) = \frac{e^x - e^{-x}}{e^x + e^{-x}} \quad (5)$$

The candidate memory cell G_t is used to update the value of the unit state C_t, and its calculation formula is as follows:

$$C_t = F_t \text{ e } C_{t-1} + I_t \text{ e } G_t \quad (6)$$

In Formula (6), t represents the current time step, t − 1 represents the previous time step, and e represents the Hadamard product.

2.3. Output Gate

The output gate calculates the output h_t of the entire processing unit according to the value of the state variable C_t and the function values of X_t and h_{t-1} after activation [33]. The calculation formulas of activation vectors O_t and h_t are as follows:

$$O_t = \sigma(W_{xo}X_t + h_{t-1}W_{ho} + b_o) \quad (7)$$

$$h_t = O_t \text{ e } \tanh(C_t) \quad (8)$$

In Equations (7) and (8), W_{xo} and W_{ho} are the weight coefficients of the gating unit, b_o is the bias vector, and h_{t-1} is the hide state for the previous time step.

2.4. Multi-Hidden-Layer LSTM Structure

A standard LSTM model typically includes an input layer, an output layer, and multiple hidden layers, as shown in Figure 3. With the increase in hidden layers, the structure becomes more complex, and the nonlinear mapping relationship between the data samples used for processing becomes more complex. However, this does not mean that the more hidden layers there are, the better the prediction of the structure for the unknown data because the number of hidden layers has a positive correlation with the iteration time, and with the increase in the number of hidden layers, the calculation time will increase exponentially. In addition to the above structure, there is a fully connected (FC) layer before the output of the model data, which is located between the hidden layer and the output layer. The final multiple hidden layers are connected by the FC layer to the target output layer to construct the desired output features [34,35].

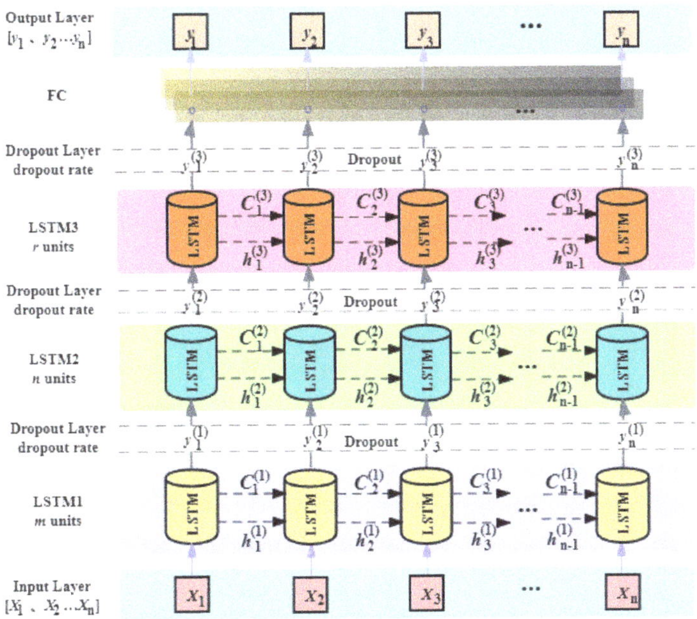

Figure 3. Multilayer LSTM structure model diagram.

Furthermore, a dropout layer can also be added after each LSTM hidden layer, and its value is usually between 0.2 and 0.5, effectively avoiding the occurrence of overfitting during model training [34,36]. Overfitting is a common problem in deep learning fields [37]. The key idea of the dropout layer is to randomly interrupt the mapping relationship between some data during the training process and discard a part of the data at a certain dropout rate. For these very reasons, the computing speed and computational efficiency of an LSTM model are greatly enhanced.

3. Prediction Method of the Wind-Induced Strain Response of Lightning Rod Structures Based on LSTM

Because the strain response between each measurement point of the lightning rod structure has a certain degree of nonlinear mapping relationship [38–40], the LSTM network has strong applicability for addressing this problem. Therefore, this paper adopts the LSTM network to repair and predict the missing or defective data of wind-induced dynamic response in the process of testing or monitoring lightning rod structures under different wind speeds. The specific process is shown in Figure 4.

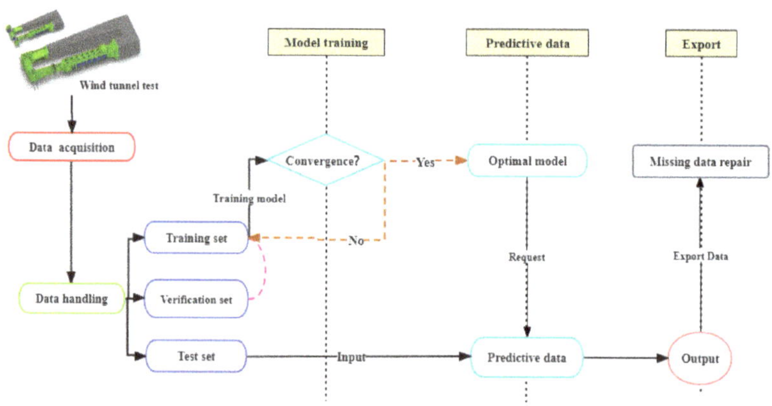

Figure 4. Flow chart of LSTM prediction and analysis of the strain value of the lightning rod structure measurement point.

First, the parameters of the LTSM model are optimized by using the strain time-history data of each measurement point of the aeroelastic model of the lightning rod structure obtained from the wind tunnel test, and the most suitable model structure is determined. Then, normal data are used to repair the disease data or "missing" data to ensure the integrity of the data chain. The detailed step descriptions are as follows:

Step 1: Classify the collected dynamic strain test data of 10 measurement points on the strong axis and weak axis of the lightning rod structure. Then, take the first 70% of the data of some measurement points as the training set and the last 30% of the data as the test set. Here, the test set data are treated as "missing" data.

Step 2: Perform noise reduction processing on the collected dataset to reduce the interference of the noise data collected during the test on capturing the nonlinear mapping relationship between the strains of each measurement point. At the same time, the data are normalized to decimals between 0 and 1 to speed up model iteration.

Step 3: Input a part of the data into the model for parameter-seeking training and determine the unknown parameters such as the number of hidden layers, the number of hidden units, and the number of iterations of the model so that the model can reach the optimal structure for the next step of data fitting and prediction.

Step 4: Input the dynamic strain data of the lowest two points of the lightning rod aeroelastic model structure into the sequence $X_1 = (x_1, x_2, x_3, \ldots, x_T)$, $X_2 = (x_1, x_2, x_3, \ldots, x_T)$ and set the dynamic strain sequence of the point to be measured as $Y = (y_1, y_2, y_3, \ldots, y_n)$, that is, using many-to-one model construction. Then, the first 70% of the data (i.e., $0.7T \times (M + 1)$) of these sequences are used to form a multidimensional matrix $[X_1, X_2, \ldots, Y]^T$ to perform the machine learning process. Here, M is the number of X vectors, and T is the number of data contained in each column of X vectors.

Step 5: After the model has captured the nonlinear relationship between X_1, $X_2 \ldots$ and Y to a considerable extent, the error loss image of the training process is used to judge whether the training is complete and whether the number of iterations is sufficient. If the training error loss image no longer decreases, go to step 6. If the image is still not completely decreased within the limited number of iterations, the number of iteration steps needs to be increased, and step 4 is repeated.

Step 6: Input the last 30% of strain data into the model to predict the Y sequence and then denormalize it to obtain the final predicted value of Y.

Step 7: Take the obtained predicted value Y as the new input value X_3, together with the previous input values X_1 and X_2, to compose a new input matrix $[X_1, X_2, X_3]^T$. The dynamic strain value of the next adjacent measurement point is used as new "missing" data to predict the value of Y, and steps S2 to S5 are repeated.

Step 8: Repeat steps S2 to S7 to predict the time-history data of the "missing" strain response data of the lightning rod structure.

By comparing the lightning rod strain response time-history data obtained in the above steps with the strain response time-history data actually measured by the wind tunnel test, the accuracy of the prediction results can be judged. On this basis, the average relative error of multiple prediction results is used as the final evaluation indicator for the prediction model.

4. Example Verification

4.1. Wind Tunnel Test of the Lightning Rod Aeroelastic Model

To study the wind-induced vibration response characteristics of frame–lightning rod structures, the research group designed and carried out a wind tunnel test of the scaled aeroelastic model of typical frame lightning rods [1]. Through the wind tunnel test and fiber grating (FBG) strain measurement technology, the strain response time-history of the lightning rod structure aeroelastic model under different wind speeds and different wind directions was measured.

The central wind tunnel during the test is a series double test section return/DC boundary layer wind tunnel with a height 3.0 m and length 24.0 m. According to the design theory of aeroelastic models, combined with the structure type of the lightning rods, the similarity ratios that should be satisfied in the design of the lightning rod aeroelastic model include the geometric ratio, mass ratio, Froude number, Cauchy number, dimensionless frequency, and damping ratio. The similar design parameters used in the aeroelastic model wind tunnel test are shown in Table 1, and the arrangement of one of the models and the strain measurement points of the FBG sensors is shown in Figure 5.

Table 1. Similarity parameters for the aeroelastic model wind tunnel test (n = 1:10).

Similarity Parameter	Geometric Ratio λ_L	Mass Ratio λ_M	Wind Speed Ratio λ_U	Flexural Rigidity Ratio λ_{EI}	Frequency Ratio λ_f	Damping Ratio λ_ζ
Similitude ratio	n	n^3	$n^{0.5}$	n^5	$n^{-0.5}$	1
Values	1:10	1:1000	1:3.16	1:100,000	3.16:1	1

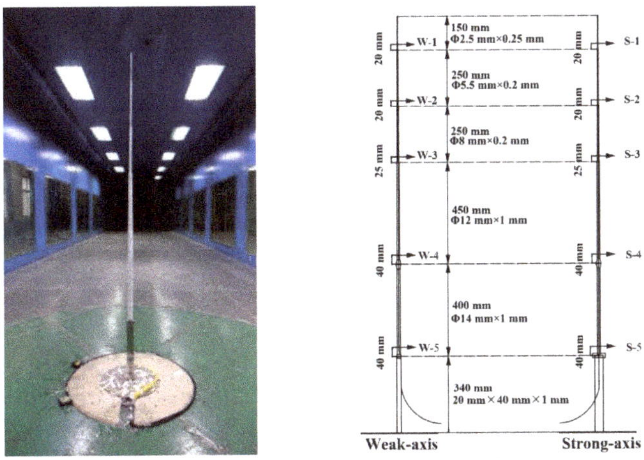

(a) (b)

Figure 5. Aeroelastic model and strain measurement points. (**a**) Aeroelastic model. (**b**) Arrangement of the FBG sensor strain measurement points.

The lightning rod aeroelastic model consists of five sections of rods with different specifications, which are defined as the first section, the second section, the third section, the fourth section and the fifth section from top to bottom. Measurement points are arranged at the center of each segment bottom to measure the strain values of the corresponding measurement points. The side corresponding to the smaller stiffness of the rod is called the weak axis, which is represented by W. The side with the larger stiffness of the rod is called the strong axis, which is represented by S. The W-axis measurement points from top to bottom are W-1, W-2, W-3, W-4, and W-5, and the S-axis measurement points from top to bottom are S-1, S-2, S-3, S-4, and S-5, respectively, as shown in Figure 5b.

Due to the slender and highly flexible structure of lightning rod aeroelastic models, light-weight and small-volume sensors are more suitable for wind tunnel tests to avoid excessive influence on the dynamic response of the model and minimize the measurement error. Therefore, FBG strain sensors are adopted in the experiment to measure the strain value at different measurement points of the structure. Figure 6 shows the location of specific FBG measurement points on the model and the FBG demodulator used in the test.

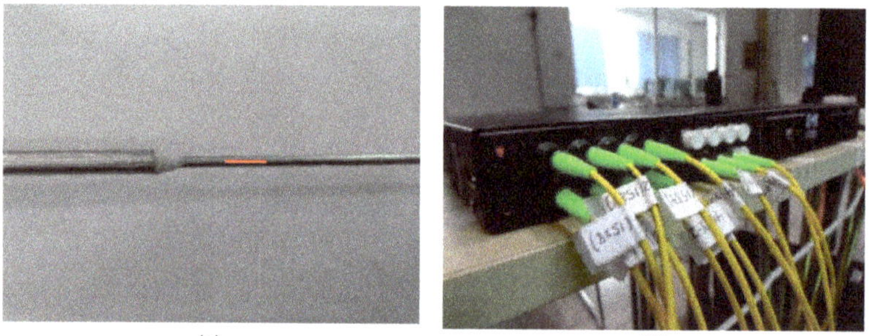

(**a**) (**b**)

Figure 6. FBG measurement point and FBG demodulator. (**a**) FBG measurement point (red). (**b**) FBG demodulator.

During the test, the definition of the wind direction angle is shown in Figure 7. The 0° wind direction angle is the case where the windward side is along the W-axis, and the 90° wind direction angle is the case where the windward side is along the S-axis. Under a 0° wind direction angle, the along-wind vibration response measurement points are located on the W-axis, and the crosswind vibration response measurement points are located on the S-axis. The wind speeds during the wind tunnel test are 5 m/s, 6 m/s, 8 m/s, and 10 m/s. According to the similarity ratio of the scaled model, the actual wind speeds of the corresponding lightning rod structure are 15.81 m/s, 18.97 m/s, 25.30 m/s, and 31.62 m/s. The above wind speed range not only includes the main wind speed range when the actual lightning rods work but also includes the extreme wind speed conditions that may occur in practice. For the convenience of guiding engineering practice, in the subsequent analysis, the wind speeds considered in this study are expressed in accordance with the actual wind speeds.

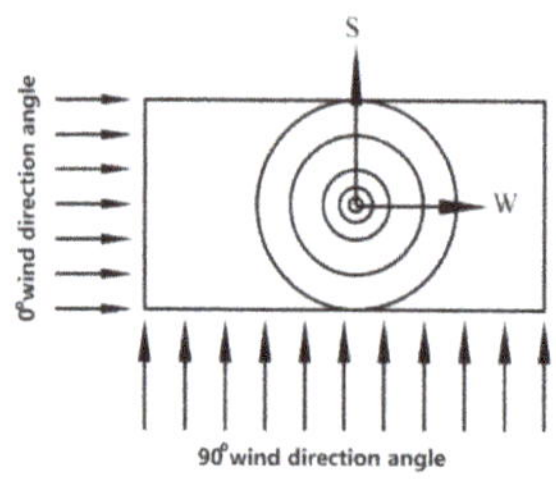

Figure 7. Schematic diagram of the lightning rod wind direction angle.

4.2. Selection and Processing of Data Samples

The strain values at the ten measurement points under the action of four different wind speeds (i.e., 15.81 m/s, 18.97 m/s, 25.30 m/s, 31.62 m/s) were sampled. The sampling frequency and time were 250 Hz and 40 s, respectively. There were 40 samples in total, and each sample contained 10,000 strain values. The first 70% of all the samples was used to train the LTSM network model. The remaining 30% of the data in the sample set of measurement points 1 to 3 were regarded as the "missing" data to test the LTSM network model. The evaluation indices [41,42] of the model were the root mean square error (RMSE), the mean absolute error (MAE), the variance accounted for (VAF) and the refined Willmott index (RWI). When the values of VAF and RWI are close enough to 1 and the values of RMSE and MAE are close enough to 0, the LSTM model can be considered excellent. The formulas for calculating the abovementioned indices were as follows:

$$\mathrm{RMSE} = \sqrt{\frac{1}{n}\sum_{\mathrm{i}=1}^{\mathrm{n}} (y_{\mathrm{i}} - y_{\mathrm{it}})^2} \tag{9}$$

$$\mathrm{MAE} = \frac{1}{n}\sum_{\mathrm{i}=1}^{\mathrm{n}} |y_{\mathrm{i}} - y_{\mathrm{it}}| \tag{10}$$

$$\mathrm{VAF} = 1 - \frac{\mathrm{var}(y_{\mathrm{i}} - y_{\mathrm{it}})}{\mathrm{var}(y_{\mathrm{i}})} \tag{11}$$

$$\mathrm{RWI} = 1 - \frac{\sum\limits_{\mathrm{i}=1}^{\mathrm{n}} |y_{\mathrm{i}} - y_{\mathrm{it}}|}{2\sum\limits_{\mathrm{i}=1}^{\mathrm{n}} |y_{\mathrm{i}} - y_{\mathrm{n}}|} \tag{12}$$

In Formulas (9) to (12), n is the number of test samples, y_{i} is the measured value at time i, y_{it} is the predicted value at time i, *var* represents variance, and y_{n} is the average value of the measured samples.

Meanwhile, the data in the training set and test set can also be normalized to improve the prediction accuracy and prediction speed. The calculation formula for normalization is:

$$x_{\mathrm{norm}} = \frac{x - m_{\mathrm{ean}}}{v_{\mathrm{ariance}}} \tag{13}$$

In Formula (13), m_{ean} is the mean of the input samples, and v_{ariance} is the variance of the input samples.

Take, as an example, the lightning rod under the action of wind speed 25.30 m/s. A total of 30,000 data points were selected from the test values of measurement points S-3, S-4, and S-5 on the strong axis to train and optimize the LTSM network model parameters. The data of measurement points S-4 and S-5 are used as the input data, and the data of measurement point S-3 are used as the output data. The measured time-history strain values of the abovementioned measurement points during the wind tunnel test are demonstrated in Figure 8.

Figure 8 shows that the time-history data fluctuate greatly in the first 5 s after the start of the test, and the difference is obvious from the data after that. To make the optimization of the model parameters more accurate and to minimize the adverse effect of noise data on the predictive results, the measured data of the first 5 s are removed, and the data from 5 s to 45 s are used for simulation and prediction in the subsequent analysis. The updated time-history data are shown in Figure 9.

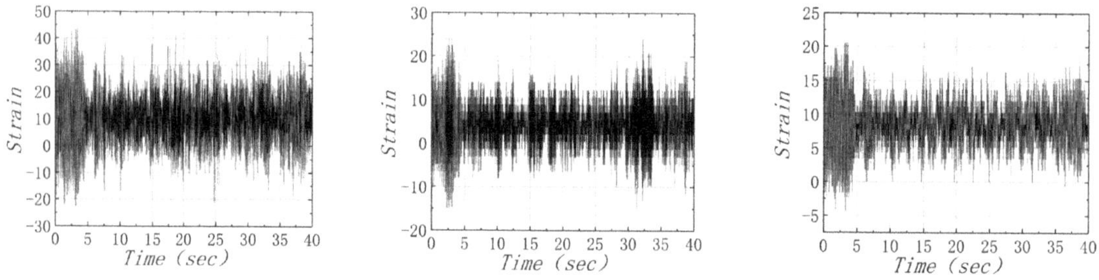

(a) (b) (c)

Figure 8. Measured strain time-history diagram of the measurement points of S-3, S-4, S-5. (**a**) Measured strain values at measurement point S-3. (**b**) Measured strain values at measurement point S-5. (**c**) Measured strain values at measurement point S-5.

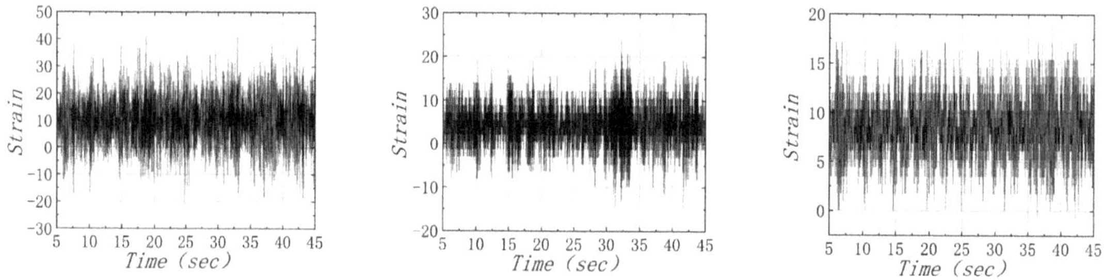

(a) (b) (c)

Figure 9. Updated strain time-history diagram of the measurement points of S-3, S-4, S-5 after noise elimination. (**a**) Updated strain values at measurement point S-3. (**b**) Updated strain values at measurement point S-5. (**c**) Updated strain values at measurement point S-5.

4.3. Determination of Model Structure and Parameters

MATLAB 2018a deep learning tools were used to build the predictive models. Due to the large number and the high dimensions of the measured data, a dropout layer is adopted to prevent the model from overfitting during the learning process. During analysis, the parameters of the LSTM model are first determined, and then the influence of different LSTM layers and hidden units on the prediction results is studied to find the optimal model parameters, and the basic control variable method is used for analysis.

To ensure that the model fully captures the nonlinear mapping relationship between the data, the number of iterations is set to 1000, and the initial learning rate is 0.005. The range of the number of hidden layers is set to 1 to 7, and the range of the number of hidden units is set to between 50 and 600, with a value interval of 50. The two independent variables (i.e., number of hidden layers and number of hidden units) are simulated, and the final fitting result of the three-dimensional surface graph is shown in Figure 10a. At the same time, to determine the most suitable number of iterations, the root mean square error (RMSE) of the test set during the training process of the LSTM neural network is counted, and its descending curve is shown in Figure 10b.

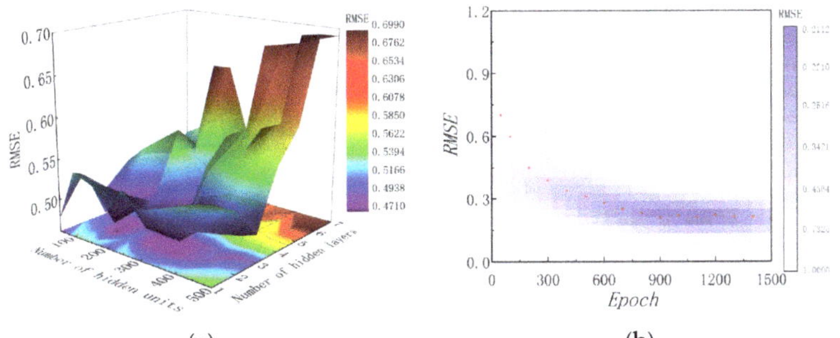

Figure 10. RMSE values for different parameters. (**a**) RMSE surface graph under different parameters. (**b**) Changing of RMSE value during the learning process.

From Figure 10a, it can be observed that the RMSE values change with the number of hidden layers and the number of hidden units. When the number of hidden layers is 3, the RMSE value is stable between 0.45 and 0.5, and the error is small compared to other combinations of hidden layers and hidden units. Figure 10b indicates that when the number of iterations is approximately 1000, the RMSE gradually tends to be stable. Since the calculation time of the model is positively correlated with the number of hidden layers, the number of hidden units and the number of iterations, to obtain the best balance between the calculation time and the calculation results, the optimal number of hidden units is set to 250, and the number of iterations is set to 1000.

The LSTM neural network model established in this paper includes an input layer for receiving input data; three LSTM layers for modeling the data; three dropout layers, with the dropout rate set to 0.2 to prevent data from being overfitted; and an FC layer for dimensional transformation of the output data.

The Adam optimizer is used in the neural network training process, the threshold activation function is the σ function, the output activation function is the *tanh* function, and the initial learning rate is 0.005. The maximum number of iterations is set to 1000. The dynamic learning rate is adopted, and the learning rate is reduced by half after every 500 training iterations. At the same time, the weight parameters of the LSTM model are normalized to prevent data from being overfitted. The parameter for normalization is set to 0.01. At this point, the neural network parameter training is complete, and the final optimal network parameters are presented in Table 2.

Table 2. LSTM network parameters.

Layer	Type	Activation Function	Output Shape
Input	InputLayer	-	(none, $n, f+1$)
LSTM1	LSTM	tanh	(none, n, 250)
Dropout1	dropout	-	0.2
LSTM2	LSTM	tanh	(none, n, 250)
Dropout2	dropout	-	0.2
LSTM3	LSTM	tanh	(none, n, 250)
Dropout3	dropout	-	0.2
FC	Dense	Linear	(none, $s \times f$)

Note: n is the length of the input time series; s is the length of the output time series; f is the number of degrees of freedom for prediction.

4.4. Analysis of Wind Vibration Response of Lightning Rod Structure under the Action of Wind Speed 25.30 m/s

The first four orders of modal frequencies of the lightning rod aeroelastic model structure are shown in Table 3. To find the weak parts of the lightning rod structure and identify the positions where the sensors are more likely to damage and cause data loss or drift during the test process, a Fourier transform of the strain time-history under 25.3 m/s wind speed and 0 degree wind direction angle is performed. The along-wind and crosswind vibration strain power spectra of the four measurement points on the upper part of the structure are shown in Figures 11 and 12, respectively.

Table 3. The first four orders of vibration mode frequencies of the lightning rod aeroelastic model.

Modal Number	Aeroelastic Model Natural Frequency (Hz)	
	W-Axis	S-Axis
1st order	7.95	8.97
2nd order	17.48	19.23
3rd order	43.41	47.49
4th order	64.02	71.84

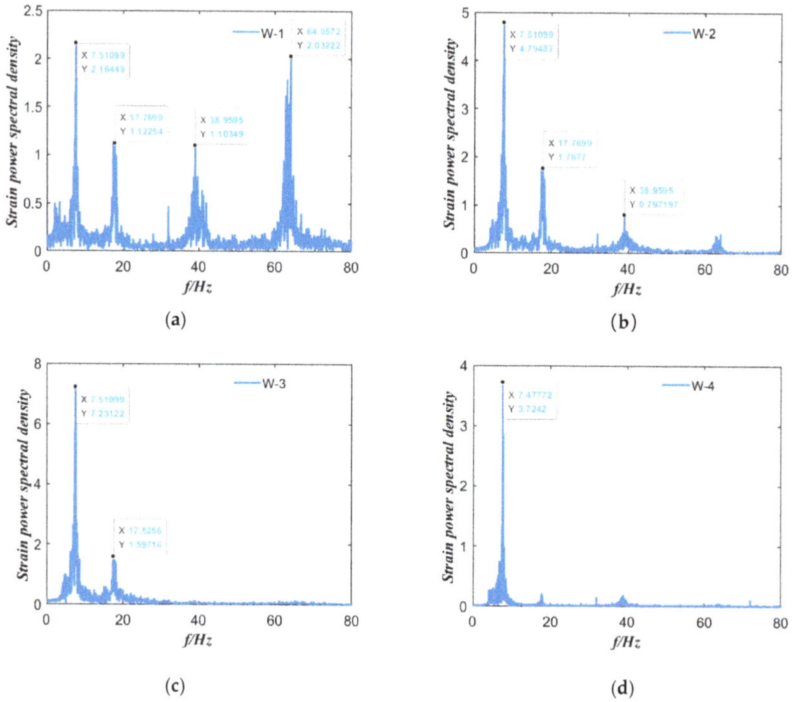

Figure 11. Along-wind vibration strain power spectrum of each measurement point of the lightning rod under a wind speed of 25.3 m/s. (**a**) Along-wind strain power spectrum at measurement point W-1. (**b**) Along-wind strain power spectrum at measurement point W-2. (**c**) Along-wind strain power spectrum at measurement point W-3. (**d**) Along-wind strain power spectrum at measurement point W-4.

Figure 11 shows that the along-wind vibration characteristic of the upper part near the W-1 measurement point of the lightning rod is the most complex; its vibration response contains the mode components of the first four orders, with corresponding frequencies of 7.51099 Hz, 17.7699 Hz, 38.9595 Hz, and 64.0572 Hz. The vibration response of the structure near measurement point W-2 mainly includes the first-, second- and third-order mode components, while the vibration response of the structure near measurement point W-3 is dominated by the contribution of the first- and second-order modes. By and large, the high-frequency vibration response of the lightning rod structure mainly occurs in the upper part of the structure, the vibration is jointly controlled by multiple modes of vibration, and the vibration amplitude becomes stronger with increasing wind speed, while the vibration of the lower part of the structure is basically dominated by the contribution of the fundamental vibration mode. Therefore, the sensors at the upper part of the lightning rod structure are more prone to damage, resulting in data loss or abnormality.

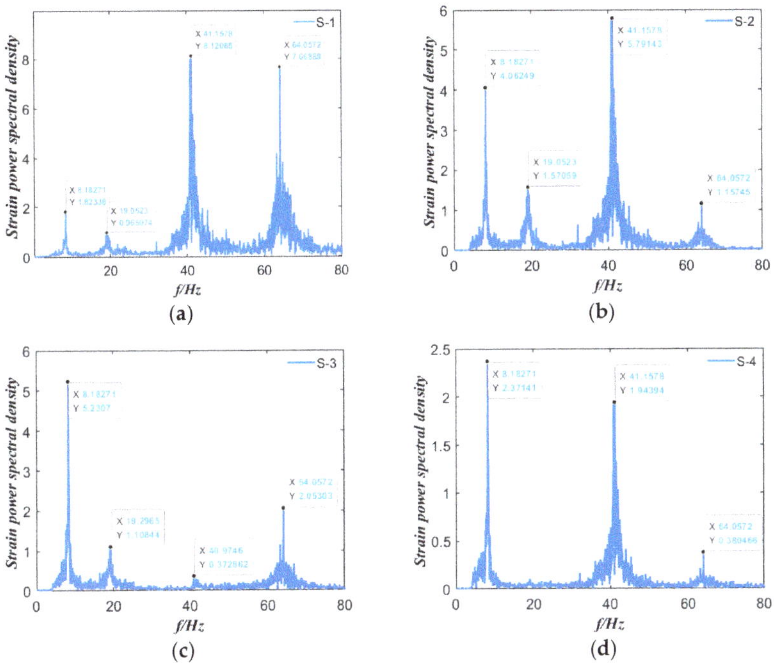

Figure 12. Crosswind vibration strain power spectrum of each measurement point of the lightning rod under a wind speed of 25.3 m/s. (**a**) Strain power spectrum at measurement point S-1. (**b**) Strain power spectrum at measurement point S-2. (**c**) Strain power spectrum at measurement point S-3. (**d**) Strain power spectrum at measurement point S-4.

Figure 12 also shows that the crosswind vibration characteristic of the structure near the upper measurement points is relatively complex, and its vibration response contains mode components of the first four orders. The corresponding modal frequencies are 8.18271 Hz, 19.0523 Hz, 41.1578 Hz, and 64.0572 Hz. Under a wind speed of 25.30 m/s, the high-frequency vibration of the lightning rod near measurement point S-1 is the most severe, which also indicates the position where exceptional monitoring data are prone to occur during the test.

4.5. Prediction Results of the Strain Response Data of the Lightning Rod

Based on the LSTM neural network model trained in the previous sections, the along-wind and crosswind strain response data of measurement points 1 to 5 of the lightning rod under different wind speeds are adopted to predict the "missing" strain response. Here, the measured strain response data of measurement points 4 and 5 are used as input data, and the strains of measurement points 1, 2 and 3 are treated as output data. Using the simulation method proposed in the above sections, the along-wind and crosswind strain response data are predicted, and the results are shown in Figures 13–18.

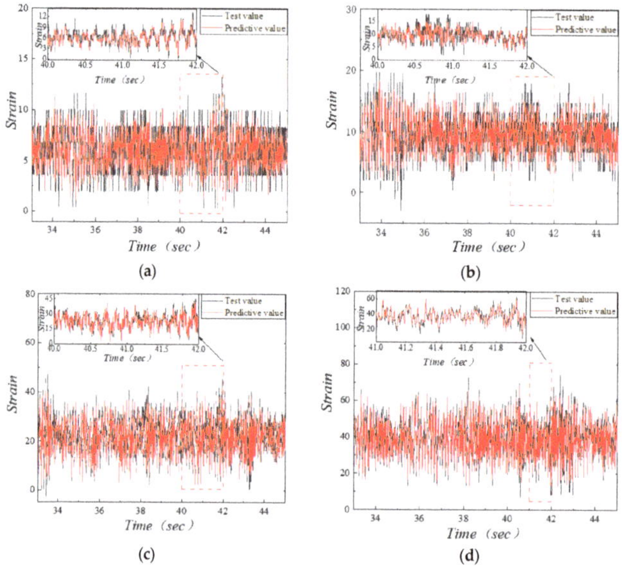

Figure 13. Comparison of the prediction results and test values of along-wind strain response at W-1 measurement point of the lightning rod under different wind speeds. (**a**) Strain values at measurement point W-1 under a wind speed of 15.81 m/s. (**b**) Strain values at measurement point W-1 under a wind speed of 18.97 m/s. (**c**) Strain values at measurement point W-1 under a wind speed of 25.30 m/s. (**d**) Strain values at measurement point W-1 under a wind speed of 31.62 m/s.

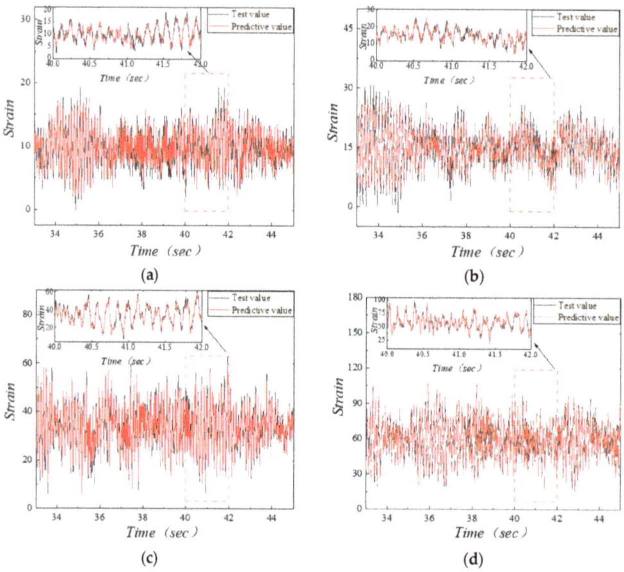

Figure 14. Comparison of the prediction results and test values of along-wind strain response at W-2 measurement point of the lightning rod under different wind speeds. (**a**) Strain values at measurement point W-2 under a wind speed of 15.81 m/s. (**b**) Strain values at measurement point W-2 under a wind speed of 18.97 m/s. (**c**) Strain values at measurement point W-2 under a wind speed of 25.30 m/s. (**d**) Strain values at measurement point W-2 under a wind speed of 31.62 m/s.

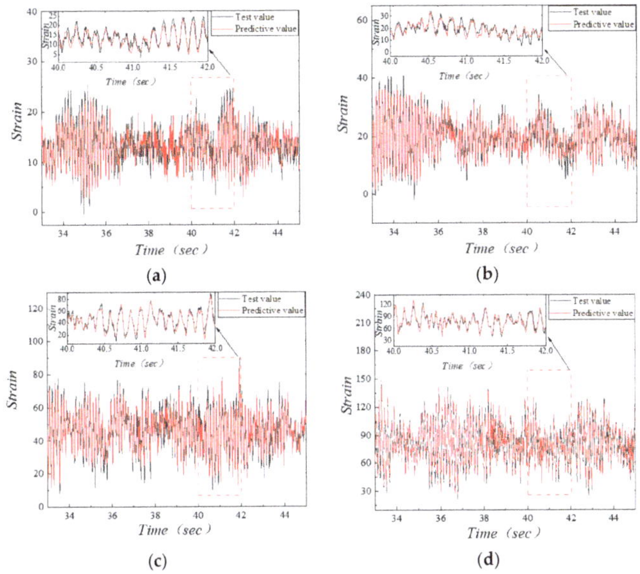

Figure 15. Comparison of the prediction results and test values of along-wind strain response at W-3 measurement point of the lightning rod under different wind speeds. (**a**) Strain values at measurement point W-3 under a wind speed of 15.81 m/s. (**b**) Strain values at measurement point W-3 under a wind speed of 18.97 m/s. (**c**) Strain values at measurement point W-3 under a wind speed of 25.30 m/s. (**d**) Strain values at measurement point W-3 under a wind speed of 31.62 m/s.

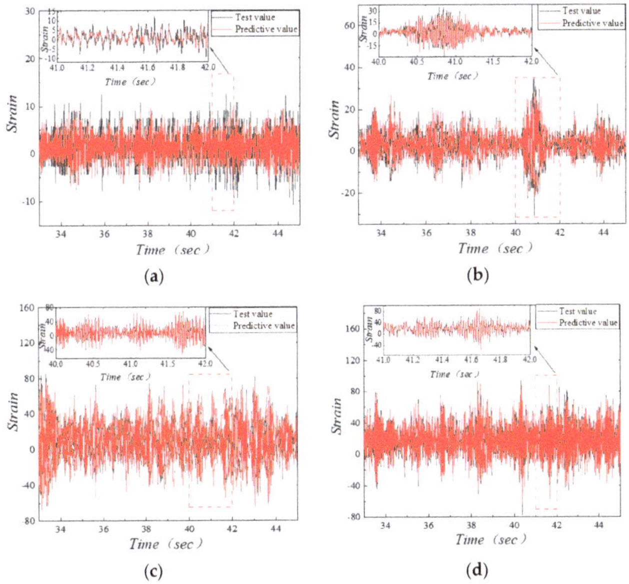

Figure 16. Comparison of the prediction results and test values of crosswind strain response at S-1 measurement point of the lightning rod under different wind speeds. (**a**) Strain values at measurement point S-1 under a wind speed of 15.81 m/s. (**b**) Strain values at measurement point S-1 under a wind speed of 18.97 m/s. (**c**) Strain values at measurement point S-1 under a wind speed of 25.30 m/s. (**d**) Strain values at measurement point S-1 under a wind speed of 31.62 m/s.

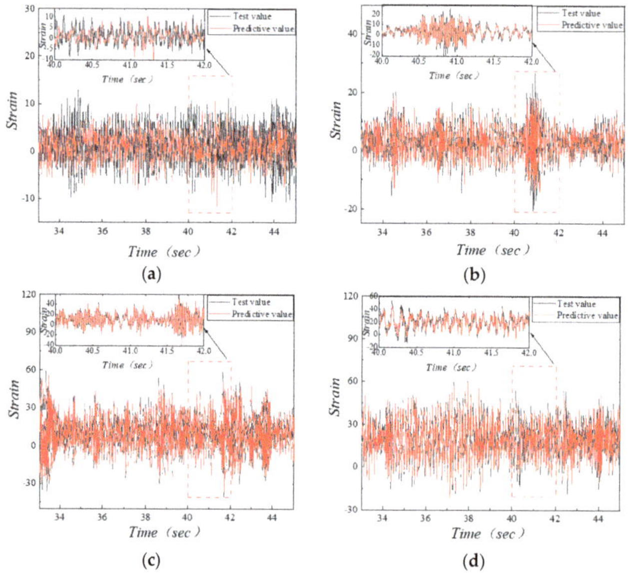

Figure 17. Comparison of the prediction results and test values of crosswind strain response at S-2 measurement point of the lightning rod under different wind speeds. (**a**) Strain values at measurement point S-2 under a wind speed of 15.81 m/s. (**b**) Strain values at measurement point S-2 under a wind speed of 18.97 m/s. (**c**) Strain values at measurement point S-2 under a wind speed of 25.30 m/s. (**d**) Strain values at measurement point S-2 under a wind speed of 31.62 m/s.

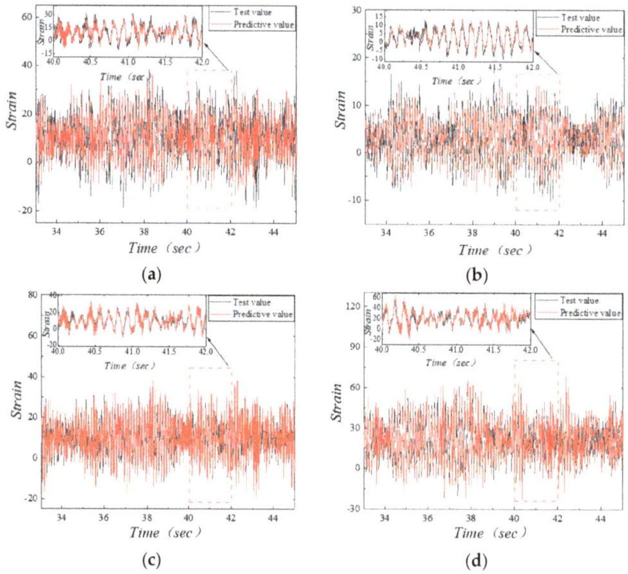

Figure 18. Comparison of the prediction results and test values of crosswind strain response at S-3 measurement point of the lightning rod under different wind speeds. (**a**) Strain values at measurement point S-3 under a wind speed of 15.81 m/s. (**b**) Strain values at measurement point S-3 under a wind speed of 18.97 m/s. (**c**) Strain values at measurement point S-3 under a wind speed of 25.30 m/s. (**d**) Strain values at measurement point S-3 under a wind speed of 31.62 m/s.

Figures 13–18 show that the strain time-history responses at each measurement point predicted by the LSTM model are in good agreement with the measured values by the wind tunnel test, regardless of whether under frequently encountered wind speed or extreme wind speed conditions, indicating that the proposed method in this paper has high reliability and stability.

In practical engineering, the vibration amplitude of a lightning rod usually increases with increasing wind speed, which may cause damage to the sensors on the upper part of the structure under extreme wind speed conditions, resulting in measured data abnormities or broken data chains. Therefore, it is more meaningful to accurately predict the vibration response of the structure under the condition of high wind speed.

4.6. Correlation Analysis of the Predictive Data under Extreme Wind Speed Conditions

The above analysis shows that the structural vibration response is most complex and strongest at measurement points W-1 and S-1 under the extreme wind speed 31.62 m/s. The sensors in these positions are most likely to fail in actual engineering, resulting in an incomplete response data chain. To identify whether the proposed prediction method can meet the needs of engineering applications, a correlation analysis of the predictive data under extreme wind speed conditions is performed in this study, and the results are shown in Figure 19. Here, R^2 is the coefficient of determination, and the range of its value is 0 to 1. Moreover, the closer the value of R^2 is to 1, the better the data prediction effect.

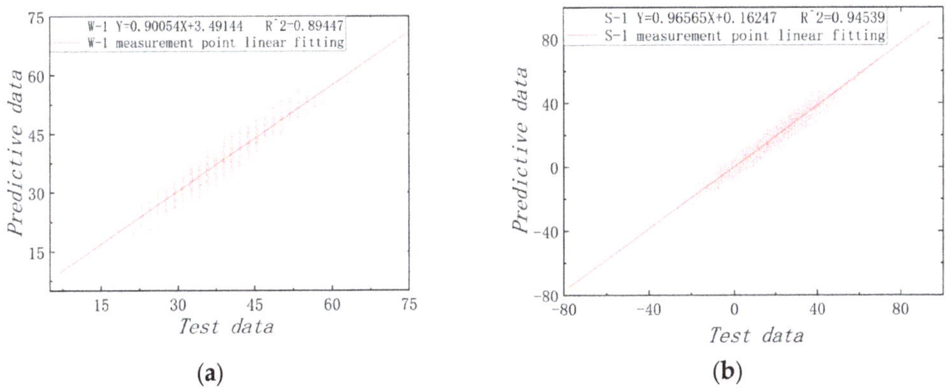

Figure 19. Scatter plot and linear fitting diagram of the prediction strain results at measurement points W-1 and S-1 under a wind speed of 31.62 m/s. (**a**) Scatter plot and linear fitting diagram of the predicted data at measurement point W-1. (**b**) Scatter plot and linear fitting diagram of the predicted data at measurement point S-1.

It can be seen from Figure 19 that even under extreme wind speed conditions, the coefficient of determination of the predicted data at the most unfavorable position of the lightning rod (i.e., measurement point 1) is approximately 0.9, and the predicted and measured values are evenly distributed on both sides of the fitting curve. There is no abnormal drift point, indicating that the prediction ability of the model in this paper has a high guarantee rate and can meet the actual engineering needs. At the same time, with the help of the unique data feature importance analysis function of the random forest method, the weight coefficients of different input data features for the prediction results can be obtained. When predicting the "missing" data at monitoring point 1, four sets of measured data from measurement point 2 to measurement point 5 were used, and the final analysis results are shown in Figure 20.

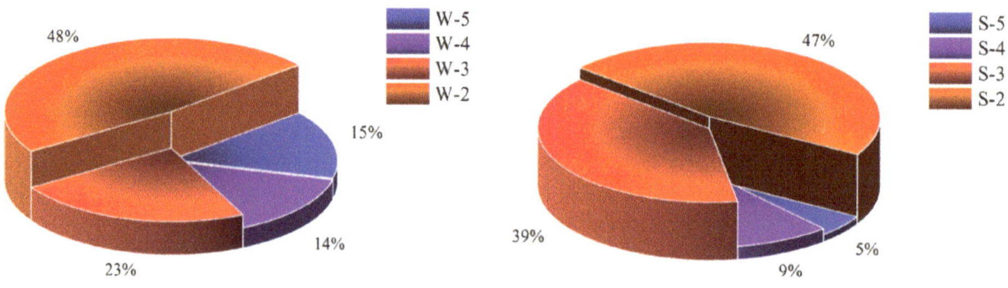

(**a**) (**b**)

Figure 20. Weight coefficients of the measured data of each measurement point for the predicted results at measurement point 1 under a wind speed of 31.62 m/s. (**a**) Weight coefficients of the measured data at measurement points W-2, W-3, W-4, and W-5 with respect to W-1. (**b**) Weight coefficients of the measured data at measurement points S-2, S-3, S-4, and S-5 with respect to S-1.

Figure 20 shows that when predicting the strain response of the lightning rod at measurement point S-1, the measured data at measurement point S-2 play a decisive role in the predictive results at measurement point S-1, and its weight coefficient reaches 0.48. The influence of the data of measurement point S-3 on the prediction results of measurement point S-1 is second only to that of measurement point S-2, and its weight coefficient is 0.23. The influence of the data of measurement point S-5 and measurement point S-4 on the prediction results is basically similar, and their weight coefficients are approximately 0.15. In general, the measured response data of the monitoring points that are closer to the predicted measurement point have a greater impact on the response of the predicted measurement point, while the measured response data of monitoring points that are far from the predicted measurement point have less impact on the response of the predicted measurement point and are considered ignorable. To compare the precision of the prediction results of each working condition analyzed in this paper more intuitively, the RMSE, MAE, RWI and VAF of the prediction results corresponding to each working condition are also analyzed, and the results are shown in Figures 21–24.

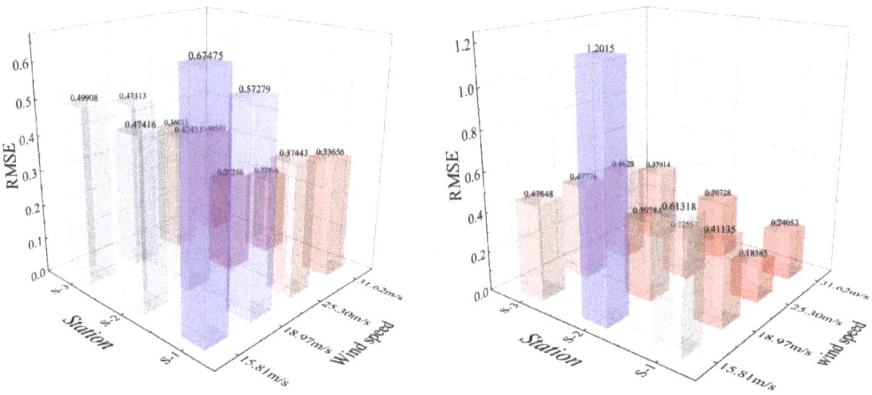

(**a**) (**b**)

Figure 21. RMSE values of each measurement point under different wind speeds. (**a**) RMSE values of each measurement point on the W-axis. (**b**) RMSE values of each measurement point on the S-axis.

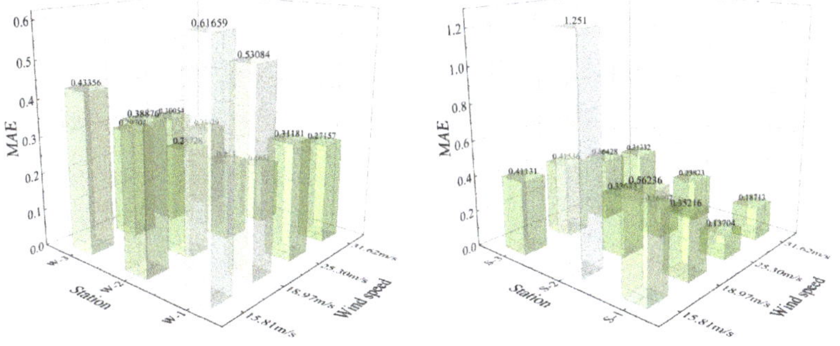

(a) (b)

Figure 22. MAE values of each measurement point under different wind speeds. (**a**) MAE values of each measurement point on the W-axis. (**b**) MAE values of each measurement point on the S-axis.

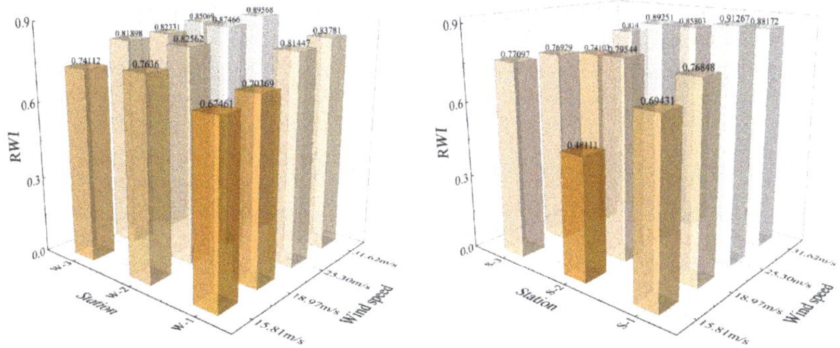

(a) (b)

Figure 23. RWI values of each measurement point under different wind speeds. (**a**) RWI values of each measurement point on the W-axis. (**b**) RWI values of each measurement point on the S-axis.

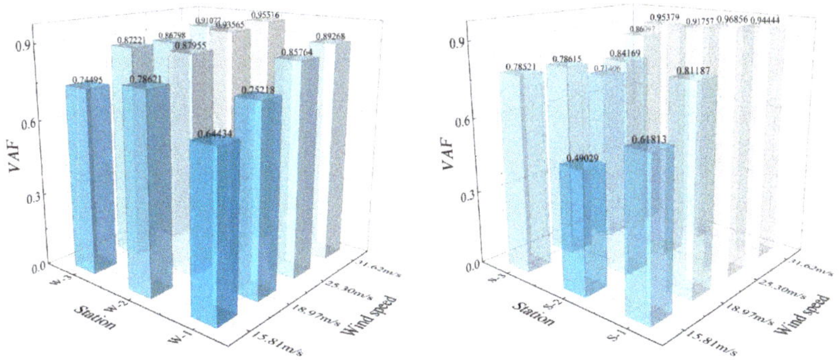

(a) (b)

Figure 24. VAF values of each measurement point under different wind speeds. (**a**) VAF values of each measurement point on the W-axis. (**b**) VAF values of each measurement point on the S-axis.

Figures 21 and 22 show that within the wind speed range analyzed in this paper, the RMSE and MAE values of the prediction results of each measurement point on the W-axis of the lightning rod show a negative correlation as a whole with the increase in wind speed, and under the same wind speed conditions, the RMSE and MAE values of each measurement point are relatively stable.

Figures 23 and 24 indicate that within the wind speed range analyzed in this paper, the RWI and VIF values of each measurement point of the lightning rod structure gradually increase with increasing wind speed and finally tend to 1. On the S-axis side of the lightning rod, when the wind speed is 25.30 m/s, the prediction precision of the response results of measurement point S-1 is the highest. For other measurement points on the S-axis, the prediction precision can still reach an excellent level, which shows that the LSTM model built in this paper has a certain degree of generalization capability.

5. Conclusions

This paper proposes a method based on an LSTM neural network to predict the "missing" response data of lightning rod structures under wind-induced vibration. The prediction of the "missing" strain value of each measurement point in the along-wind and crosswind directions of the lightning rod under different wind speeds is carried out, and the random forest method is used to analyze the correlation relationship between the predicted data and the measured data under extremely high wind speed conditions. The main conclusions are as follows:

1. The high-frequency and complex response of the lightning rod structure in the along-wind direction and the crosswind direction mainly occurs near measurement point 1 on the upper part of the structure. Problems such as loss or abnormality of monitoring data are prone to occur. Therefore, it is important to focus on this measurement point and prepare for missing data prediction during testing and monitoring.
2. Under the normal working wind speed range of the lightning rod structure, regardless of whether frequently encountered wind speed or extreme wind speed, the strain responses of the measurement point predicted by the LSTM model are in good agreement with the measured values of the wind tunnel test. Even under the case of an extreme wind speed of 31.62 m/s, the values of RMSE, MAE, R^2, RWI and VAF are 0.24053, 0.18213, 0.94539, 0.88172 and 0.94444, respectively, which are within the acceptable range, indicating that the LSTM method can better capture the nonlinear mapping relationship between the strains of each measurement point and has high reliability and stability.
3. In the structural vibration response prediction, measurement point 2 plays a decisive role in the prediction result at measurement point 1, while the influence of measurement points 5 and 4 on the prediction results is almost negligible. In general, the measured response data of the monitoring points that are closer to the predicted measurement point have a greater impact on the response of the predicted measurement point, while the measured response data of monitoring points that are far from the predicted measurement point have less impact on the response of the predicted measurement point and are considered ignorable.
4. Finally, it should be noted that it is necessary to perform more complex processing on the noise data to make the prediction results more accurate and meet engineering requirements.

Author Contributions: Conceptualization, G.Z. and M.Z.; methodology, G.Z., K.X. and Y.W.; software, G.Z., H.Q. and M.Z.; validation, K.X., Y.W. and G.Z.; formal analysis, K.X., Y.W. and H.Q.; investigation, Y.W. and M.Z.; resources, H.Q. and M.Z.; data curation, K.X.; writing—original draft preparation, G.Z., K.X. and Y.W.; writing—review and editing, G.Z., H.Q. and M.Z. All authors have read and agreed to the published version of the manuscript.

Funding: This research was supported by the Program for Innovative Research Team (in Science and Technology) at the University of Henan Province (Grant No. 23IRTSTHN006) and the Cultivating Fund Project for Young Teachers of Zhengzhou University (Grant no. JC21539028).

Institutional Review Board Statement: Not applicable.

Informed Consent Statement: Not applicable.

Data Availability Statement: Data are contained within this article.

Conflicts of Interest: The authors declare no conflict of interest.

References

1. Zhao, G.; Li, J.; Zhang, M.; Yi, Y. Experimental study on the bearing capacity and fatigue life of lightning rod structure joints in high-voltage substation structures. *Thin-Walled Struct.* **2022**, *175*, 109282. [CrossRef]
2. Guojun, D.; Lei, G.; Manling, D. Break Fault Analysis and Countermeasure of Frame Lighting Rod. *Henan Electr. Power* **2015**, *4*, 6–9.
3. Tao, S. A Cause Analysis of Fracture Failure with Architecture Type Lightning Rod. *Insul. Surge Arresters* **2017**, *2*, 49–52.
4. Jian, W.; Heming, D. The Deformation Analysis and Countermeasures on the Frame Lightning Rod of 750 kV Substation in the Wind Damage Areas. *Insul. Surge Arresters* **2017**, *2*, 14–18.
5. Gomez-Cabrera, A.; Escamilla-Ambrosio, P.J. Review of Machine-Learning Techniques Applied to Structural Health Monitoring Systems for Building and Bridge Structures. *Appl. Sci.* **2022**, *12*, 10754. [CrossRef]
6. Ye, X.W.; Jin, T.; Yun, C.B. A review on deep learning-based structural health monitoring of civil infrastructures. *Smart Struct. Syst.* **2019**, *24*, 567–585.
7. Rucevskis, S.; Kovalovs, A.; Chate, A. Optimal Sensor Placement in Composite Circular Cylindrical Shells for Structural Health Monitoring. *J. Phys. Conf. Ser.* **2023**, *2423*, 12021. [CrossRef]
8. Luo, Y.; Mei, Y.; Shen, Y.; Yang, P.; Jin, L.; Zhang, P. Field measurement of temperature and stress on steel structure of the National Stadium and analysis of temperature action. *J. Build. Struct.* **2013**, *34*, 24–32.
9. Amezquita-Sanchez, J.P.; Valtierra-Rodriguez, M.; Adeli, H. Wireless smart sensors for monitoring the health condition of civil infrastructure. *Sci. Iran.* **2018**, *25*, 2913–2925.
10. Sabato, A.; Niezrecki, C.; Fortino, G. Wireless MEMS-based accelerometer sensor boards for structural vibration monitoring: A review. *IEEE Sens. J.* **2016**, *17*, 226–235. [CrossRef]
11. Sarkar, S.; Reddy, K.K.; Giering, M. Deep learning for structural health monitoring: A damage characterization application. In Proceedings of the Annual Conference of the PHM Society, Denver, CO, USA, 3–6 October 2016; Volume 8.
12. Oh, B.K.; Glisic, B.; Kim, Y.; Park, H.S. Convolutional neural network-based data recovery method for structural health monitoring. *Struct. Health Monit.* **2020**, *19*, 1821–1838. [CrossRef]
13. Muin, S.; Mosalam, K.M. Structural health monitoring using machine learning and cumulative absolute velocity features. *Appl. Sci.* **2021**, *11*, 5727. [CrossRef]
14. Yuen, K.V.; Kuok, S.C. Ambient interference in long-term monitoring of buildings. *Eng. Struct.* **2010**, *32*, 2379–2386. [CrossRef]
15. Cross, E.J.; Manson, G.; Worden, K.; Pierce, S.G. Features for damage detection with insensitivity to environmental and operational variations. *Proc. R. Soc. A Math. Phys. Eng. Sci.* **2012**, *468*, 4098–4122. [CrossRef]
16. Sarmadi, H.; Karamodin, A. A novel anomaly detection method based on adaptive Mahalanobis-squared distance and one-class kNN rule for structural health monitoring under environmental effects. *Mech. Syst. Signal Process.* **2020**, *140*, 106495. [CrossRef]
17. Padil, K.H.; Bakhary, N.; Abdulkareem, M.; Li, J.; Hao, H. Nonprobabilistic method to consider uncertainties in frequency response function for vibration-based damage detection using artificial neural network. *J. Sound Vib.* **2020**, *467*, 115069. [CrossRef]
18. Bao, Y.; Tang, Z.; Li, H.; Zhang, Y. Computer vision and deep learning–based data anomaly detection method for structural health monitoring. *Struct. Health Monit.* **2019**, *18*, 401–421. [CrossRef]
19. Tang, Z.; Chen, Z.; Bao, Y.; Li, H. Convolutional neural network-based data anomaly detection method using multiple information for structural health monitoring. *Struct. Control Health Monit.* **2019**, *26*, e2296. [CrossRef]
20. Avci, O.; Abdeljaber, O.; Kiranyaz, S.; Hussein, M.; Inman, D.J. Wireless and real-time structural damage detection: A novel decentralized method for wireless sensor networks. *J. Sound Vib.* **2018**, *424*, 158–172. [CrossRef]
21. Hochreiter, S.; Schmidhuber, J. Long short-term memory. *Neural Comput.* **1997**, *9*, 1735–1780. [CrossRef]
22. Li, S.; Li, S.; Laima, S.; Li, H. Data-driven modeling of bridge buffeting in the time domain using long short-term memory network based on structural health monitoring. *Struct. Control Health Monit.* **2021**, *28*, e2772. [CrossRef]
23. Sherstinsky, A. Fundamentals of recurrent neural network (RNN) and long short-term memory (LSTM) network. *Phys. D Nonlinear Phenom.* **2020**, *404*, 132306. [CrossRef]
24. Iparraguirre-Villanueva, O.; Alvarez-Risco, A.; Salazar, J.L.H.; Beltozar-Clemente, S.; Zapata-Paulini, J.; Yáñez, J.A.; Cabanillas-Carbonell, M. The Public Health Contribution of Sentiment Analysis of Monkeypox Tweets to Detect Polarities Using the CNN-LSTM Model. *Vaccines* **2023**, *11*, 312. [CrossRef] [PubMed]
25. Smagulova, K.; James, A.P. A survey on LSTM memristive neural network architectures and applications. *Eur. Phys. J. Spec. Top.* **2019**, *228*, 2313–2324. [CrossRef]
26. Bakhshi Ostadkalayeh, F.; Moradi, S.; Asadi, A.; Nia, A.M.; Taheri, S. Performance Improvement of LSTM-based Deep Learning Model for Streamflow Forecasting Using Kalman Filtering. *Water Resour. Manag.* **2023**, 1–17. [CrossRef]
27. Chen, X.; Chen, W.; Dinavahi, V.; Liu, Y.; Feng, J. Short-Term Load Forecasting and Associated Weather Variables Prediction Using ResNet-LSTM Based Deep Learning. *IEEE Access* **2023**, *11*, 5393–5405. [CrossRef]
28. Graves, A.; Schmidhuber, J. Framewise phoneme classification with bidirectional LSTM and other neural network architectures. *Neural Netw.* **2005**, *18*, 602–610. [CrossRef]

29. Liu, H.-J.; Shan, W.-F.; Geng, G.-Z. Earthquake Prediction Using Dissolved Radon Data in Long and Short Term Memory Network. *Sci. Technol. Eng.* **2020**, *20*, 4029–4035.
30. Gers, F.A.; Schmidhuber, J.; Cummins, F. Learning to forget: Continual prediction with LSTM. *Neural Comput.* **2000**, *12*, 2451–2471. [CrossRef]
31. Nasir, J.; Aamir, M.; Haq, Z.U.; Khan, S.; Amin, M.Y.; Naeem, M. A new approach for forecasting crude oil prices based on stochastic and deterministic influences of LMD Using ARIMA and LSTM Models. *IEEE Access* **2023**, *11*, 14322–14339. [CrossRef]
32. De Pater, I.; Mitici, M. Developing health indicators and RUL prognostics for systems with few failure instances and varying operating conditions using a LSTM autoencoder. *Eng. Appl. Artif. Intell.* **2023**, *117*, 105582. [CrossRef]
33. Roux, A.; Changey, S.; Weber, J.; Lauffenburger, J.-P. LSTM-Based Projectile Trajectory Estimation in a GNSS-Denied Environment. *Sensors* **2023**, *23*, 3025. [CrossRef] [PubMed]
34. Zhang, R.; Chen, Z.; Chen, S.; Zheng, J.; Büyüköztürk, O.; Sun, H. Deep long short-term memory networks for nonlinear structural seismic response prediction. *Comput. Struct.* **2019**, *220*, 55–68. [CrossRef]
35. Byeon, W.; Breuel, T.M.; Raue, F.; Liwicki, M. Scene labeling with lstm recurrent neural networks. In Proceedings of the IEEE Conference on Computer Vision and Pattern Recognition, Boston, MA, USA, 7–12 June 2015; pp. 3547–3555.
36. Long, X.; Ding, X.; Li, J.; Dong, R.; Su, Y.; Chang, C. Indentation Reverse Algorithm of Mechanical Response for Elastoplastic Coatings Based on LSTM Deep Learning. *Materials* **2023**, *16*, 2617. [CrossRef]
37. Long, B.; Tan, F.; Newman, M. Forecasting the Monkeypox Outbreak Using ARIMA, Prophet, NeuralProphet, and LSTM Models in the United States. *Forecasting* **2023**, *5*, 127–137. [CrossRef]
38. Xie, X.; Luo, Y.; Zhang, N. Miss data reconstruction in stress monitoring of steel spatial structures using neural network techniques. *Spat. Struct.* **2019**, *25*, 38–44.
39. Han, J.; Pak, W. Hierarchical LSTM-Based Network Intrusion Detection System Using Hybrid Classification. *Appl. Sci.* **2023**, *13*, 3089. [CrossRef]
40. Xu, Z.; Chen, J.; Shen, J.; Xiang, M. Recursive long short-term memory network for predicting nonlinear structural seismic response. *Eng. Struct.* **2022**, *250*, 113406. [CrossRef]
41. Raja, M.N.A.; Shukla, S.K. Predicting the settlement of geosynthetic-reinforced soil foundations using evolutionary artificial intelligence technique. *Geotext. Geomembr.* **2021**, *49*, 1280–1293. [CrossRef]
42. Demirbay, B.; Bayram, D.; Uur, A. A bayesian regularized feed-forward neural network model for conductivity prediction of ps/mwcnt nanocomposite film coatings. *Appl. Soft Comput.* **2020**, *96*, 106632. [CrossRef]

buildings

MDPI

Article

Bayesian RC-Frame Finite Element Model Updating and Damage Estimation Using Nested Sampling with Nonlinear Time History

Kunyang Wang [1], Yukihide Kajita [2,*] and Yaoxin Yang [3]

[1] Division of Structural and Earthquake Engineering, Department of Civil Engineering, Graduate School of Engineering, Kyushu University, Fukuoka 8190385, Japan
[2] Department of Civil and Structural Engineering, Faculty of Engineering, Kyushu University, Fukuoka 8190385, Japan
[3] YJK Building Software, Beijing 100013, China
* Correspondence: ykajita@doc.kyushu-u.ac.jp

Abstract: This paper proposes a Bayesian RC-frame finite element model updating (FEMU) and damage state estimation approach using the nonlinear acceleration time history based on nested sampling. Numerical RC-frame finite element model (FEM) parameters are selected through nested sampling, and their probability density is estimated using nonlinear time history. In the first step, we estimate the error standard deviation and select the FEM parameters that are required to be updated by FEMU. In the second step, we estimate the probability density of the selected parameters and realize the FEMU through the resampling method and kernel density estimation (KDE). Additionally, we propose a damage state estimate approach, which is a derivative method of the FEMU sample. The numerical results demonstrate that the proposed approach is reliable for the Bayesian FEMU and damage state estimation using nonlinear time history.

Keywords: Bayesian model updating; structural health monitoring; nested sampling; Bayesian model selection; finite element model; nonlinear model; damage degree estimation

1. Introduction

The finite element model (FEM) has been widely used in the engineering field, particularly in civil engineering. The role of FEM is to predict or calculate the relevant response of structures. Approaches such as incremental dynamics analysis (IDA) and pushover have been developed from FEM for predicting the structural response and damage in accidents, particularly in seismic incidents [1–4]. Evidently, FEM is not the same as the actual structure. Errors, such as noise and material properties in the FEM and the actual structure, can result in incorrect results of model calculations. Therefore, it is important to refine the FEM based on the collected structural responses, which is termed finite element model updating (FEMU). FEMU, as a part of the structural health monitoring and model updating method, has been developed in recent decades along with many other structural health monitoring methods [5–10].

Model updating methods are generally divided into deterministic and nondeterministic methods that consider errors. Deterministic methods, such as Machine learning and Kalman Filter, do not consider the effects of errors; these approaches have been successfully applied in updating some simple linear and nonlinear models [11–13]. The key to a deterministic model updating method is to modify the model and match the results to the collected response data. However, in the case of complex structures, the difference and error between the FEM and the actual structure may lead to incorrect results; thus, we need to consider the error and difference between the FEM and the actual engineering structure when performing FEMU. Bayesian methods have proven to be a successful non-deterministic approach to model updating considering errors. It has been successfully used for updating many linear and nonlinear models [14–18].

While performing model updating, we always encounter complex Bayesian posterior distribution problems, which are challenging to solve; particularly, when we use high-dimensional parameters, time-history response etc. To solve this problem, Beck [19] proposed the use of Markov chain Monte Carlo (MCMC) sampling for model updating in 2002, and the related methods have been fully developed in recent years [20–25]. In fact, it remains challenging to use these related MCMC methods to perform model updating when we encounter high-dimensional parameter problems. Therefore, it is important to find a way of reducing the number of identified parameters and using more efficient sampling methods for model updating, especially in FEMU.

In previous studies, in order to solve high-dimensional and complex equation problems in model updating based on Bayesian methods, they were more likely to use high-efficiency sampling methods immediately or use complex methods to simplify numerical models, such as model selection methods [26], individually. Different from other sampling methods already been used in model updating based on Bayesian methods. This paper proposed an approach that combined model selection and rapidly sampled from the posterior based on a nested sampling method. The proposed approach not only realized the number of parameter reductions but also accurately estimated the probability distribution of the nonlinear model.

The nested sampling method proposed by Skilling [27] is completely different from the MCMC method, and it has proven to be more than five times more efficient than the MCMC method [28]. J.Speagle [29] created a package called Dynesty to make it easier to implement the complex nested sampling method.

The proposed method in this paper is realized by changing the value of the stop criterion in nested sampling to estimate parameter distribution in two steps. At first, using minimal values of the stop criterion and the number of live points to do initial sampling; analyzing and reducing the number of parameters with convergence curves. Then, using normal values of the stop criterion and the number of live points to do sampling for the simplified Bayesian equation. In this way, nested sampling will successfully be used in simplifying the numerical Bayesian model and probability distribution estimation for FEMU.

This paper realized a 2D RC-frame FEMU and parameter selection using nonlinear time history with nested sampling. Then, the probability distributions of the selected parameters were estimated using resampling.

2. Theory Background

2.1. Bayesian Method Based on Nonlinear Time History

In the Bayesian method, the basic formula is as follows [30]:

$$\textbf{\textit{Posterior}} = \textbf{\textit{Prior}} \times \textbf{\textit{Likelihood/Evidence}} \quad (1)$$

In the equation used in FEMU, **Posterior** is the distribution of the structural parameters, and its specific description is as follows:

$$p\left(\boldsymbol{\theta}|\tilde{\mathbf{d}},\mathbf{M}\right) = \textbf{\textit{Posterior}} \quad (2)$$

where $\boldsymbol{\theta}$ is a set of parameters of the structure and $\tilde{\mathbf{d}}$ is the measured data vector we collected from the structure. In this proposed method, $\tilde{\mathbf{d}}$ is the acceleration time history of the structure and **M** is the given model. **Prior** is established through prior knowledge of the engineering structures, which is usually determined using historical data and engineering experience. In FEMU, **Prior** is generally treated as a uniform distribution [31,32]:

$$\textbf{\textit{Prior}} = \pi(\boldsymbol{\theta}) = c \tag{3}$$

where c is a constant, which is determined using the range of $\boldsymbol{\theta}$.

In Equation (1), **Likelihood** is established using the measured data vector and output data vector of the Numerical Model. For a given model **M**, the error $\boldsymbol{e}$ between the real collected and output from the Finite Element Model differs from time to time:

$$\boldsymbol{e} = \widetilde{y} - y(\boldsymbol{\theta}|\mathbf{M}) \tag{4}$$

where $\boldsymbol{e}$ is the error between the real data $\widetilde{y}$, which is collected from the structure and simulated acceleration; $y(\boldsymbol{\theta}\,|\mathbf{M})$, which is outputted from the numerical model. Generally, the error vector is assumed as a normal distribution whose Mean = 0 and Variance = σ [26]. Therefore, likelihood can be derived as below:

$$\textbf{\textit{Likelihood}} = L(\boldsymbol{\theta}) = exp\left[\sum -\frac{||y(\boldsymbol{\theta}|M) - \widetilde{y}||^2}{2\sigma^2}\right] \tag{5}$$

Evidence is used to perform model selection [33,34]. This paper proposes another approach to performing model selection, where it is generally treated as a constant [26].

From the above formulas, the **Posterior** probability distribution can be derived as:

$$p\left(\boldsymbol{\theta}|\,\widetilde{\mathbf{d}},\mathbf{M}\right) \propto exp\left[\sum -\frac{||y(\boldsymbol{\theta}|\mathbf{M}) - \widetilde{y}||^2}{2\sigma^2}\right] \tag{6}$$

Solving this equation using appropriate sampling methods, we can get the probability distribution of the parameters.

2.2. Nested Sampling

Nested sampling is a sampling method proposed by Skilling, which is often used in astronomy to solve high-dimensional Bayesian problems. There are three basic steps to obtain samples through nested sampling [28]:

1. "Slicing" the posterior into many simpler distributions.
2. Sampling from each of those in turn.
3. Re-combining the results afterwards.

Because step one converts a high-dimensional posterior to a one-dimensional posterior, which makes it easier to solve the high-dimensional problem, such as FEMU, using nested sampling.

In some cases, nested sampling is used to estimate Evidence for Bayesian model selection [32]. However, in the FEMU of Civil Engineering, it would take a significant amount of time to select an appropriate Bayesian model using estimating evidence. This paper proposes another approach to perform model selection (parameters selection) using nested sampling but without using evidence.

2.2.1. Basic Overview of Nested Sampling

For the nested sampling approach, the key is to use one other parameter instead of all the true parameters. The method is shown below:

$$Z = \int_{\Omega_\theta} L(\boldsymbol{\theta})\pi(\boldsymbol{\theta})d\boldsymbol{\theta} = \int_0^1 L(\mathbf{X})d\mathbf{X} \tag{7}$$

$$Z = \int_{\Omega_\theta} c{\cdot}L(\boldsymbol{\theta})d\boldsymbol{\theta} = \int_0^1 L(\mathbf{X})d\mathbf{X} \tag{8}$$

J. Speagle [28] has listed many mathematical approaches to achieve the transformation of Equation (8). The prior distribution integral is defined as:

$$X = X(\lambda) = \int_{L(\boldsymbol{\theta})\geq\lambda} \pi(\boldsymbol{\theta})d\boldsymbol{\theta} = \int_{L(\boldsymbol{\theta})\geq\lambda} cd\boldsymbol{\theta} \tag{9}$$

$$X_i(\lambda_{i-1}) = \int_{L(\boldsymbol{\theta})\geq\lambda_{i-1}} cd\boldsymbol{\theta} \tag{10}$$

where $\lambda_i = L(\boldsymbol{\theta}_i) \propto p\left(\boldsymbol{\theta}|\,\tilde{\mathbf{d}},\mathbf{M}\right)$, and it is the value of likelihood at the i-th time iteration. To calculate the value of $\mathbf{Z}$ faster, the k-th iteration can be simplified as below:

$$Z_k = \sum_{i=1}^{k}(X_{i-1} - X_i)L_i \tag{11}$$

where L_i is the i-th time iteration value of $L(\boldsymbol{\theta})$. The iteration process gives:

$$\lambda_i > \lambda_{i-1} \tag{12}$$

As λ increases, X decreases from 1 to 0:

$$\lim_{i\to\infty} \lambda_i = L_{\max} \tag{13}$$

$$0 < X_N < \ldots < X_2 < X_1 \tag{14}$$

2.2.2. Stopping Criterion

Unlike MCMC, nested sampling sets the number of iteration step to stop the sampling loop, nested sampling stops the loops by controlling the value of Z. The stopping criterion is as follows:

$$\Delta \ln Z_k = \ln(Z_k) - \ln(Z_{k-1}) < \varepsilon \tag{15}$$

To obtain a full distribution curve, that is mostly [26–28], the stopping criterion was set as:

$$\varepsilon = 10^{-3}(K-1) + 10^{-2} \tag{16}$$

If ε is set as an infinite small value in the loop and the number of "live points" is set very little, the iteration steps will increase significantly. From Equations (13) and (14), we can easily derive:

$$\lambda_j \approx \lambda_{j-1} \approx L_{max} \tag{17}$$

where j is the number of the last time iterate step. It is a process similar to the maximum likelihood estimation [35].

2.2.3. Sampling Flow

The details of the nested sampling algorithm flow used in this study are as given in [28]:

1. Draw K "live" points $\boldsymbol{\theta}_K = \{\boldsymbol{\theta}_1, \ldots, \boldsymbol{\theta}_k\}$ from the prior $\pi(\boldsymbol{\theta})$, live points distribution is the same as prior. In this paper, because prior $\pi(\boldsymbol{\theta})$ is a uniform distribution, samples will be selected randomly.
2. Compute the minimum likelihood L_{min} among the current set of live points. Record it as L_1, accumulate $\mathbf{Z}$, and record these K "live" points into samples.
3. Add a new point $\boldsymbol{\theta}'$, which is subject to the constraint $L(\boldsymbol{\theta}') \geq L_{min}$, and replace the point of L_{min} in step 2. Treat the new set of "live" points as $\boldsymbol{\theta}_K^*$.
4. Compute whether it meets the stopping criterion. If it does, end this flow. If it does not, continue this flow.
5. Replace the original $\boldsymbol{\theta}_K$ by $\boldsymbol{\theta}_K^*$ in step 1, and go back to step 1.

2.2.4. Re-Combine Samples

As shown in Equation (17), if we set $\varepsilon = 0.01$ and $\sigma = 0.2$ [26], we can easily obtain the real value of the error's variance σ, similar to the maximum likelihood estimation. Then, set ε as Equation (16) and the previously calculated real value of σ into the nested sampling again. After the completion of all the previous processes, samples of all the parameters will be obtained.

Unlike the MCMC, the samples recorded using nested sampling cannot immediately calculate the probability distribution through the Kernel density estimation [20,36] (KDE). These samples are required to be resampled using resampling methods [30,37]. In this paper, we use the systematic resampling method to reconstruct the samples collected from nested sampling. [30]

2.2.5. Comparison of Nested Sampling and MCMC in Efficiency

As mentioned before, nested sampling has more advantages in solving complex Bayesian problems, which MCMC could not. Moreover, because nested sampling does not have a "Burin-in" process, the efficiency of nested sampling is much more than the efficiency of the MCMC method. We also make a comparison by using nested sampling and MCMC to sample a standard Cauchy distribution. The comparison of different methods in sampling efficiency is analyzed using the sample mean; the result is shown below

Because the target distribution is a standard Cauchy distribution, the mean value of the sample points should be close to zero and belong to the iteration process. This means if the sample mean of the iteration process is stabilized to zero more rapidly, then this method shows that this property is the more efficient method. As shown in Figure 1, the sample mean of nested sampling has already stably converged to zero at the 100-iteration step, but the sample mean of MCMC has not converged to zero even at the 500-iteration step. The result shows that nested sampling is five times more efficient than the MCMC method.

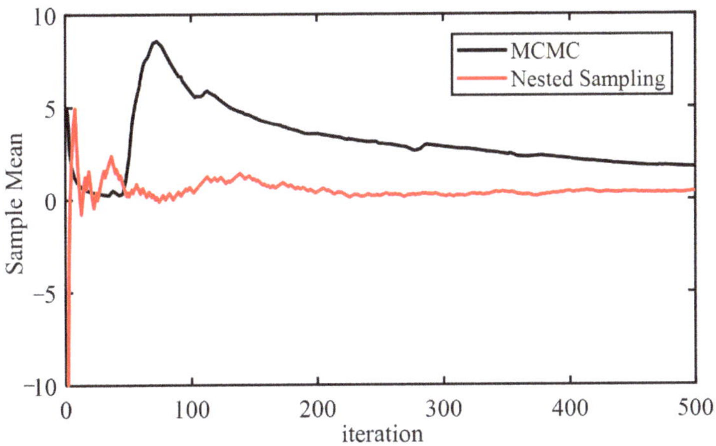

Figure 1. Sample Mean in Iteration Process.

3. Numerical Example

3.1. Two-Dimensional RC-Frame Finite Element Model

For a defined Finite Element Model, if the shape and section are defined and unchangeable, the only property of the structure that can be updated is the material stress–strain model parameters. Unlike steel structures, RC-frame structures have more materials and a more complex stress–strain curve. To prove the advantage of the proposed approach. In this study, a 2D RC-Frame FEM was created in Opensees as an example [38,39].

Figure 2a shows an example of the shape of a 2D RC frame. Figure 2b shows the column and beam sections that built this entire finite element model. Figure 2c,d show how the column and beam sections were meshed by the core and cover concrete in Opensees. The compressive strength of concrete is 30 MPa and the tension strength of steel is 400 MPa. In the column and beam sections, confined and unconfined concrete materials properties follow the Mander stress–strain model [40], as shown in Figure 3.

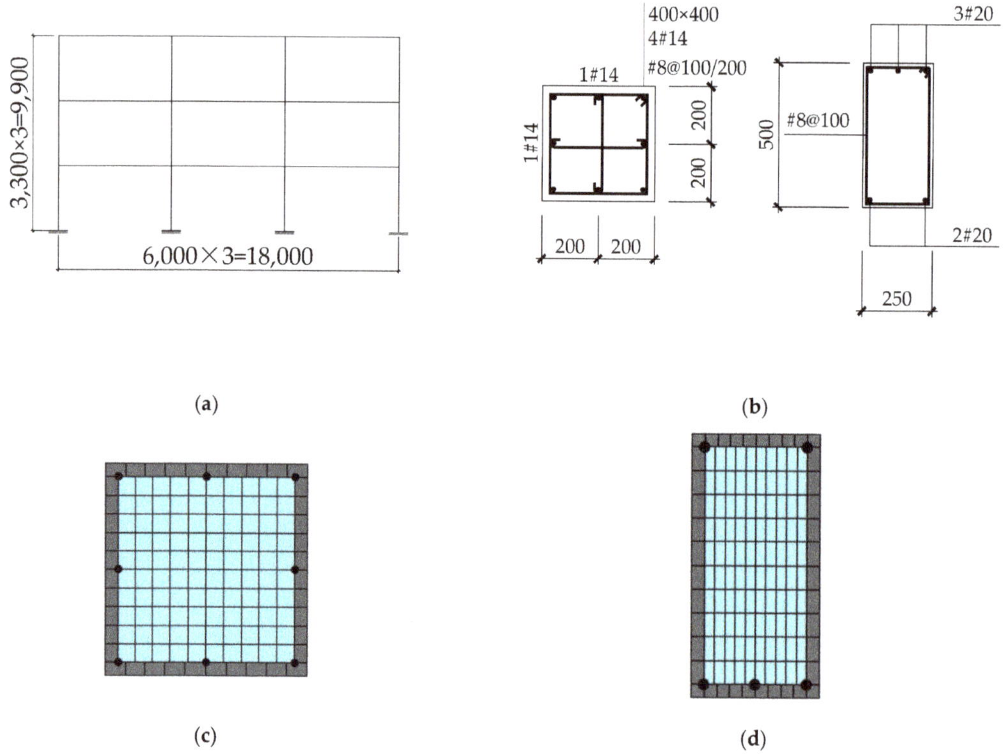

Figure 2. Two-Dimensional RC-Frame Finite Element Structure (unit: mm). (**a**) RC-Frame Structure; (**b**) Column and Beam sections; (**c**) Mesh of Column Section; (**d**) Mesh of Beam Section.

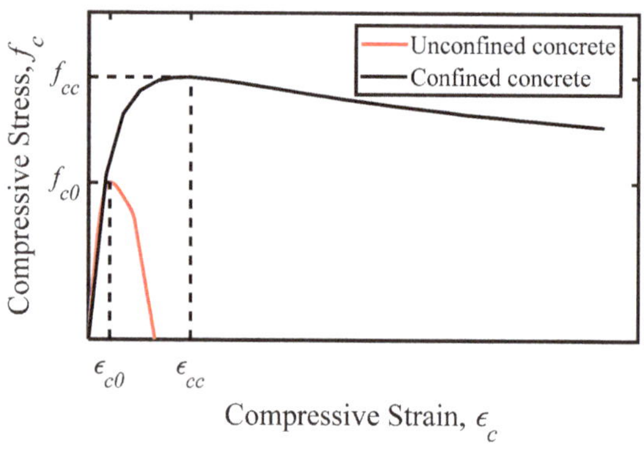

Figure 3. Mander Stress–strain Model.

The relationship between the parameters in Figure 3 is:

$$f_c = \frac{f_{cc} x r}{r - 1 + x^r} \tag{18}$$

In the Equation (18):

$$x = \frac{\varepsilon_c}{\varepsilon_{cc}} \tag{19}$$

$$\varepsilon_{cc} = \varepsilon_{c0}\left[1 + 5\left(\frac{f_{cc}}{f_{c0}} - 1\right)\right] \tag{20}$$

$$f_{cc} = f_{c0}\left[-1.254 + 2.254\sqrt{1 + \frac{7.94 f_1{}'}{f_{c0}}} - 2\frac{f_1{}'}{f_{c0}}\right] \tag{21}$$

$$r = \frac{E_c}{E_c - E_{sec}} \tag{22}$$

$$E_c = e_c\sqrt{f_{c0}} \tag{23}$$

$$E_{sec} = \frac{f_{cc}}{\varepsilon_{cc}} \tag{24}$$

In the above equations, f_{c0} is the unconfined concrete peak stress, ε_{c0} is the unconfined concrete peak strain, f_{cc} and ε_{cc} are the confined concrete peak stress and strain, respectively; E_c is the initial elastic modulus of concrete, E_{sec} is the secant modulus at the peak stress point; and $f_1{}'$ is the effective restraint stress of the hoop reinforcement, which depends on the shape and strength of hoops:

$$f_1{}' = k_e f_1 \tag{25}$$

$$f_1 = \frac{1}{2}\rho_s f_{yh} \tag{26}$$

For rectangular hoops:

$$k_e = \frac{\left(1 - \sum_{i=1}^{n} \frac{\omega_i'}{6 b_c d_c}\right)\left(1 - \frac{s'}{2b_c}\right)\left(1 - \frac{s'}{2d_c}\right)}{1 - \rho_{cc}} \tag{27}$$

In the above equations, ρ_{cc} is the reinforcement rate of the longitudinal reinforcement in the core area of the hoop constraint; s' is the net distance of the hoop; ω_i' is the net distance of the *i*-th longitudinal reinforcement; and b_c and d_c are the distances between the centerlines of the hoops along the two directions of the constraint concrete section.

After performing the above calculations, these parameters are input to concrete 02 (a stress–strain rule in Opensees) to build the confined concrete materials.

For these reinforced concrete sections, steel materials properties follow the Giuffré-Menegotto-Pinto stress–strain model [40] (Steel 02 called in Opensees). Giuffré-nMenegotto-Pinto stress–strain model has four main parameters, f_y, E, R, and *Ratio*, where f_y and E are the yield strength stress and initial elastic modulus of steel. R is the parameter control for the transition from elastic to plastic branches, and Ratio is the value of plastic elastic modulus to the initial elastic modulus. Generally, they are set through a suggested value [41].

Derived from these previous Equations (18)–(27), to build a finite element model, the initial values of vector $\boldsymbol{\theta}$ should be known.

The flowchart in Figure 4 shows how the FEM is built

$$\boldsymbol{\theta} = \left\{f_{c0}, \varepsilon, f_{yh}, \varepsilon_{cu}, e_c, F_y, E, Ratio\right\} \tag{28}$$

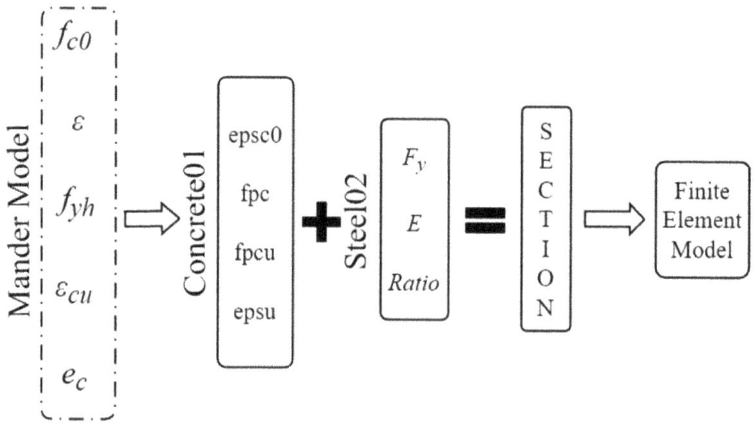

Figure 4. FEM Build Flowchart.

3.2. *Structural Parameter Identification based on Nested Sampling*

As shown in Figure 5, the input seismic recorder is I-ELC180 ground motion (obtained from the 1940 Imperial Valley earthquake at the Array #9 station). To prove the proposed method as much as possible, amplify the I-ELC180 ground motion PGA to 0.1 g, 0.3 g and input them into the FEM separately to create two different damage state cases (Case.1 PGA = 0.1 g, Case.2 PGA = 0.3 g). To simulate the acceleration collection in real engineering structures. The response acceleration data of the structural top story is recorded, and 20% root mean square (rms) [42] white noise is added to the acceleration to simulate collected responses in reality. It is easy to derive that, unlike the real structure, all the errors of FEMU in the numerical examples are oriented from the added white noise.

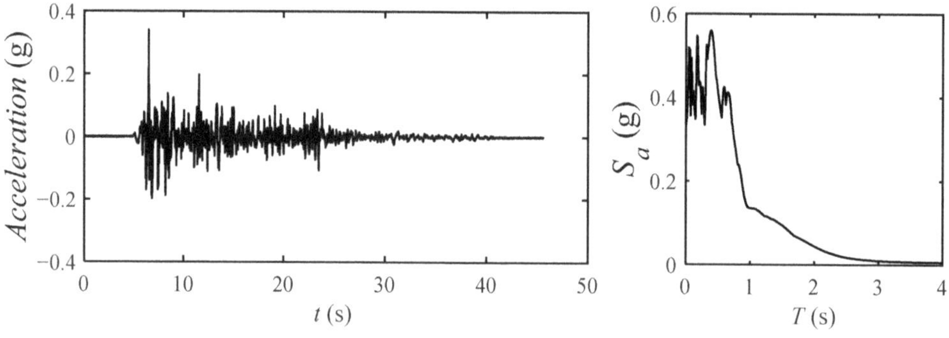

Figure 5. EL-Centro Ground Motion.

3.2.1. Initial Sampling and Structural Parameters Selection

The simulated acceleration in Figure 5 is input to Equation (6), a large enough sample range, covering the parameter distributions range, is set to prior. The primary sampling is started using the method proposed in Section 2. The primary sampling process of different cases is shown in Figure 6.

In Figure 6a, it is evident that the parameters, such as ε_0, f_{c0}, and E_0, have significant convergence in the iterative process, implying that the relationships between ε_0, f_{c0} and E_0 with the collected acceleration are strong. That is because, under the seismic force with PGA = 0.1 g, acceleration is barely collected with nonlinearity. In Figure 6b, it is evident that parameters, such as f_{c0}, ε_0, F_y, and E_0, are apparently gradually converging with the iteration; however, because the acceleration is nonlinear, the samples of f_{yh}, $\varepsilon_{cu}/\varepsilon_0$, e_c, and *Ratio* are different. Parameters such as F_y affect the nonlinear time history. If we still insist on updating these parameters without a significant relationship, it will not only lead to inefficient sampling but may also produce incorrect results. Therefore, in FEMU, appropriate parameters should be selected for updating.

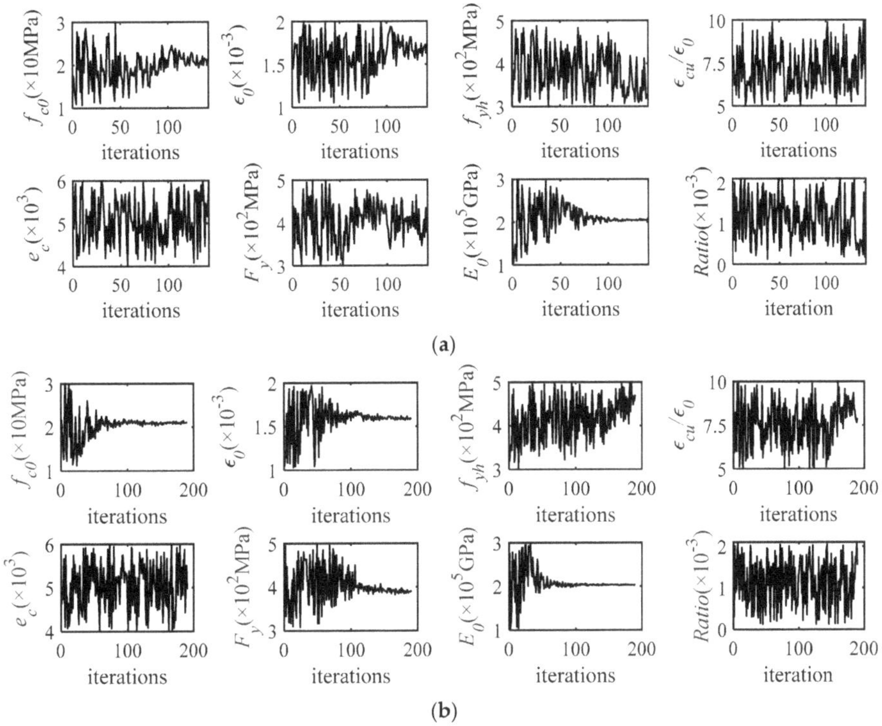

Figure 6. Primary Sampling Process of Different Cases. (**a**) PGA = 0.1 g; (**b**) PGA = 0.3 g.

This paper recommends only sampling the parameters that are closely related to the collected information, such as ε_0, f_{c0}, and E_0 in Case 1. For the other parameters without close relationships, this paper recommends setting them as a suggested value.

The suggested value is defined using the historical data and engineering experience of the structure. For example, in this paper, the concrete which built the FEM was C30, and the Steel was HRB400. These suggested values in the FEM are shown in Table 1.

Table 1. Suggested Value of Different Parameters.

Parameters	Suggested Value
f_{c0} ($\times$10 MPa)	2.01
ε_0 ($\times 10^{-3}$)	1.64
f_{yh} ($\times 10^2$ MPa)	4.00
$\varepsilon_{cu}/\varepsilon_0$	7.00
e_c ($\times 10^3$)	5.00
F_y ($\times 10^2$ MPa)	4.00
E_0 ($\times 10^5$ GPa)	2.06
Ratio ($\times 10^3$)	1.00

As described in Section 2, the primary sample process is similar to MLE. Therefore, we can record the last iteration's acceleration response as y_N, and substitute it into Equation (4) to calculate the standard deviation of error.

As mentioned in Section 2, considering the error as normal distribution and computing standard deviation, the standard deviation is estimated, which is shown below along with the true value.

3.2.2. Secondary Sampling

The selected parameters in the primary sampling are input into Equation (6). The stopping criterion is set as Equation (16). Then, sampling the modified equation as secondary sampling. The proposed approach is used to calculate the distribution of different parameters.

The results are as follows:

As shown in Figure 7, nested sampling successfully estimates the parameter probability density of the parameters and provides probabilistic FEM updating. To prove the reliability of the solution in different cases, this paper converted the error between the most likely estimated parameters in Figure 7 with true values.

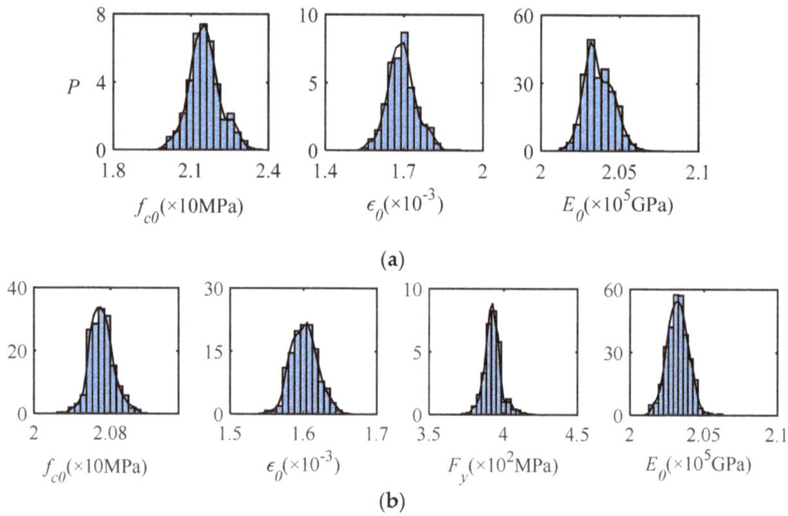

Figure 7. Probability distribution of the parameters for different cases. (**a**) PGA = 0.1 g; (**b**) PGA = 0.3 g.

As shown in Tables 2 and 3, the sampling results have a small error of no more than 6% with these true values, implying that the results of the proposed method are reliable.

Table 2. Comparison of the estimate and true value of the standard deviation.

STD	Estimated	True	Error (%)
σ(PGA = 0.1 g)	0.0695	0.0697	0.29
σ(PGA = 0.3 g)	0.2537	0.2535	0.08

Table 3. Comparison of the estimate and true parameters.

Case	Parameters	Estimate	True	Error (%)
PGA = 0.1 g	$f_{c0}(\times 10\text{ Mpa})$	2.01	2.13	5.97
	$\varepsilon_0(\times 10^{-3})$	1.64	1.66	1.22
	$E_0(\times 10^5\text{ Gpa})$	2.06	2.03	1.46
PGA = 0.3 g	$f_{c0}(\times 10\text{ Mpa})$	2.01	2.07	2.99
	$\varepsilon_0(\times 10^{-3})$	1.64	1.61	1.83
	$F_y(\times 10^2\text{ Mpa})$	4.00	3.95	1.25
	$E_0(\times 10^5\text{ Gpa})$	2.06	2.04	0.97

3.3. Damage State Estimation

The purpose of FEM updating is to accurately estimate the performances and damage state under an earthquake.

Mostly, the damage degree of the RC frame buildings is accomplished through the damage index as other structures. In general, the damage index can estimate the seismic damage degree of structural components and the whole body quantitatively. In recent years, a lot of calculated methods of damage index in long-term research have been proposed for different structures [43–45]. Actually, it is difficult to choose an adequate method to calculate the damage index, which can capture the damage level of structures using a single value. In this paper, because the model in the present research is simple, the seismic damage of the structure was expressed in the form of the Maximum Inter-Story Drift Ratio (MIDR), which is considered a useful damage index to estimate the RC Frame structural seismic damage [46].

In the case of RC frame building using MIDR to estimate the damage degree, Masi [47] proposed a relationship between MIDR and damage degree. As shown in Table 4 below, the damage state can be determined using the structural MIDR.

Table 4. Relation between the MIDR and damage state [48].

MIDR (%)	<0.25	0.25–0.50	0.50–1.00	1.00–1.50	>1.50
Degree of damage	Null	Slight	Moderate	Heavy	Destruction

Similar to the parameters, the output responses of the FEM also have probability density. By inputting the ground motion to the FEM samples recorded in nested sampling and collecting the output MIDR of different models in samples, the distribution of the MIDR can be obtained. According to the probability density of the MIDR, the damage state of a structure's underground motion can be obtained by integrating the probability density of the MIDR.

The MIDR probability densities of the 2D RC Frame Structure in different ground motion cases are shown below.

As shown in Figure 8, the MIDR estimated in Case 1 (PGA = 0.1 g) is in a null-damage-state MIDR range, and for Case 2 (PGA = 0.3 g), the estimated MIDR is in the Moderate-damage-state MIDR range; therefore, the estimated damage state probability is completely 100% Null and Moderate in different cases after the integral.

It's easy for us to obtain the real Inter-Story Drift Ratio of the defined finite element model in different cases by Opensees. The output of the real IDR of the structure in different cases is shown in Figure 9 below.

As shown in Figure 9, the real max inner-story drift ratio and damage state in different cases are easily classified, and the results are shown in Table 5 below.

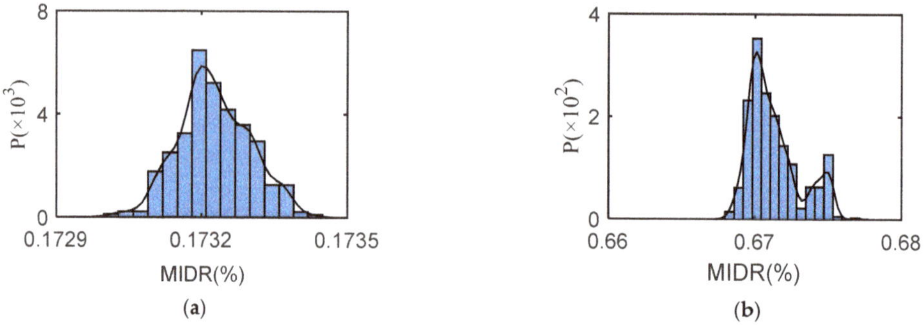

Figure 8. Estimation of MIDR. (**a**) Probability Distribution of MIDR (PGA = 0.1 g); (**b**) Probability Distribution of MIDR (PGA = 0.3 g).

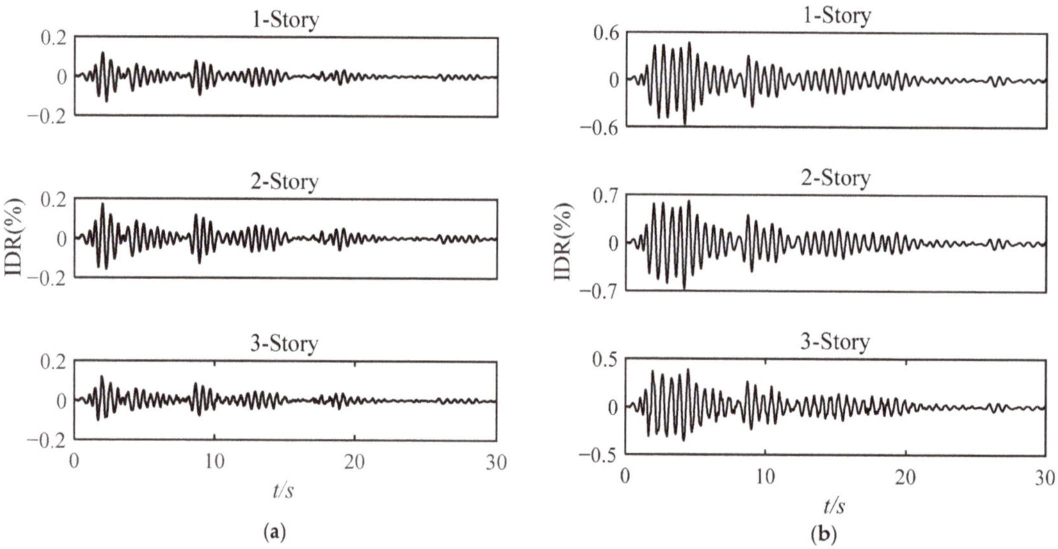

Figure 9. Structural Real Performances in Different Cases. (**a**) IDR of Each Story (PGA = 0.1 g); (**b**) IDR of Each Story (PGA = 0.3 g).

Table 5. Structural Real MIDR and Damage State.

Case	MIDR	Damage State
PGA = 0.1 g	0.17	Null
PGA = 0.3 g	0.67	Moderate

As we can see by comparing Figure 8 and Table 5, the real damage degree of the RC Frame FEM is null and moderate, which is as same as the estimated damage state. This implies that the proposed approach can successfully estimate the degree of structural global damage.

Furthermore, we can also calculate the MIDR probability distribution for each story using these output samples in nested sampling. Because the global damage degree of the structure, in the case of PGA = 0.1 g is null, there is no need to do further local damage degree analysis for the structure. Obviously, in the case of PGA = 0.1 g, the damage degree of each story in the structure is null. In the case of PGA = 0.3 g, the MIDR probability distribution of each story is calculated and shown below.

As shown in Figure 10, In the case of PGA = 0.3 g, we can deduce that the second story of the structure suffered the most serious damage compared with other stories. The estimated damage degree of the first story is the same as the second story, both are moderate damage. The third story of the structure estimate MIDR is in the range of 0.25–0.50, so the estimated damage degree of the third story is slight damage. This is the same as shown in Figure 9b.

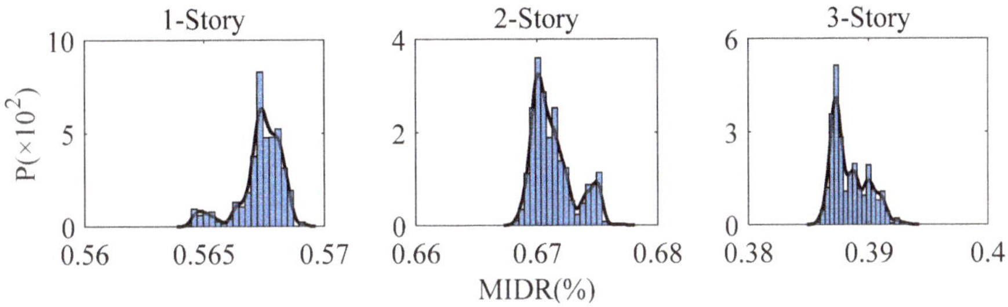

Figure 10. MIDR Probability of Each Story.

Derived from the result, the proposed method can accurately locate the damage location and the damage state. In this result, it had not shown many probabilistic properties in estimating the structural damage state. This is because the error in the numerical example is too low to expand the range of sample distribution. This reason leads to an estimated results probability of 100%.

4. Conclusions

This paper proposed a FEMU approach by using nested sampling with nonlinear time history and its application in structural damage estimation. Different from other common sampling methods, the major advantage of the proposed method is that it can combine model selection with estimate probability distribution by changing the stop criterion.

The results from the example in different damage cases show that the nested sampling is reliable in the number of parameters' reduction and selection without calculated evidence in different situations. This will help simplify high-dimensional space Bayesian problems in the future. Moreover, the results also show that the method could be used to estimate probability distributions by using the nonlinear time history and nonlinear models such as FEM, which is more advanced than other methods that are based on the Markov chain Monte Carlo approach.

We have also presented a method for damage probability estimation by using the samples created using nested sampling. It provided a new method to estimate damage state and location in probability.

The reliability of the proposed approach has been demonstrated by the numerical example in different cases. Because a large number of algorithms perform well in numerical examples but cannot be used in engineering, a real engineering structural example is still required to prove its application in the future. Future work will consider using the proposed method in the shaking table test and more complex structures, which have more parameters to estimate.

Author Contributions: Methodology, K.W.; Software, K.W. and Y.Y.; Investigation, K.W.; Writing—original draft, K.W.; Writing—review & editing, Y.K.; Funding acquisition, Y.K. All authors have read and agreed to the published version of the manuscript.

Funding: This research received no external funding.

Data Availability Statement: https://github.com/WanghealthKunyang/Bayesian-Samples-of-Nested-Sampling/issues.

Conflicts of Interest: The authors declare no conflict of interest.

References

1. Chopra, A.K.; Goel, R.K. A modal pushover analysis procedure for estimating seismic demands for buildings. *Earthq. Eng. Struct. Dyn.* **2002**, *31*, 561–582. [CrossRef]
2. Vamvatsikos, D.; Cornell, C.A. Applied incremental dynamic analysis. *Earthq. Spectra* **2004**, *20*, 523–553. [CrossRef]
3. Mander, J.B.; Dhakal, R.P.; Mashiko, N.; Solberg, K.M. Incremental dynamic analysis applied to seismic financial risk assessment of bridges. *Eng. Struct.* **2007**, *29*, 2662–2672. [CrossRef]
4. Farrar, C.R.; Duffey, T.A. Bridge modal properties using simplified finite element analysis. *J. Bridge Eng.* **1998**, *3*, 38–46. [CrossRef]
5. Friswell, M.; Mottershead, J.E. *Finite Element Model Updating in Structural Dynamics*; Springer Science & Business Media: BerlinHeidelberg, Germany, 1995; p. 5.
6. Imregun, M.; Visser, W.J.; Ewins, D.J. Finite element model updating using frequency response function data: I. Theory and initial investigation. *Mech. Syst. Signal Process.* **1995**, *9*, 187–202. [CrossRef]
7. Jaishi, B.; Ren, W.X. Structural finite element model updating using ambient vibration test results. *J. Struct. Eng.* **2005**, *131*, 617–628. [CrossRef]
8. Marwala, T. *Finite-Element-Model Updating Using Computational Intelligence Techniques: Applications to Structural Dynamics*; Springer: London, UK, 2010; pp. 225–231.
9. Ren, W.X.; Chen, H.B. Finite element model updating in structural dynamics by using the response surface method. *Eng. Struct.* **2010**, *32*, 2455–2465. [CrossRef]
10. Abedin, M.; y Basalo, F.J.D.C.; Kiani, N.; Mehrabi, A.B.; Nanni, A. Bridge load testing and damage evaluation using model updating method. *Eng. Struct.* **2022**, *252*, 113648. [CrossRef]
11. Li, W.; Yong, H.; Zikai, X. Machine Learning-Based Probabilistic Seismic Demand Model of Continuous Girder Bridges. *Adv. Civ. Eng.* **2022**, *2022*, 3867782. [CrossRef]
12. Atalla, M.J.; Inman, D.J. On model updating using neural networks. *Mech. Syst. Signal Process.* **1998**, *12*, 135–161. [CrossRef]
13. Levin, R.I.; Lieven, N.A.J. Dynamic finite element model updating using neural networks. *J. Sound Vib.* **1998**, *210*, 593–607. [CrossRef]
14. Astroza, R.; Alessandri, A.; Conte, J.P. A dual adaptive filtering approach for nonlinear finite element model updating accounting for modeling uncertainty. *Mech. Syst. Signal Process.* **2019**, *115*, 782–800. [CrossRef]
15. Behmanesh, I.; Moaveni, B.; Lombaert, G.; Papadimitriou, C. Hierarchical Bayesian model updating for structural identification. *Mech. Syst. Signal Process.* **2015**, *64*, 360–376. [CrossRef]
16. Beck, J.L.; Katafygiotis, L.S. Updating models and their uncertainties. I: Bayesian statistical framework. *J. Eng. Mech.* **1998**, *124*, 455–461. [CrossRef]
17. Li, X.; Kurata, M. Probabilistic updating of fishbone model for assessing seismic damage to beam–column connections in steel moment-resisting frames. *Comput. Aided Civ. Infrastruct. Eng.* **2019**, *34*, 790–805. [CrossRef]
18. Lam, H.F.; Yang, J.H.; Beck, J.L. Bayesian operational modal analysis and assessment of a full-scale coupled structural system using the Bayes-Mode-ID method. *Eng. Struct.* **2019**, *186*, 183–202. [CrossRef]
19. Jensen, H.A.; Millas, E.; Kusanovic, D.; Papadimitriou, C. Model-reduction techniques for Bayesian finite element model updating using dynamic response data. *Comput. Methods Appl. Mech. Eng.* **2014**, *279*, 301–324. [CrossRef]
20. Beck, J.L.; Siu-Kui, A. Bayesian Updating of Structural Models and Reliability using Markov Chain Monte Carlo Simulation. *J. Eng. Mech.* **2002**, *128*, 380–391. [CrossRef]
21. Ching, J.; Muto, M.; Beck, J.L. Bayesian Linear Structural Model Updating using Gibbs Sampler with Modal Data. In Proceedings of the 9th International Conference on Structural Safety and Reliability, Rome, Italy, 19–23 June 2005; Millpress: Rotterdam, The Netherlands, 2005; pp. 2609–2616.
22. Cheung, S.H.; Beck, J.L. Bayesian Model Updating Using Hybrid Monte Carlo Simulation with Application to Structural Dynamic Models with Many Uncertain Parameters. *J. Eng. Mech.* **2009**, *135*, 243–255. [CrossRef]
23. Boulkaibet, I.; Mthembu, L.; Marwala, T.; Friswell, M.I.; Adhikari, S. Finite element model updating using the separable shadow hybrid Monte Carlo technique. *Top. Modal Anal. II* **2014**, *8*, 267–275.
24. Boulkaibet, I.; Mthembu, L.; Marwala, T.; Friswell, M.I.; Adhikari, S. Finite element model updating using the shadow hybrid Monte Carlo technique. *Mech. Syst. Signal Process.* **2015**, *52–53*, 115–132. [CrossRef]
25. Lam, H.F.; Yang, J.H.; Au, S.K. Bayesian model updating of a coupled-slab system using field test data utilizing an enhanced Markov chain Monte Carlo simulation algorithm. *Eng. Struct.* **2015**, *102*, 144–155. [CrossRef]
26. Zhao, Y.; Gong, M.; Zuo, Z.; Gao, Y. Bayesian estimation approach based on modified SCAM algorithm and its application in structural damage identification. *Struct. Control Health Monit.* **2021**, *28*, e2654. [CrossRef]
27. Yuen, K.-V. *Bayesian Methods for Structural Dynamics and Civil Engineering*; John Wiley & Sons: New York, NY, USA, 2010.
28. Skilling, J. Nested sampling. AIP Conference Proceedings. *Am. Inst. Phys.* **2004**, *735*, 395–405.
29. Keeton, C.R. On statistical uncertainty in nested sampling. *Mon. Not. R. Astron. Soc.* **2011**, *414*, 1418–1426. [CrossRef]
30. Speagle, J.S. Dynesty: A dynamic nested sampling package for estimating Bayesian posteriors and evidences. *Mon. Not. R. Astron. Soc.* **2020**, *493*, 3132–3158. [CrossRef]
31. Jaynes, E.T. Prior probabilities. *IEEE Trans. Syst. Sci. Cybern.* **1968**, *4*, 227–241. [CrossRef]

32. Xin, Y.; Hao, H.; Li, J.; Wang, Z.C.; Wan, H.P.; Ren, W.X. Bayesian based nonlinear model updating using instantaneous characteristics of structural dynamic responses. *Eng. Struct.* **2019**, *183*, 459–474. [CrossRef]
33. Beck, J.L.; Yuen, K.V. Model selection using response measurements: Bayesian probabilistic approach. *J. Eng. Mech.* **2004**, *130*, 192–203. [CrossRef]
34. Cao, T.; Zeng, X.; Wu, J. An arithmetic example study of nested sampling algorithm for groundwater model evaluation. *Hydrogeol. Eng. Geol.* **2017**, *44*, 69–76. (In Chinese)
35. Myung, I.J. Tutorial on maximum likelihood estimation. *J. Math. Psychol.* **2003**, *47*, 90–100. [CrossRef]
36. Terrell, G.R.; Scott, D.W. Variable kernel density estimation. *Ann. Stat.* **1992**, *20*, 1236–1265. [CrossRef]
37. Douc, R.; Cappé, O. Comparison of resampling schemes for particle filtering Ispa 2005. In Proceedings of the 4th International Symposium on Image and Signal Processing and Analysis, Zagreb, Croatia, 15–17 September 2005; IEEE: Piscataway, NJ, USA, 2005; pp. 64–69.
38. McKenna, F. OpenSees: A framework for earthquake engineering simulation. *Comput. Sci. Eng.* **2011**, *13*, 58–66. [CrossRef]
39. Yang, Y.; Yang, Y.Q.; Zhao, Y.N. Comparative analysis of seismic capacity of RC frame structures based on different damage models. *Earthq. Eng. Eng. Vib.* **2020**, *40*, 118–126. (In Chinese)
40. Mander, J.B.; Priestley, M.J.N.; Park, R. Theoretical stress-strain model for confined concrete. *J. Struct. Eng.* **1988**, *114*, 1804–1826. [CrossRef]
41. Filippou, F.C.; Popov, E.P.; Bertero, V.V. *Effects of Bond Deterioration on Hysteretic Behavior of Reinforced Concrete Joints*; Report EERC 83–19; Earthquake Engineering Research Center, University of California: Berkeley, CA, USA, 1983.
42. Gasparini, D.A.; DebChaudhury, A. Dynamic response to nonstationary nonwhite excitation. *J. Eng. Mech. Div.* **1980**, *106*, 1233–1248. [CrossRef]
43. Chen, G.; Wang, Z.; Zuo, X.; Du, X.; Gao, H. Shaking table test on the seismic failure characteristics of a subway station structure on liquefiable ground. *Earthq. Eng. Struct. Dyn.* **2013**, *42*, 1489–1507. [CrossRef]
44. Zhang, S.; Gaohui, W. Effects of near-fault and far-fault ground motions on nonlinear dynamic response and seismic damage of concrete gravity dams. *Soil Dyn. Earthq. Eng.* **2013**, *53*, 217–229. [CrossRef]
45. Zhong, Z.; Wang, Z.; Zhao, M.; Du, X. Structural damage assessment of mountain tunnels in fault fracture zone subjected to multiple strike-slip fault movement. *Tunn. Undergr. Space Technol.* **2020**, *104*, 103527. [CrossRef]
46. Naeim, F. *The Seismic Design Handbook*, 2nd ed.; Kluwer Academic: Boston, MA, USA, 2011.
47. Masi, A.; Vona, M.; Mucciarelli, M. Selection of natural and synthetic accelero- grams for seismic vulnerability studies on reinforced concrete frames. *J. Struct. Eng.* **2011**, *137*, 367–378. [CrossRef]
48. Morfidis, K.; Kostinakis, K. Seismic parameters' combinations for the optimum prediction of the damage state of R/C buildings using neural networks. *Adv. Eng. Softw.* **2017**, *106*, 1–16. [CrossRef]

buildings

MDPI

Article

Bayesian Model-Updating Implementation in a Five-Story Building

Oscar D. Hurtado [1,*], Albert R. Ortiz [1], Daniel Gomez [1] and Rodrigo Astroza [2]

1 School of Civil Engineering and Geomatics, Universidad del Valle, Cali 760032, Colombia; albert.ortiz@correounivalle.edu.co (A.R.O.); daniel.gomez@correounivalle.edu.co (D.G.)
2 Facultad de Ingeniería y Ciencias Aplicadas, Universidad de los Andes, Santiago 7620001, Chile; rastroza@miuandes.cl
* Correspondence: oscar.david.hurtado@correounivalle.edu.co

Abstract: Simplifications and theoretical assumptions are usually incorporated into the numerical modeling of structures. However, these assumptions may reduce the accuracy of the simulation results. This problem has led to the development of model-updating techniques to minimize the error between the experimental response and the modeled structure by updating its parameters based on the observed data. Structural numerical models are typically constructed using a deterministic approach, whereby a single best-estimated value of each structural parameter is obtained. However, structural models are often complex and involve many uncertain variables, where a unique solution that captures all the variability is not possible. Updating techniques using Bayesian Inference (BI) have been developed to quantify parametric uncertainty in analytical models. This paper presents the implementation of the BI in the parametric updating of a five-story building model and the quantification of its associated uncertainty. The Bayesian framework is implemented to update the model parameters and calculate the covariance matrix of the output parameters based on the experimental information provided by modal frequencies and mode shapes. The main advantage of this approach is that the uncertainty in the experimental data is considered by defining the likelihood function as a multivariate normal distribution, leading to a better representation of the actual building behavior. The results showed that this Bayesian model-updating approach effectively allows a statistically rigorous update of the model parameters, characterizing the uncertainty and increasing confidence in the model's predictions, which is particularly useful in engineering applications where model accuracy is critical.

Keywords: modal analysis; Bayesian inference; parametric uncertainty; probabilistic model updating; full-scale testing; finite element modeling

1. Introduction

Commonly, the mechanical behavior of infrastructure systems is determined through analytical methods, experimental testing, and field observations. In structural engineering, modeling uses mathematical and computational techniques to simulate the mechanical behavior under different load conditions. Deterministic mathematical models are widely adopted to accurately represent the behavior of real structures while avoiding excessive computational costs. These models are favored for their ease of manipulation; however, their accuracy in representing the real behavior of built structures is often limited [1,2]. Therefore, the model-updating methodology attempts to decrease the gap between the constructed structural system and its structural model's behavior [3].

The Finite Element Method (FEM) has been extensively applied in structural engineering [4]. The traditional approach to estimating the parameters of an FE model is based on a unique result for a given set of inputs, no matter how many times it is re-calculated. This deterministic modeling approach limits the information obtained from experimental data, since the uncertainty associated with each structural parameter is not included in the analysis [5]. The uncertainty might arise from a combination of factors, including lack of knowledge, material variability, load changes, measurement errors, and modeling assumptions [6–8]. Over the years, significant contributions have been made towards incorporating uncertainty in structural models and their parameters, aiming to update building code models [9].

The uncertainty may be accounted for using probabilistic methods when statistical distributions are incorporated, providing more realistic estimations of the structural response. Inferential statistics proposes an alternative solution to the deterministic approach by generating models where the results are given probabilistically. Thus, it is possible to know how accurate the model outputs are, draw conclusions based on percentages, and make decisions to develop safer designs [10]. Parametric updating of structural models often implements the Bayesian over other probabilistic approaches due to the incorporation of prior knowledge. This technique provides more accurate parameter estimation by accounting for uncertainty and updating beliefs as new information is obtained. Additionally, it provides a systematic framework for model selection and is able to handle complex and nonlinear models. Thus, valuable experimental information should be obtained to feed the models to be updated. Many structural identification methods typically rely on a deterministic approach, which involves experimental measurement of properties and estimating unknown or non-measurable model parameters by minimizing an error function based on these measurements. With this deterministic approach, the information obtained from tests is limited to the measurements taken and their accuracy, and there is no consideration of uncertainty or variability in the measurements. On the other hand, probabilistic approaches such as Bayesian updating take into account uncertainty in the model parameters and incorporate prior knowledge, which may lead to more accurate and robust parameter estimation.

Ambient vibration testing is a useful and non-destructive method for measuring the response of a structure, allowing the estimation of modal parameters. The Bayesian Inference (BI) approach proposes a methodology to quantify the uncertainty associated with the model and its input parameters. The information obtained from experimental tests is augmented using a Bayesian-approached model to quantify the uncertainty in the data [11]. However, implementing Bayesian model updating with FE models is still challenging. Updating the model parameters using Bayesian methods requires evaluating functions multiple times, which might be computationally expensive when dealing with large FE models. Nonetheless, the development of efficient computational methods and algorithms continues to advance the application of Bayesian model updating in the field of structural engineering [12,13].

This paper aims to implement the BI framework for the parametric updating of a structural FE model based on experimental modal properties. The methodology proposes quantifying the model's uncertainty taking into account different kinds of observations using a multivariate normal likelihood function. The reference structure for implementing the developed method is a full-scale five-story reinforced concrete building. Specialized software for performing Bayesian updating of a complex model is utilized. Furthermore, the software allows cloud-based computational solutions and other techniques to make the updating process of this computationally expensive model feasible and valuable.

This specific research holds a remarkable significance in the field due to the incorporation of two innovative approaches: the utilization of test data and the implementation of iterative calculation of the covariance matrix. The research conducted by Loyola (2018) [14] highlights that the architecture, engineering, construction, and operation (AECO) industry lags behind in harnessing the potential of big data compared to other sectors. With limited examples and a lack of practical application, the use of data and Bayesian models in this study brings a significant contribution to decision-making processes in building design. Furthermore, the comparison between the iterative calculation of the covariance matrix and the use of an identity matrix as covariance matrix in the paper represents a significant disclosure within the field, contributing valuable insights and paving the way for further developments.

The organization of this paper is as follows. The Introduction provides an overview of the research topic and an explanation of its significance. Section 2 presents the Bayesian inference framework. Section 3 describes the experimental program, including the characteristics of the reference structure and data collection. In Section 4, the implementation of Bayesian inference in the parametric updating of the reference structure model using identified modal properties is presented. The results are presented in Section 5, which involves assessing the behavior of various parametric uncertainties and evaluating the predictive performance of the updated structural model. Finally, the main findings are discussed and the potential implications of the implementation of this framework are considered in the Conclusions.

2. BI in Finite Element Model Updating (FEMU)

A comprehensive understanding of the BI methodology employed in this study requires specific concepts explained in [15–17]. Uncertainty in modeling complex systems may be classified as epistemic or aleatory. Epistemic uncertainty arises from a lack of knowledge and might be reduced through improved data and modeling, while aleatory uncertainty results from inherent variability or randomness and cannot be reduced. Therefore, it is the epistemic error that we aim to reduce through the implementation of BI in structural models. BI is based on Bayes' theorem [18], which is defined as:

$$P(A \mid B) = \frac{P(B \mid A) \cdot P(A)}{P(B)} \tag{1}$$

where $P(A|B)$ is the conditional probability of event A given event B has occurred, $P(B|A)$ is the conditional probability of event B given event A has occurred, $P(A)$ is the prior probability of event A, $P(B)$ is the prior probability of event B.

Bayes' theorem is based on the fundamental principle where laws of probability guide rational belief. In BI, the probabilities represent degrees of belief rather than frequencies or long-term averages. This methodology provides a framework for updating beliefs as more experimental evidence or information becomes available, starting with prior beliefs and using Bayes' rule to derive a posterior probability distribution. Therefore, BI provides a flexible methodology for complex modeling through data and making inferences [19]. Bayesian methods have been widely used in engineering for model updating and uncertainty quantification due to their ability to handle various challenges, such as missing data, hierarchical structures, nonlinear relationships, and model uncertainty [20–23]. In particular, BI provides a flexible and robust approach to statistical modeling, making it ideal for complex engineering problems. The most common form of Bayes' theorem in this context is defined as:

$$P(\theta|D, M_j) = \frac{P(D|\theta, M_j) \cdot P(\theta|M_j)}{P(D)} \tag{2}$$

where $P(\theta|D, M_j)$ is the posterior Probability Density Function (PDF) of the parameter vector θ, for the model M_j, given the evidence $P(D)$. $P(\theta|M_j)$ is the a priori probability distribution, representing the prior belief of the parameters. $P(D|\theta, M_j)$ is the conditional probability where the evidence D is fulfilled given the θ parameters evaluated in the M_j model. The marginal likelihood $P(D)$, is a normalizing constant that ensures an area equal to one when the posterior distribution is integrated.

The marginal likelihood in BI becomes more complex as the number of model parameters and data points increase, making it computationally expensive. Furthermore, the complexity of the likelihood function employed in the model can also impact the marginal likelihood. Since it does not affect the shape of the posterior distribution, it is common to work with the unnormalized posterior distribution which is proportional to the product of the prior distribution and the likelihood function, as shown in Zhang and Feissel (2011) [24]:

$$P(\theta|D, M_j) \propto P(D|\theta, M_j) \cdot P(\theta|M_j). \tag{3}$$

2.1. Likelihood Function

In BI, the likelihood function is a critical component representing the probability of observing a particular data set given a group of model parameters. To formulate the likelihood function, it is necessary to assume a probabilistic relationship between the model predictions and the experimental data, accounting for both aleatory and epistemic uncertainty, as noted by Argyris et al. (2020) [25]. Depending on the characteristics of the data, different prediction error equations might be used for each type of parameter. The likelihood function is a multivariate normal distribution to account for multiple observations of the same structure and better characterize the associated errors beyond the updated parameters. It is well suited for this purpose since it may represent multiple outputs, each with its own mean and covariance within a single distribution [26,27]. Thus, the likelihood function $\phi(x)$ is defined as follows:

$$\phi(x) = \left(\frac{1}{2\pi}\right)^{p/2} |\Sigma|^{-1/2} e^{\left(-\frac{1}{2}(x-\mu)^{\intercal}\Sigma^{-1}W(x-\mu)\right)} \tag{4}$$

where μ and x are the experimental and model data, respectively. The likelihood function takes the maximum value when the vector x equals the vector μ. p is the dimension of the normal distribution. The weight matrix W assigns different values to the obtained information from the tests and designates relative importance to different parts of the model or measurement data. The weight matrix may be used to reflect confidence levels in different parts of the data or different accuracy levels in measurements. Thus, the Bayesian model-updating approach may provide a more informed estimation of the model parameters, considering the weight matrix's measurements and information. The covariance matrix Σ is also a fundamental component of the multivariate normal distribution used as the likelihood function in BI. It characterizes the degree of variability and the relationship between the different variables in the dataset. In Bayesian model-updating problems, the covariance matrix plays a fundamental role in capturing the uncertainty associated with the measurements and the model parameters. A precise and accurate covariance matrix estimation is necessary to obtain reliable results and make informed decisions based on the posterior distribution. It is defined as

$$\Sigma = \begin{bmatrix} \sigma_1^2 & \sigma_{1,2} & \cdots & \sigma_{1,n} \\ \sigma_{2,1} & \sigma_2^2 & \cdots & \sigma_{2,n} \\ \vdots & \vdots & \ddots & \vdots \\ \sigma_{n,1} & \sigma_{n,2} & \cdots & \sigma_n^2 \end{bmatrix}_{n \times n} \tag{5}$$

where the diagonal elements of the covariance matrix σ_i^2 represent the variances of each variable i in the dataset, while the off-diagonal elements of the covariance matrix $\sigma_{i,j}$ represent the covariances between pairs of variables i and j in the dataset. The size n is equal to the number of variables in the dataset. $|\Sigma|$ is the determinant of the covariance matrix, and Σ^{-1} is the inverse of the covariance matrix called the precision matrix, which is also updated.

Identity Matrix as a Covariance Matrix

This paper presents two approaches for incorporating the covariance matrix into the likelihood function. The first approach involves computing the covariance matrix for each iteration, using a Numpy function called `numpy.cov()`; accessed on 1 April 2023, which may result in a more precise estimation of the posterior distribution by considering the unique characteristics of the data. The other approach involves using an identity matrix as the covariance matrix.

Assuming an identity matrix as the covariance matrix is a common practice when there is a lack of prior knowledge about the correlations between model output parameters or when there is no reason to believe that these correlations are non-zero [28,29]. By assuming uncorrelated parameters with equal uncertainty, the updating algorithm is simplified, and the identity matrix may also serve as a starting point for more complex covariance structures if needed [30]. However, this approach assumes equal variances of observations across all parameters, which is not always true [31]. This may lead to suboptimal or biased parameter estimates, and in such scenarios, a more intricate covariance matrix may be necessary to incorporate the varying variances of observations.

When deciding whether to compute the covariance matrix for each sample or use an identity matrix, multiple factors must be taken into consideration. In situations with a small sample size, where the variability of the experimental data is low, it is generally suggested to compute the covariance matrix for each sample to obtain a more precise estimate of the posterior distribution [32]. However, in cases where using an identity matrix may still provide reasonable results offering computational efficiency, it is a commonly used alternative [33]. Ultimately, the choice between these or another options depends on the specific needs of the problem being tackled, such as the desired precision of the posterior distribution and the available computational resources.

3. Experimental Program

3.1. Test Building

The experimental structure consists of a full-scale five-story concrete building tested at the University of California, San Diego [34]. Based on the coordinate system illustrated in Figure 1a, the structure consisted of three column and wall axes aligned in the longitudinal direction (Y direction), and two column and wall axes oriented in the transverse direction (X direction). The reinforced concrete was poured in situ with a floor area of 6.6 × 11 m and a mezzanine height of 4.27 m, resulting in a total height of 21.34 m from the top of the foundation (zero elevation) to the deck [34]. Figure 1 shows the bare and built structure.

(**a**) (**b**)

Figure 1. Five-story test building. (**a**) Bare structure without non-structural elements, (**b**) Complete structure.

The building had six identical reinforced concrete columns (660 × 460 mm) with a longitudinal steel ratio (ρ) of 1.42% and a precast welded mesh of 12 mm to 102 mm tie rods as transverse reinforcement. Each reinforced concrete slab, 0.2 m thick, was designed with two-way reinforcement at the top and bottom, and incorporated perforations to enable the installation of various building services, including plumbing, electrical wiring, fire sprinklers, sensors, and camera cables. Additionally, the slabs were configured with two large openings to accommodate an elevator and a stairwell. Two walls placed in the Y-direction bound the elevator span and are made of reinforced concrete with a 0.15 m thick reinforcing mesh. These walls also provide additional transversal and torsional stiffness; thus, one of Y-axis spans of the building was transversely braced at all floor levels with 32 mm diameter steel rods anchored to the concrete slabs above and below. A summary of the compressive strength and elastic modulus of concrete for a selected number of cylinders is presented in Table 1 [34].

Table 1. Summary results for compressive concrete cylinder tests before seismic testing.

Element	Average f'_c [MPa]	Average E_c [GPa]
Columns and walls	57.2	32.6
Slabs and beams	51.7	33.1

3.2. Test and Data Collection

A series of dynamic tests, which included white noise, pulse, and earthquake motions, were performed on the structure's base in the horizontal X direction using the shake table. In addition, Ambient Vibration Tests (AVT) were conducted during the construction of the test building and during the base excitation testing phases to identify the system. The test building was subjected to 13 earthquake motion tests, 31 low amplitude white noise base excitation tests, and 45 pulse-like base excitation tests using the NEES@UCSD shake table [34].

The natural frequencies and damping ratios associated with ten structural mode shapes were identified by Pantoli et al. (2016) [35] using AVT data. Two output-only system identification methods, Data-Driven Stochastic Subspace Identification (SSI-DATA) and Natural Excitation Technique combined with Eigensystem Realization Algorithm (NExT-ERA), were utilized for estimating the modal properties of the building. Both methods assume broad-band and stationary excitation for the AVT data [36]. Among the ten identified mode shapes, the first three modes and associated frequencies from SSI-DATA were selected in this study for the Bayesian updating of the parameters. The identified frequencies for the mode shapes are 1.91 Hz, 1.89 Hz, and 2.66 Hz, corresponding to the first, second, and third modes, respectively. These mode shapes were classified as longitudinal, transverse+torsional, and torsional, respectively.

Astroza et al. (2016) [36] utilized the Modal Assurance Criterion (MAC) value to determine that the identified mode shapes from each method were not significantly different, implying that any alternative method would not have substantially altered the results. The identified modal properties provided crucial input for further model updating and accurate structural assessment of the building.

4. Methodology

The general methodology for implementing the Bayesian parametric update in the model of the five-story building is illustrated in Figure 2. This diagram provides a comprehensive overview of the steps involved in the methodology, which will be further elaborated upon in the subsequent subsections, with a particular emphasis on the critical steps.

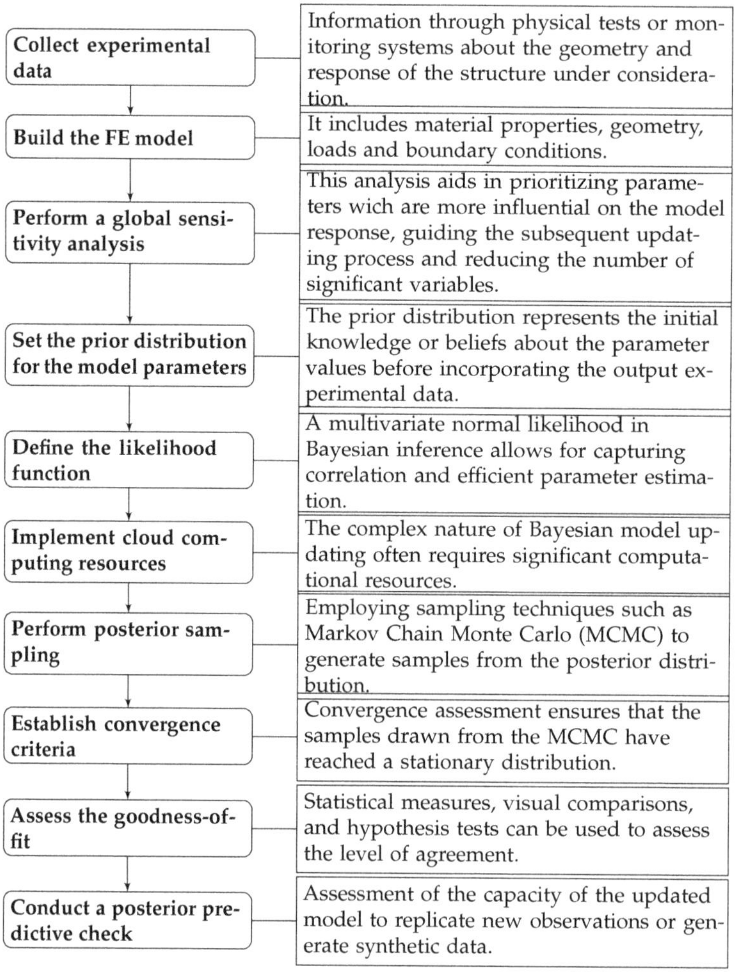

Figure 2. Workflow of the implemented methodology for Bayesian parametric update in the model of the five-story building.

4.1. Modeling of the Structure

The model presented in this study is built upon the foundation of a previously developed model by Gutierrez (2020) [37]. The building is modeled in Opensees using a FE program in Python called Openseespy. Figure 3 shows a graphical representation of the 3D FE model developed in Openseespy. The structure is first discretized into elements and nodes. Then, the elements' shape, type, boundary conditions, and dimensions are defined. The model consists of 20 input parameters, such as the moduli of elasticity E for beams, columns, slabs, and walls, and two output parameters, which are the modal coordinates and frequencies of the first three mode shapes. The selection of these parameters is made based on the availability of experimental data during the course of this study. This choice is guided by the aim of utilizing the most pertinent and credible information from the experimental test. Beams and columns are modeled using the *elasticBeamColumn* element. The *ShellMITC4* element is used for slabs and walls, which are joined with approximately 750 nodes. The moduli of elasticity are defined as variable parameters in the model. The beam elements are separated into two groups on each floor: beams on the X-axis and Y-axis. Each group has the same E per level and a cross-sectional area of 0.22 m^2. The slabs of each floor are discretized into 184 elements, where each shell element has an elastic section with a depth of 0.20 m. The 0.19 m side walls are modeled with 80 shell elements, reaching from the base of the building to the roof. The truss members are modeled with a fixed E using the *UniaxialMaterial* element, where steel properties are implemented as an elastic material. These truss elements are included in the model as fictitious members with strain-rate effects, making them suitable to include damping in the system [34].

The shear modulus (G) for each of the mentioned structural elements is calculated indirectly using a function that incorporates Poisson's ratio (ν) and the modulus of elasticity (E) (See Equation (6)). The value of ν is set as constant and defined as 0.16 following the work of Pearson (1999) [38].

$$G = \frac{E}{2(1+\nu)} \tag{6}$$

A remarkable aspect to consider is that the utilized calibration model may not have fully accounted for the presence of stiff zones in the joints. In practice, joints introduce stiffness to the structural system due to the connections and interactions between elements. However, in the calibration model, these stiffening effects might not have been explicitly included, leading to an underestimation of the overall stiffness.

In order to address the absence of stiff zones in the joints, the calibration process indirectly increased the moduli of elasticity for beams and columns. By raising the moduli of elasticity, the model effectively incorporates the stiffening effects that would be present in the joints, although in an indirect manner.

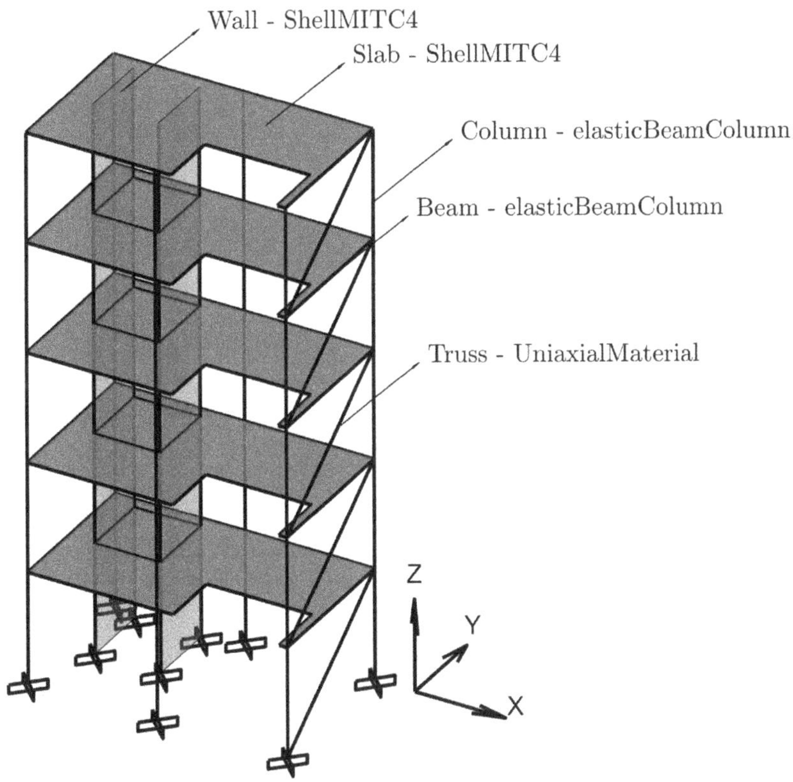

Figure 3. Finite element model developed in Openseespy.

4.2. Global Sensitivity Analysis

As explained in the previous section, the FE model is defined with 20 input parameters, which are the moduli of elasticity for the structural elements. Before carrying out the updating process, it is important to investigate which parameters have the most significant effects on the response that will be subsequently used for estimation. Thus, a one-at-a-time (OAT) local sensitivity analysis is performed by varying each parameter by ±5% and ±10% of its corresponding nominal value. Then, the variation effects on the model outputs are subsequently studied. The responses obtained by perturbing each model's parameter (i.e., keeping the others fixed) are compared with the response obtained with the nominal values of each parameter (Figure 4).

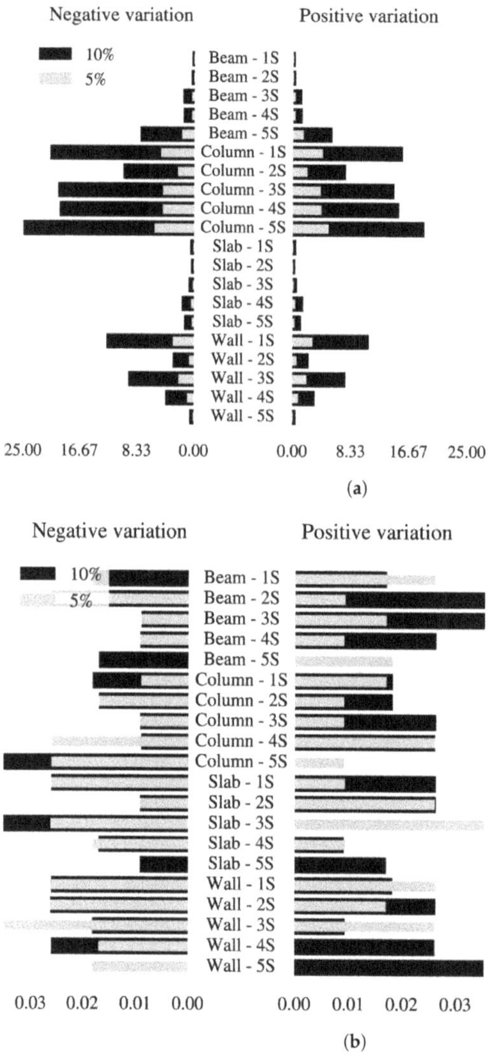

Figure 4. Results of the OAT sensitivity analysis. Note that the number accompanying the letter *S* indicates the story where the structural elements are. (**a**) Changes (%) on natural frequencies regarding variations on the moduli of elasticity, (**b**) Changes (%) on mode shapes regarding variations on the moduli of elasticity.

Figure 4a shows that the most sensitive parameters for the frequencies correspond to the moduli of elasticity of columns, while the least sensitive parameters correspond to the moduli of elasticity of slabs. However, in the case of mode shapes (Figure 4b), there is no structural element whose modulus of elasticity parameter is particularly sensitive. The values presented in Figure 4a,b are obtained through the calculation of the mean squared error (MSE) using the vector of experimental and model values of the mode shapes and frequencies. By utilizing this approach, we may assess the level of agreement between the experimental and model values, providing a comprehensive evaluation of the accuracy and performance of the model in capturing the experimental data. The sensitivity analysis results only provide information on the relative importance of the model parameters on the building's first three natural frequencies and mode shapes. However, it does not mean that a higher preponderance is given to some structural elements when performing the model updating. Hence, a direct comparison of the influence of input parameters in the model is feasible, particularly in relation to the moduli of elasticity of specific beams such as the second and third level beams (Beam 3S and Beam 4S, respectively). These beams exhibit a similar influence on both mode shapes and associated frequencies, allowing for a meaningful evaluation of their respective contributions. Although the amplitude of the shape modes does have an influence on the results depicted in Figure 4, it is not the decisive factor in the regrouping of variables. This is primarily due to the error calculation method employed in the analysis. These results define a regrouping of the input parameters to make more efficient use of the available computational resources. The regrouping is performed with the idea that all parameters included in a subgroup have a single associated Probability Density Function (PDF). As shown in Table 2, seven groups are defined and divided based on their PDFs.

Table 2. Parameter regrouping for Bayesian updating of the structural model.

Group	Element	Previous Nomenclature	New Nomenclature
1	Beam—1st Story Beam—2nd Story	Ebeam1 Ebeam2	Ebeam1
2	Beam—3rd Story Beam—4th Story Beam—5th Story	Ebeam3 Ebeam4 Ebeam5	Ebeam2
3	Column—1st Story Column—2nd Story	Ecol1 Ecol2	Ecol1
4	Column—3rd Story Column—4th Story Column—5th Story	Ecol3 Ecol4 Ecol5	Ecol2
5	Slab—1st Story Slab—2nd Story Slab—3rd Story Slab—4th Story Slab—5th Story	Eslab1 Eslab2 Eslab3 Eslab4 Eslab5	Eslab1
6	Wall—1st Story Wall—2nd Story	Ewall1 Ewall2	Ewall1
7	Wall—3rd Story Wall—4th Story Wall—5th Story	Ewall3 Ewall4 Ewall5	Ewall2

4.3. Bayesian Updating Algorithm

4.3.1. Likelihood Function Estimation

As the likelihood function serves as a bridge between prior knowledge and new information, the correct likelihood function depends on the type of data being analyzed and the nature of the estimated parameters. There are various forms that the likelihood function may take, such as normal distributions and gamma distributions, each tailored to a specific problem. The multivariate normal distribution is chosen because it may adequately capture the relationships between multiple observations, making it well-suited for modeling complex systems. The multivariate normal distribution also has well-understood properties for propagation errors and estimating uncertainty, making it a useful tool for Bayesian model updating. Following the definition made in Equation (4), the vectors x and μ are expressed with the following configuration:

$$\mu = \begin{bmatrix} f_{1-\text{exp}} \\ f_{2-\text{exp}} \\ f_{3-\text{exp}} \\ \Phi_{1,1-\text{exp}} \\ \vdots \\ \Phi_{1,20-\text{exp}} \\ \Phi_{2,1-\text{exp}} \\ \vdots \\ \Phi_{2,20-\text{exp}} \\ \Phi_{3,1-\text{exp}} \\ \vdots \\ \Phi_{3,20-\text{exp}} \end{bmatrix}_{63\times 1} \quad x = \begin{bmatrix} f_{1-\text{mod}} \\ f_{2-\text{mod}} \\ f_{3-\text{mod}} \\ \Phi_{1,1-\text{mod}} \\ \vdots \\ \Phi_{1,20-\text{mod}} \\ \Phi_{2,1-\text{mod}} \\ \vdots \\ \Phi_{2,20-\text{mod}} \\ \Phi_{3,1-\text{mod}} \\ \vdots \\ \Phi_{3,20-\text{mod}} \end{bmatrix}_{63\times 1} \tag{7}$$

where f_i represents the frequency of the mode shape i, while $\Phi_{i,j}$ denotes the modal coordinate j of the mode shape i. The subscripts exp and mod indicate the experimental and model values, respectively.

The weight matrix W is defined in this study as:

$$W = \begin{bmatrix} 1 & 0 & 0 & 0 & 0 & 0 & \dots & 0 \\ 0 & 1 & 0 & 0 & 0 & 0 & \dots & 0 \\ 0 & 0 & 1 & 0 & 0 & 0 & \dots & 0 \\ 0 & 0 & 0 & \frac{1}{20} & 0 & 0 & \dots & 0 \\ 0 & 0 & 0 & 0 & \frac{1}{20} & 0 & \dots & 0 \\ 0 & 0 & 0 & 0 & 0 & \frac{1}{20} & \dots & 0 \\ \vdots & \vdots & \vdots & \vdots & \vdots & \vdots & \ddots & \vdots \\ 0 & 0 & 0 & 0 & 0 & 0 & \dots & \frac{1}{20} \end{bmatrix}_{63\times 63} \tag{8}$$

where the first three values in the diagonal of the matrix contain the weights of the frequency values associated with the first three mode shapes, and the weights of the modal coordinates for the first three modes are in the following 60 values. Since there are 20 modal coordinates for each mode, the respective values are normalized by a factor of $1/20$, while the associated frequencies are multiplied by a factor of 1, thus giving each mode the same weight as its corresponding frequency.

Following Equation (5), the covariance matrix measures the degree of interdependence between output variables in a dataset. In the diagonal elements, σ_i^2, with $i \leq 3$, corresponds to the variance of the frequency associated with the i mode shape, and $i > 3$ corresponds to the variance of respective modal coordinates. The off-diagonal elements are the covariance between a pair of output variables, whether frequencies or mode shapes.

4.3.2. Posterior Sampling

The model updating is performed using Python together with the following libraries: *Arviz* for the exploratory analysis of Bayesian models, *Numpy* for the implementation of mathematical functions and linear algebra operators, *Bilby* for parameter estimation using BI [39], *Matplotlib* for plotting, *Scipy* as a statistical package (probability distributions, correlation functions, etc.), *Emcee* as a sampler of Goodman and Weare's Markov chain affine invariant [40], *Pandas* for data manipulation and analysis, and *Openseespy* as Opensees interpreter.

The numerical model updating of the structure aims to estimate the parameter set of the x vector that maximizes the likelihood function $\phi(x)$ (Equation (4)). However, it is necessary to include the MAC to ensure that the correct modal forms are being compared. To ensure that a pair of experimental and model mode shapes are comparable, it is necessary to both normalize the mode shapes and calculate the MAC value between them. Once the MAC value has been calculated, the mode shapes can be compared, and the highest MAC value indicates the most similar mode shapes. The covariance matrix is automatically estimated using the command integrated into the *Numpy* library.

The numerical model code is included within the main Bayesian updating code to save resources in data transfer in the system to be worked. The estimations are performed using a Markov chain Monte Carlo (MCMC) ensemble sampler with 142,000 samples and LogNormal distributions as priors for all parameters (Table 3). Values for the prior distributions are obtained from the experimental data available in [34]. The choice of using a LogNormal distribution is based on the work conducted by Mirza and MacGregor (1982) [41] and Nowak and Szerszen (2003) [42]. These studies involved collecting a substantial amount of data on the compressive strength of concrete, which is closely associated to the modulus of elasticity. It was observed that the data in both studies exhibited a distribution that could be reasonably approximated by a LogNormal distribution.

Table 3. Prior distributions for model updating.

Prior Distributions			
Parameters	**Type of Distribution**	**Mean [MPa]**	**Std. Dev. [MPa]**
Ebeam1	LogNormal	35,000	9700
Ebeam2	LogNormal	35,000	9700
Ecol1	LogNormal	35,000	9700
Ecol2	LogNormal	35,000	9700
Eslab1	LogNormal	35,000	9700
Ewall1	LogNormal	35,000	9700
Ewall2	LogNormal	35,000	9700

4.4. Cloud Computing as Alternative

One of the main drawbacks faced in this study is the execution time of the Bayesian updating algorithm. In this case, two variables have the most significant impact on running time: the quantity of nodes and elements of the FE model and the relatively low sensitivity between the input parameters and the model's output. Therefore, the Infrastructure-as-a-Service (IaaS) cloud computing provided by Google is used based on its scalable and cost-effective way of using Virtual Machines (VM) on demand, charging them only for the computing resources leased for a period. Moreover, Google cloud service allows integration with Google Colaboratory, an open-source Python programming tool used primarily for machine learning. Natively, Google Colaboratory uses VMs to run the notebook code on the server. The VM used in this study was the **c2-standard-8**, which focuses on ultra-high performance for processing-intensive workloads and is mainly used for workloads linked to processing. The complete model-updating simulation took about 215 h.

5. Results and Discussions

In this section, it will be presented and compared the results obtained from the two different approaches for the covariance matrix in the likelihood function: calculating the covariance matrix at each iteration and using an identity matrix as the covariance matrix, hereafter, it will be referred to as iterative-approach and identity-approach, respectively.

By comparing the results obtained from these two approaches, it is possible to assess their respective strengths and limitations in terms of accuracy, computational efficiency, and robustness. Such an analysis may help determine which approach is best suited for a particular problem and may inform the design of future studies involving Bayesian updating of structural models.

5.1. Convergence Criteria

The convergence of the Markov chains is verified using two methods: evolution of the Effective Sample Sizes (ESS) and the Markov Chain Standard Error (MCSE). The ESS measures the efficiency of Monte Carlo methods reaching the number of effective samples necessary for estimating the posterior distribution of the corresponding parameter to have enough information to guarantee a satisfactory result according to the evaluator's criterion [43]. In this case, the limit number that needs to be exceeded to guarantee convergence according to the method is 400 effective samples [44]. For both, iterative-approach and identity-approach, using the bulk-ESS and tail-ESS methods, the number of effective samples obtained in the chain far exceeds the limit, shown in Figure 5, with more than 2000 in the first case and more than 5000 in the second. In most of the chains, the iterative-approach shows a slightly faster convergence compared to the identity-approach. This may be due to the fact that the iterative-approach takes into account the specific characteristics of the data, allowing for a more precise estimation of the posterior distribution. However, the difference in convergence speed between the two approaches is so small that it may be considered negligible.

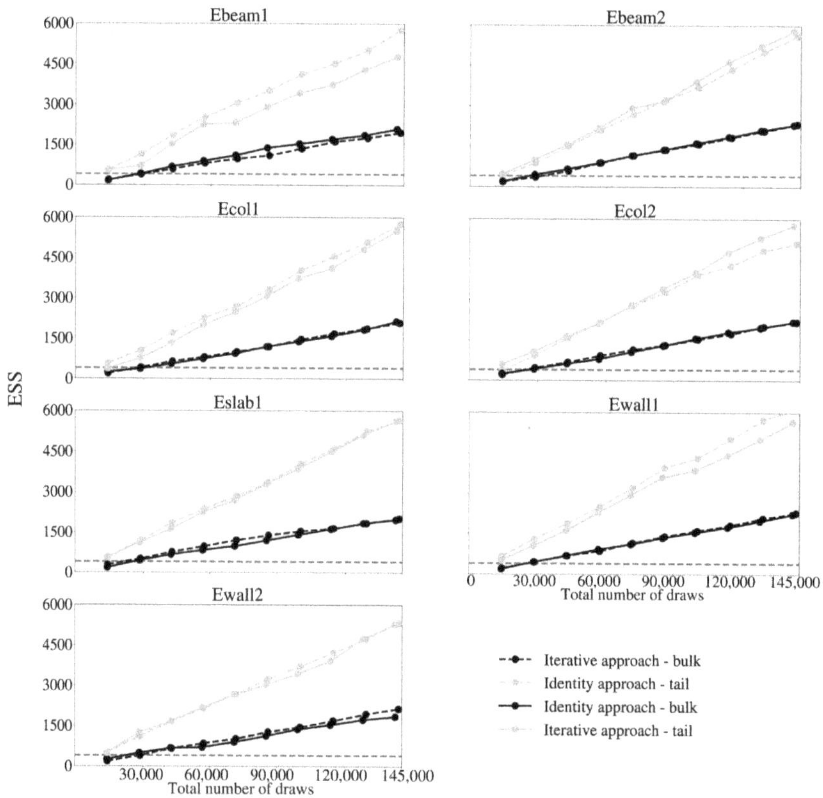

Figure 5. Evolution of effective sample size (ESS) for posterior sampling.

On the other hand, the MCSE method is defined as the standard deviation of the chains divided by their effective sample size [45]. The MCSE provides a quantitative measure of the magnitude of the estimation noise. Although the acceptable limit is also given at the discretion of the researcher, the acceptable uncertainty associated with the mean of the posterior distribution may be taken as a reference (Figure 6). The MCSE results indicate that there is no significant difference in the precision of estimating the posterior distribution of the parameters between the two approaches. Thus, both approaches may be considered equally effective in estimating the posterior distribution.

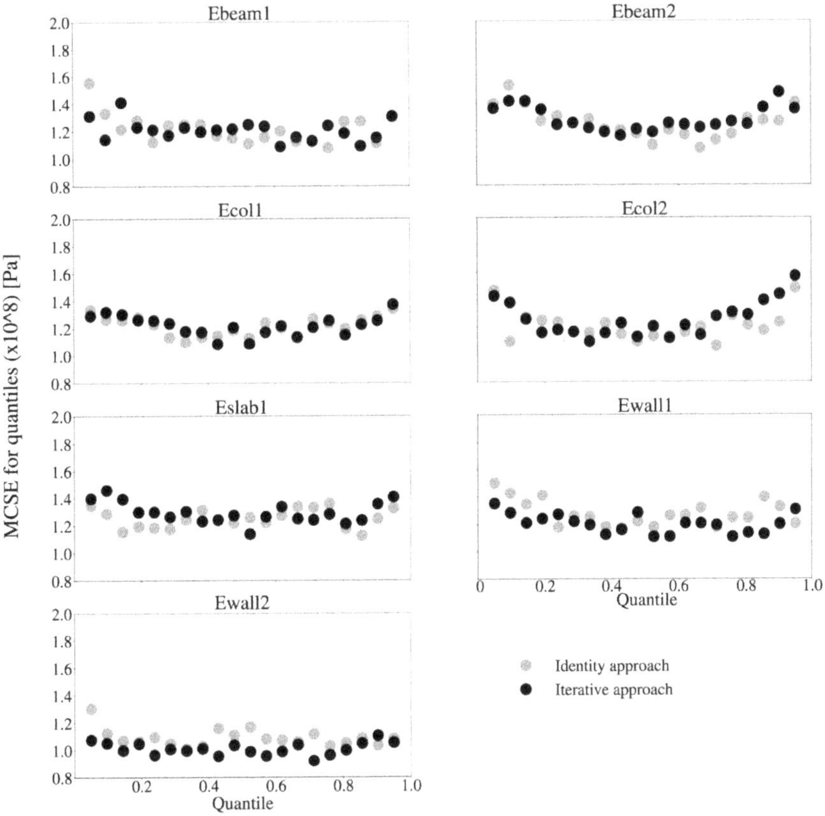

Figure 6. Markov chain standard error (MCSE) for posterior sampling.

5.2. *Numerical Evaluation of the Model Updating*

The posterior probability density allows for a detailed analysis of the model parameters, including the identification of the most likely values, the range of uncertainty, and the correlation between different parameters. Thus, Figure 7 illustrates the final distribution obtained for each parameter using both approaches. The figure provides a visual representation of the PDF generated through the respective methods and allows for a comparative analysis of the results.

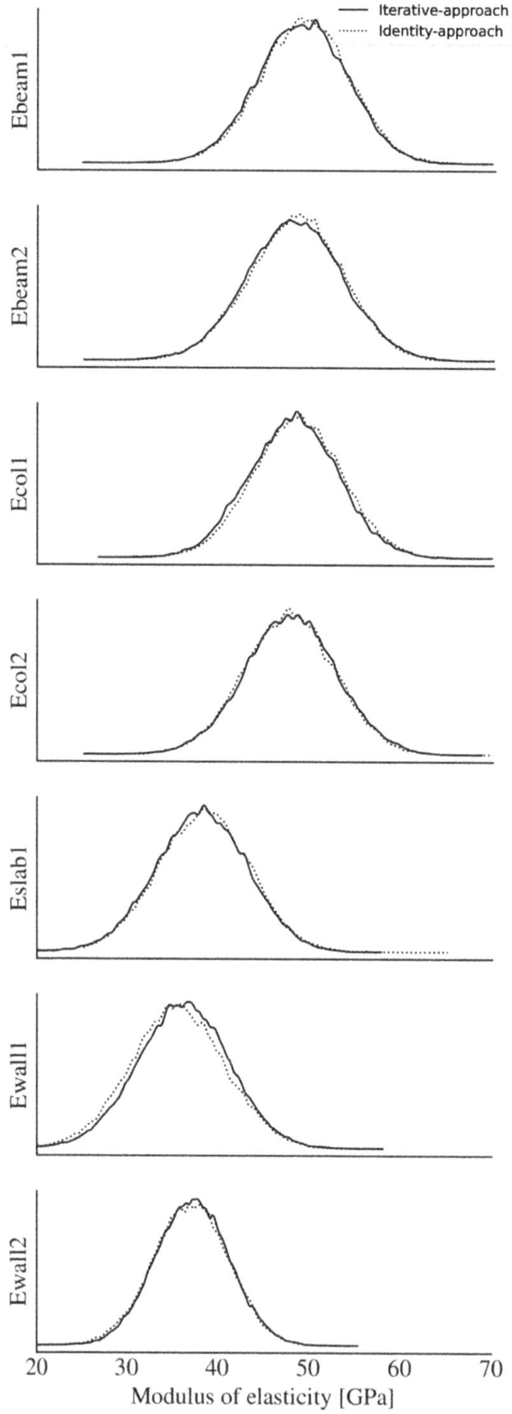

Figure 7. Posterior probability density function and tracing values.

Figure 8 illustrates that the posterior distributions obtained from both approaches exhibit significant overlap, indicating that they are almost identical. It depicts the trends and high-density intervals for each parameter where Ebeam1, Ebeam2, Ecol1, and Ecol2 have a considerably higher modulus of elasticity than Eslab1, Ewall1, and Ewall2. These plots may effectively highlight data points deviating significantly from the expected trend. The posterior distribution of the model parameters is found to be highly comparable between both approaches, indicating that the precision in estimating posterior distributions is comparable as well. Decision-makers may use this information in designing or assessing engineering models [46].

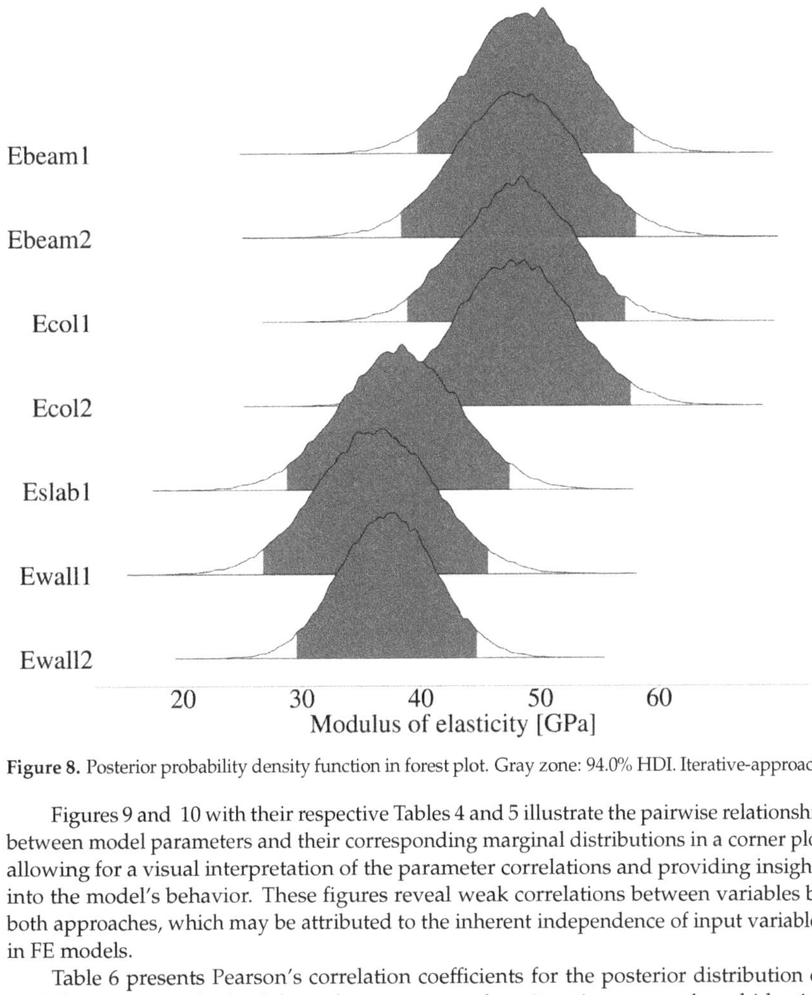

Figure 8. Posterior probability density function in forest plot. Gray zone: 94.0% HDI. Iterative-approach.

Figures 9 and 10 with their respective Tables 4 and 5 illustrate the pairwise relationship between model parameters and their corresponding marginal distributions in a corner plot, allowing for a visual interpretation of the parameter correlations and providing insights into the model's behavior. These figures reveal weak correlations between variables by both approaches, which may be attributed to the inherent independence of input variables in FE models.

Table 6 presents Pearson's correlation coefficients for the posterior distribution of model parameters obtained from the two approaches: iterative-approach and identity-approach. The above-the-diagonal values represent the correlation coefficient obtained from the iterative-approach and the below-the-diagonal values represent the correlation coefficient obtained from the identity-approach. This table provides valuable information about the correlation between the posterior distribution of model parameters obtained from the two approaches, which may be useful for understanding the impact of the covariance matrix on the estimation of the posterior distribution. The values in bold represent the largest values for a coefficient between the same parameters. As the results are mixed, it is difficult to establish a clear correlation between the parameters and either of the approaches. Initial regrouping of highly correlated parameters performed in Section 4.2 may have contributed to general weak correlations between the model final parameters.

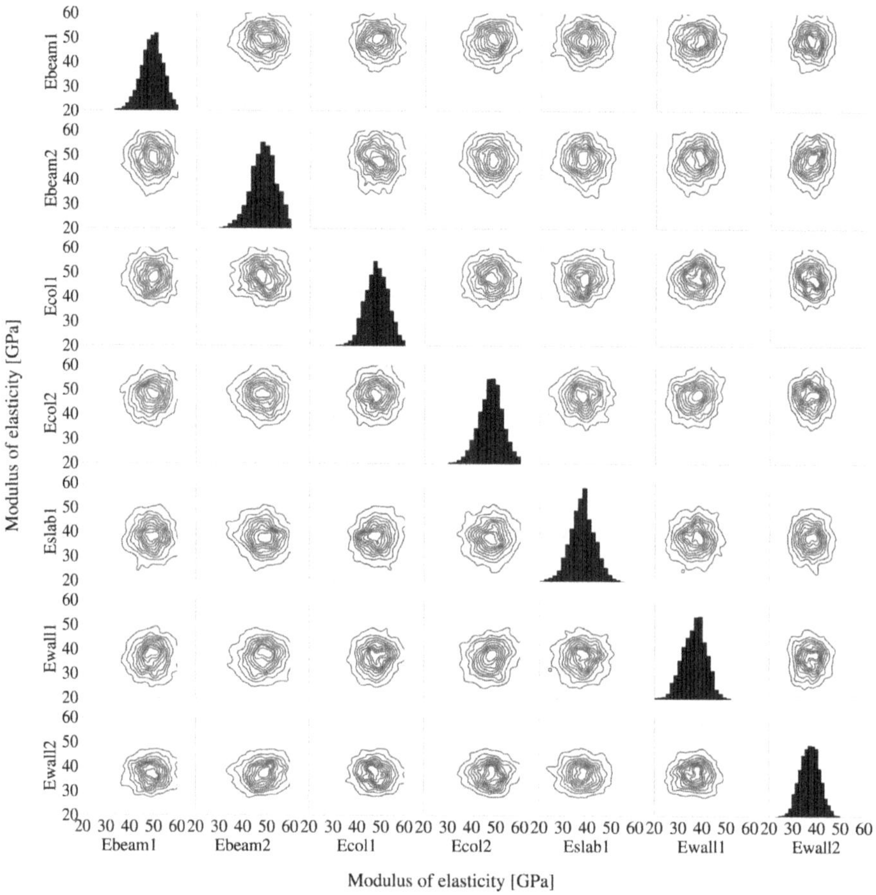

Figure 9. Corner plot of model parameters. Iterative-approach.

Table 4. Summary of posterior distribution of model parameters. Iterative-approach.

	Mean [GPa]	SD [GPa]	HDI 3% [GPa]	HDI 97% [GPa]
Ebeam1	48.93	4.87	39.75	57.99
Ebeam2	48.16	5.27	38.39	58.18
Ecol1	48.14	4.88	38.90	57.23
Ecol2	47.92	5.03	38.60	57.68
Eslab1	38.19	5.00	28.75	47.48
Ewall1	36.23	5.03	26.74	45.66
Ewall2	37.25	4.05	29.50	44.72

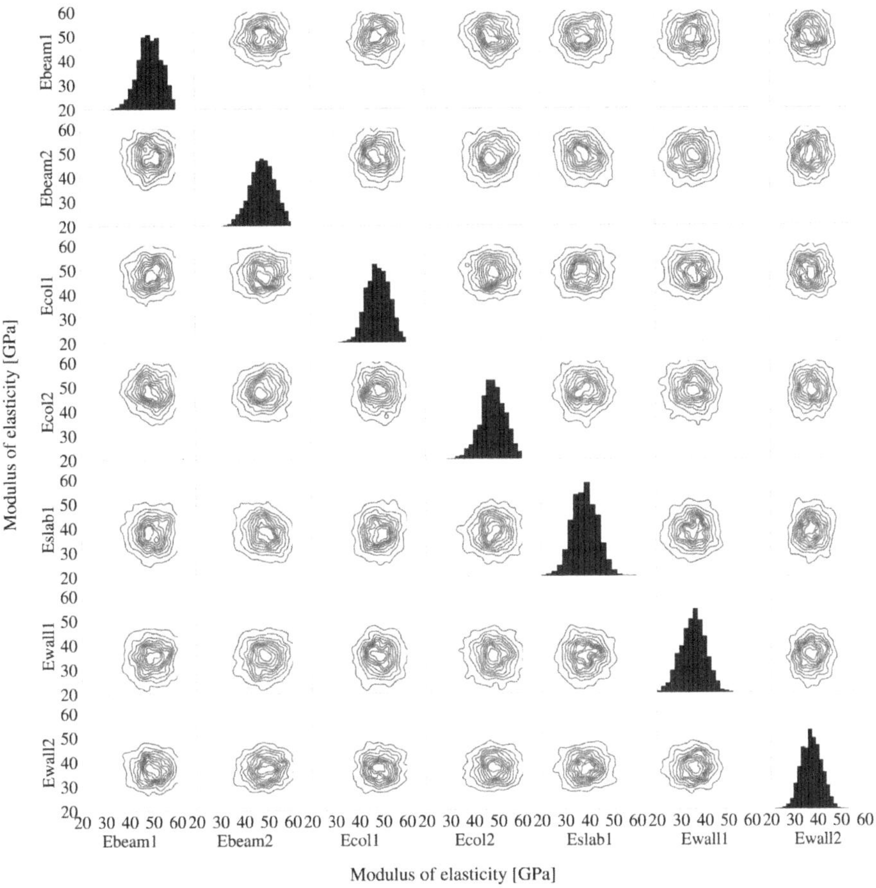

Figure 10. Corner plot of model parameters. Identity-approach.

Table 5. Summary of posterior distribution of model parameters. Identity-approach.

	Mean [GPa]	SD [GPa]	HDI 3% [GPa]	HDI 97% [GPa]
Ebeam1	49.22	4.86	40.01	58.25
Ebeam2	48.29	5.15	38.43	57.79
Ecol1	48.55	4.85	39.28	57.54
Ecol2	47.76	4.96	38.59	57.25
Eslab1	38.38	5.04	28.63	47.50
Ewall1	35.55	5.17	25.84	45.34
Ewall2	37.08	4.15	29.20	44.77

This study uses two key parameters to assess the accuracy of the structural response and the effectiveness of the model updating. First is the Posterior Predictive Check (PPC), which evaluates the agreement between observed and predicted responses. Second is the Modal Assurance Criterion (MAC), which quantifies the similarity between experimental and updated modal properties. Given the similar results obtained from both approaches in the posterior distributions, either approach will be used interchangeably to compute the results derived from the posterior distributions going forward. Figure 11a–c show the results of a subsequent prediction check in which the frequencies predicted by the model are compared with the observed data. A PPC is used to evaluate the goodness-of-fit of the model and the data, involving the generation of new frequency sets based on estimated model parameters and a comparison to the experimental data. It compares the predictions generated by the model with the actual observed values (experimental). Analyzing the PPC plot compared to experimental values helps validate the model's predictions and provides insights into its performance and reliability. The results present a good model performance relative to the experimental data used in the updating process. Specifically, the model's predictions closely match the observed data in terms of the frequencies, providing evidence of the model's improved accuracy and reliability.

Table 6. Pearson's Correlation Coefficient for posterior distribution of model parameters. Above the diagonal: Iterative-approach. Below the diagonal: Identity-approach.

	Ebeam1	Ebeam2	Ecol1	Ecol2	Eslab1	Ewall1	Ewall2
Ebeam1	1	−0.01629	0.01	0.02	0.000633	0.009928	0.007501
Ebeam2	−0.000538	1	0.016919	−0.003235	−0.005969	0.00827	0.004841
Ecol1	0.020662	0.005784	1	−0.002967	0.020706	−0.007723	0.021349
Ecol2	0.019176	0.012743	−0.007397	1	0.006645	−0.001015	−0.01976
Eslab1	0.008815	−0.02366	0.017139	0.034489	1	0.012989	0.010983
Ewall1	−0.003382	0.006759	0.026454	0.005082	0.022536	1	0.017236
Ewall2	−0.004593	0.003377	0.024154	0.010754	−0.00095	−0.011894	1

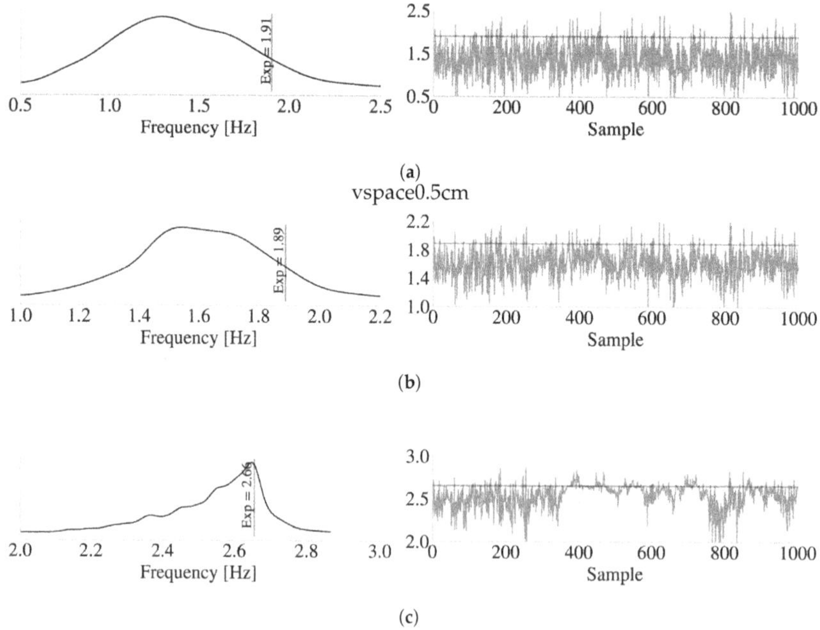

Figure 11. Posterior predictive check and tracing values. (**a**) Frequency 1, (**b**) Frequency 2, (**c**) Frequency 3.

The quality of the mode shape comparison is quantified by the MAC values, where a value of 1 indicates a perfect match of the mode shapes between the experimental and FE model. The modal coordinates and associated uncertainties are calibrated using the posterior distribution of the updated model parameters, followed by a comparison of the resulting MAC with experimentally identified data. A graphical comparison of the MAC matrix calculated for the value corresponding to the prior and posterior distribution of the modal coordinates is shown in Figure 12. The prior values correspond to the estimated values for the modal coordinates, which are based exclusively on prior information, and are not yet updated with any new data or information. The posterior values correspond to the posterior distribution of the modal coordinates, which means that the estimated values for the modal coordinates, after the Bayesian updating process, have been incorporated into the analysis. As shown in the diagonal elements of the MAC matrix, the proposed updates have significantly improved the fit of the FE model to the experimental results. The updated MAC values are closer to 1 than the initial values, indicating a closer agreement between the updated model and the experimental measurements. The good performance in the MAC fit suggests that this Bayesian model-updating approach effectively captured the uncertainties and updated the model parameters to match the experimental results.

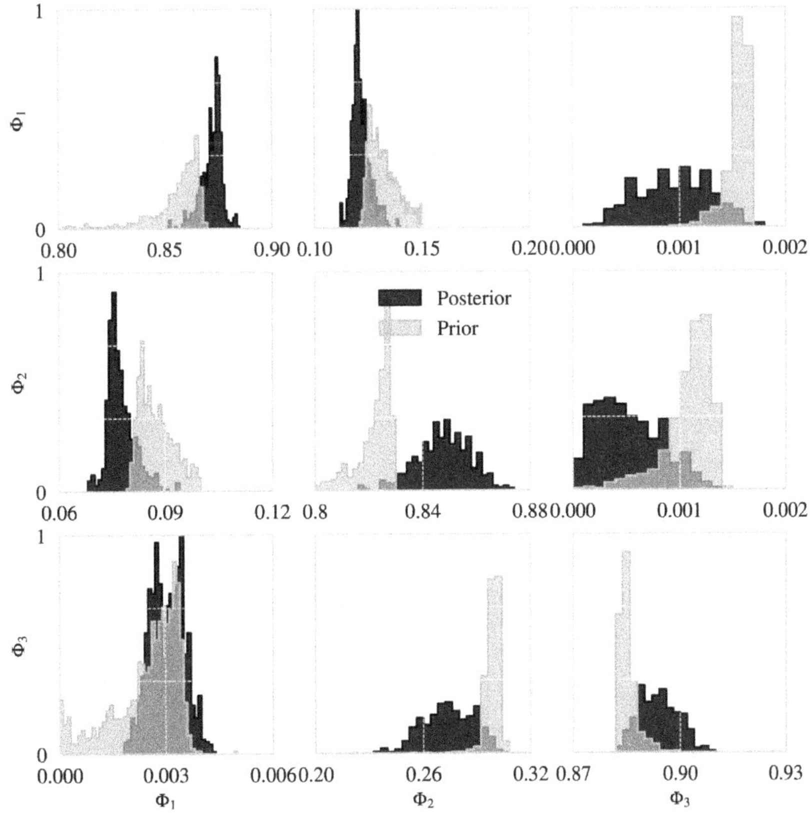

Figure 12. Prior and posterior MAC PDF.

5.3. Covariance Matrix Analysis

The covariance matrix is a type of matrix that provides insights into the correlation between frequencies and mode shapes and their associated uncertainties [27]. The shape and values of the covariance matrix reveal critical structural features such as the degree of symmetry, the presence of localized modes, and the overall complexity of the vibration patterns. Through a modification in the source code of the BI library, it is possible to save the covariance matrix for each sample. The original matrix has a size of 63×63; however, only those corresponding to the first three frequencies are shown in Figure 13. The variance (main diagonal elements) and covariance (off-diagonal elements) of the frequencies are plotted together in this plot. A higher covariance between two natural frequencies indicates a more significant correlation between the corresponding vibration modes of the structure.

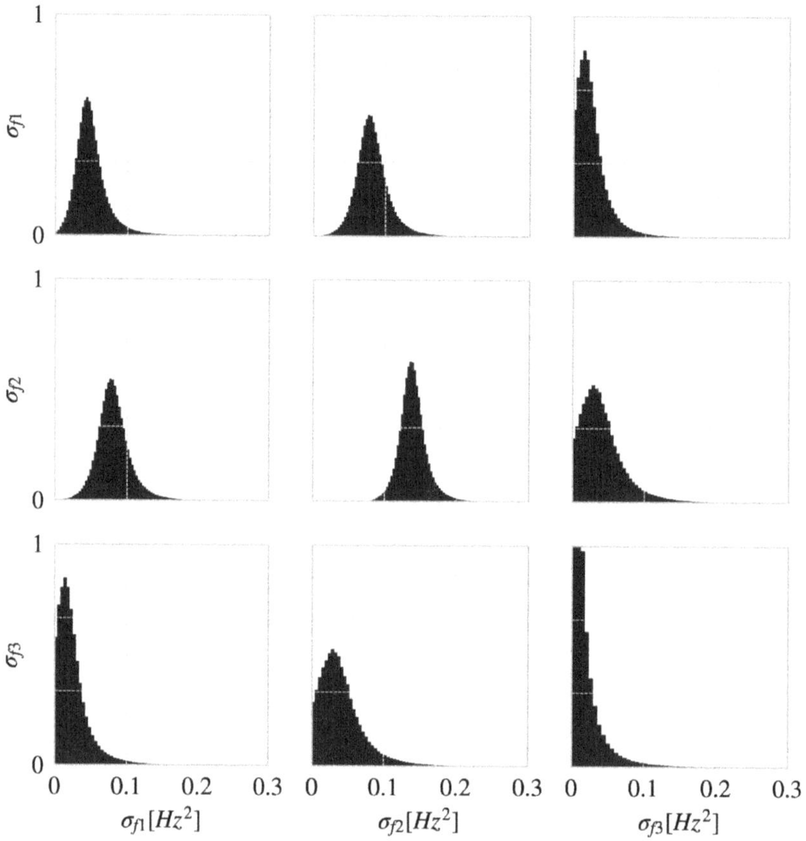

Figure 13. First three rows and columns on the Covariance matrix representing the first three frequencies of the building.

6. Conclusions

This paper implemented a methodology for updating the structural model parameters using BI. The modal properties of a five-story full-scale RC building were used to update the FE model. The parametric uncertainty quantification process and its results were explained following the proposed methodology. The updating algorithm determined the posterior probability of the FE model's parameters and calculated the covariance matrix of the observations, comparing them with model realizations. The covariance matrix between observations and the updated model allowed for identifying the error computed through a multivariate normal likelihood function.

Although, in theory, iteratively calculating the covariance matrix enabled a more comprehensive understanding of the system's behavior, both used approaches for incorporating the covariance matrix into the Bayesian model-updating framework (identity-approach and iterative-approach) resulted in similar posterior distributions and convergence rates. Therefore, in this case, both approaches may be used interchangeably without affecting the results' reliability or accuracy. This finding provides a practical advantage in terms of computational efficiency when dealing with large sample sizes.

The obtained posterior updating of the input parameters of the FE model corresponds to distributions similar to normal shapes without clear biases, i.e., approximately symmetric distributions. In all cases, the posterior distributions have a smaller standard deviation with respect to the prior distributions. Therefore, the error associated with the model and parameter's values is smaller than initially assumed. Furthermore, the way the MAC function was calculated provides valuable information about the reliability of the model update in terms of mode shapes, highlighting the importance of quantifying the uncertainty associated also with the associated frequency through PPC and its comparison with experimental data. The proposed methodology may improve the reliability of structural models and assist in decision-making for the design and assessment of structures.

Author Contributions: Conceptualization, O.D.H., A.R.O., D.G. and R.A.; methodology, O.D.H. and A.R.O.; formal analysis, O.D.H. and A.R.O.; funding acquisition, A.R.O. and D.G.; writing—original draft preparation, O.D.H.; writing—review and editing, D.G., A.R.O. and R.A. All authors have read and agreed to the published version of the manuscript.

Funding: This work was financially supported by the Universidad del Valle, Colombia (Grant No. 21185: Evaluation of structural reliability reliability of reinforced concrete elements subjected to reinforced concrete elements subjected to flexural and shear).

Acknowledgments: The authors would like to acknowledge Matthew Pitkin from University of Glasgow (Scotland) for providing suggestions in the model-updating algorithm and Mathias Gutiérrez from Universidad de los Andes (Chile) for providing information about the numerical model of the structure.

Conflicts of Interest: The authors declare no conflict of interest.

References

1. Armstrong, C.G. Modelling requirements for finite-element analysis. *Comput.-Aided Des.* **1994**, *26*, 573–578. [CrossRef]
2. Shan, S.; Wang, G.G. Survey of modeling and optimization strategies to solve high-dimensional design problems with computationally-expensive black-box functions. *Struct. Multidiscip. Optim.* **2010**, *41*, 219–241. [CrossRef]
3. Zárate, B.A.; Caicedo, J.M. Finite element model updating: Multiple alternatives. *Eng. Struct.* **2008**, *30*, 3724–3730. [CrossRef]
4. Jagota, V.; Sethi, A.P.S.; Kumar, K. Finite element method: An overview. *Walailak J. Sci. Technol. (WJST)* **2013**, *10*, 1–8.
5. Rios, J.M. Uncertainty in Structural Dynamic Models. Ph.D. Thesis, Swansea University, Swansea, UK, 2005.
6. Igusa, T.; Buonopane, S.; Ellingwood, B. Bayesian analysis of uncertainty for structural engineering applications. *Struct. Saf.* **2002**, *24*, 165–186. [CrossRef]
7. Bulleit, W.M. Uncertainty in structural engineering. *Pract. Period. Struct. Des. Constr.* **2008**, *13*, 24–30. [CrossRef]
8. Gomez, D.; Dyke, S.J.; Rietdyk, S. Structured uncertainty for a pedestrian-structure interaction model. *J. Sound Vib.* **2020**, *474*, 115237. [CrossRef]
9. Vailati, M.; Monti, G.; Khazna, M.; Realfonzo, R.; De Iuliis, M. Probabilistic seismic response analysis of existing masonry structures. In *Brick and Block Masonry*; CRC Press: Boca Raton, FL, USA, 2016; pp. 2489–2498. [CrossRef]
10. Zellner, A. Bayesian and non-Bayesian approaches to statistical inference and decision-making. *J. Comput. Appl. Math.* **1995**, *64*, 3–10. [CrossRef]
11. Briggs, A.H.; Weinstein, M.C.; Fenwick, E.A.; Karnon, J.; Sculpher, M.J.; Paltiel, A.D. Model parameter estimation and uncertainty analysis: A report of the ISPOR-SMDM Modeling Good Research Practices Task Force Working Group-6. *Med. Decis. Mak.* **2012**, *32*, 722–732. [CrossRef] [PubMed]
12. Hoang, T.V.; Matthies, H.G. An efficient computational method for parameter identification in the context of random set theory via Bayesian inversion. *Int. J. Uncertain. Quantif.* **2021**, *11*, 1–18. [CrossRef]
13. Sehgal, S.; Kumar, H. Structural dynamic model updating techniques: A state of the art review. *Arch. Comput. Methods Eng.* **2016**, *23*, 515–533. [CrossRef]
14. Loyola, M. Big data in building design: A review. *J. Inf. Technol. Constr.* **2018**, *23*, 259–284.
15. Beck, J.L.; Katafygiotis, L.S. Updating models and their uncertainties. I: Bayesian statistical framework. *J. Eng. Mech.-Proc. ASCE* **1998**, *124*, 455–462. [CrossRef]
16. Katafygiotis, L.S.; Beck, J.L. Updating models and their uncertainties. II: Model identifiability. *J. Eng. Mech.* **1998**, *124*, 463–467. [CrossRef]
17. Jaynes, E.T. *Probability Theory: The Logic of Science*; Cambridge University Press: Cambridge, UK, 2003.
18. Bayes, T. LII. An essay towards solving a problem in the doctrine of chances. By the late Rev. Mr. Bayes, FRS communicated by Mr. Price, in a letter to John Canton, AMFR S. In *Philosophical Transactions of the Royal Society of London*; Royal Society of London: London, UK, 1763; pp. 370–418.

19. Swinburne, R. Bayes' Theorem. *Mind* **2004**, *194*. [CrossRef]
20. Denison, D.G.; Holmes, C.C.; Mallick, B.K.; Smith, A.F. *Bayesian Methods for Nonlinear Classification and Regression*; John Wiley & Sons: Hoboken, NJ, USA, 2002; Volume 386.
21. MacLehose, R.F.; Hamra, G.B. Applications of Bayesian methods to epidemiologic research. *Curr. Epidemiol. Rep.* **2014**, *1*, 103–109. [CrossRef]
22. Ma, Z.; Chen, G. Bayesian methods for dealing with missing data problems. *J. Korean Stat. Soc.* **2018**, *47*, 297–313. [CrossRef]
23. Zhang, Z.; Sun, C.; Guo, B. Transfer-learning guided Bayesian model updating for damage identification considering modeling uncertainty. *Mech. Syst. Signal Process.* **2022**, *166*, 108426. [CrossRef]
24. Zhang, E.; Feissel, P.; Antoni, J. A comprehensive Bayesian approach for model updating and quantification of modeling errors. *Probabilistic Eng. Mech.* **2011**, *26*, 550–560. [CrossRef]
25. Argyris, C.; Papadimitriou, C.; Panetsos, P.; Tsopelas, P. Bayesian model-updating using features of modal data: Application to the metsovo bridge. *J. Sens. Actuator Netw.* **2020**, *9*, 27. [CrossRef]
26. Ghurye, S.; Olkin, I. A characterization of the multivariate normal distribution. *Ann. Math. Stat.* **1962**, *33*, 533–541. [CrossRef]
27. Tong, Y.L. *The Multivariate Normal Distribution*; Springer Science & Business Media: Berlin/Heidelberg, Germany, 2012.
28. Srivastava, M.S. Some tests concerning the covariance matrix in high dimensional data. *J. Jpn. Stat. Soc.* **2005**, *35*, 251–272. [CrossRef]
29. Srivastava, M.S.; Kollo, T.; von Rosen, D. Some tests for the covariance matrix with fewer observations than the dimension under non-normality. *J. Multivar. Anal.* **2011**, *102*, 1090–1103. [CrossRef]
30. Birke, M.; Dette, H. A note on testing the covariance matrix for large dimension. *Stat. Probab. Lett.* **2005**, *74*, 281–289. [CrossRef]
31. Fisher, T.J. On testing for an identity covariance matrix when the dimensionality equals or exceeds the sample size. *J. Stat. Plan. Inference* **2012**, *142*, 312–326. [CrossRef]
32. Yan, W.J.; Katafygiotis, L.S. A novel Bayesian approach for structural model updating utilizing statistical modal information from multiple setups. *Struct. Saf.* **2015**, *52*, 260–271. [CrossRef]
33. Zhang, X.; Boscardin, W.J.; Belin, T.R. Sampling correlation matrices in Bayesian models with correlated latent variables. *J. Comput. Graph. Stat.* **2006**, *15*, 880–896. [CrossRef]
34. Chen, M.C.; Pantoli, E.; Wang, X.; Astroza, R.; Ebrahimian, H.; Hutchinson, T.C.; Conte, J.P.; Restrepo, J.I.; Marin, C.; Walsh, K.D.; et al. Full-scale structural and nonstructural building system performance during earthquakes: Part I–specimen description, test protocol, and structural response. *Earthq. Spectra* **2016**, *32*, 737–770. [CrossRef]
35. Pantoli, E.; Chen, M.C.; Hutchinson, T.C.; Astroza, R.; Conte, J.P.; Ebrahimian, H.; Restrepo, J.I.; Wang, X. Landmark data set from the building nonstructural components and systems (BNCS) project. *Earthq. Spectra* **2016**, *32*, 1239–1259. [CrossRef]
36. Astroza, R.; Ebrahimian, H.; Conte, J.P.; Restrepo, J.I.; Hutchinson, T.C. Influence of the construction process and nonstructural components on the modal properties of a five-story building. *Earthq. Eng. Struct. Dyn.* **2016**, *45*, 1063–1084. [CrossRef]
37. Gutierrez, M. Bayesian Finite Element Model Parameter and Covariance Estimation. Master's Thesis, Universidad de los Andes, Santiago, Chile, 2020.
38. Persson, B. Poisson's ratio of high-performance concrete. *Cem. Concr. Res.* **1999**, *29*, 1647–1653. [CrossRef]
39. Ashton, G.; Hübner, M.; Lasky, P.D.; Talbot, C.; Ackley, K.; Biscoveanu, S.; Chu, Q.; Divakarla, A.; Easter, P.J.; Goncharov, B.; et al. BILBY: A user-friendly Bayesian inference library for gravitational-wave astronomy. *Astrophys. J. Suppl. Ser.* **2019**, *241*, 27. [CrossRef]
40. Goodman, J.; Weare, J. Ensemble samplers with affine invariance. *Commun. Appl. Math. Comput. Sci.* **2010**, *5*, 65–80. [CrossRef]
41. Mirza, S.A.; MacGregor, J.G. Probabilistic study of strength of reinforced concrete members. *Can. J. Civil Eng.* **1982**, *9*, 431–448. [CrossRef]
42. Nowak, A.S.; Szerszen, M.M. Calibration of design code for buildings (ACI 318): Part 1—Statistical models for resistance. *Struct. J.* **2003**, *100*, 377–382. [CrossRef]
43. Pinto, F.; Toledo, J.; Birrell, M.; Bazaez, R.; Hernandez, F.R.A. Uncertainty Quantification in Constitutive Models of Highway Bridge Components: Seismic Bars and Elastomeric Bearings. *Materials* **2023**, *16*, 1792. [CrossRef] [PubMed]
44. Vehtari, A.; Gelman, A.; Simpson, D.; Carpenter, B.; Bürkner, P.C. Rank-normalization, folding, and localization: An improved for assessing convergence of MCMC (with Discussion). *Bayesian Anal.* **2021**, *16*, 667–718. [CrossRef]
45. Kruschke, J. *Doing Bayesian Data Analysis: A Tutorial with R, JAGS, and Stan*; Academic Press: Waltham, MA, USA, 2014. . [CrossRef]
46. Kampenes, V.B.; Dybå, T.; Hannay, J.E.; Sjøberg, D.I. A systematic review of effect size in software engineering experiments. *Inf. Softw. Technol.* **2007**, *49*, 1073–1086. [CrossRef]

MDPI AG
Grosspeteranlage 5
4052 Basel
Switzerland
Tel.: +41 61 683 77 34

Buildings Editorial Office
E-mail: buildings@mdpi.com
www.mdpi.com/journal/buildings

www.ingramcontent.com/pod-product-compliance
Lightning Source LLC
Chambersburg PA
CBHW061835210926
51850CB00006B/901
* 9 7 8 3 7 2 5 8 4 8 5 9 1 *